Superalloys
Fundamentals and Applications

Edited by

Amir Al-Ahmed [1]

[1] Interdisciplinary Research Center for Renewable Energy and Power System (IRC-REPS), King Fahd University of Petroleum & Minerals (KFUPM), Dhahran-31261, Kingdom of Saudi Arabia

Published by **Materials Research Forum LLC**
Millersville, PA 17551, USA

Published as part of the book series
Materials Research Foundations
Volume 178 (2025)
ISSN 2471-8890 (Print)
ISSN 2471-8904 (Online)

Print ISBN 978-1-64490-368-1
eBook ISBN 978-1-64490-369-8

Distributed worldwide by

Materials Research Forum LLC
105 Springdale Lane
Millersville, PA 17551
USA
https://mrforum.com

Manufactured in the United States of America
10 9 8 7 6 5 4 3 2 1

Table of Contents

Preface

Superalloys are materials for high impact applications, like, in high-temperature, extreme mechanical and chemical environment etc. Since their first report over 100 years ago, these alloys are still being developed due to the constant need of present and future technologies. Because of their durability in higher working temperatures and high stress mechanical and corrosive conditions, superalloys are increasingly employed in most of the high-end technologies, such as space and aviation industries, energy technologies, drilling industries, military and medical technologies, and so on. However, more metallurgical breakthroughs are required to find new superalloys for the next generation technologies.

This book, "Fundamentals and Applications of Superalloys" has been designed for the readers interested in superalloys-based materials science and engineering research, as well as the essential concepts and applications that drive the shape of these materials. It contains 17 high impact chapters, covering different aspects (from fundamental to critical mechanism) of superalloys and its applications.

Finally, I would like to thankfully acknowledge all the contributing authors, publishing stuff and also the authorities, which provided required copyright permission, for their efforts and support to bring this book in its final shape. Additionally, Dr. Amir Al-Ahmed acknowledges the support provided by King Fahd University of Petroleum & Minerals during the editing phase of this book.

Superalloys: Fundamentals and Applications   Materials Research Forum LLC
Materials Research Foundations 178 (2025) 1-24   https://doi.org/10.21741/9781644903698-1

Chapter 1

Iron-based Superalloys

G.G. Flores-Rojas[1,2], B. Gómez-Lázaro[2], O. Farias-Elvira[1], F. López-Saucedo[2],
M. Rentería-Urquiza[1], E. Bucio[2], and E. Mendizábal[1]

[1]Department of Chemistry and Chemical Engineering, University Center for Exact Sciences and
Engineering, University of Guadalajara, Guadalajara 44430, Jalisco, Mexico

[2]Department of Radiation Chemistry and Radiochemistry, Institute of Nuclear Sciences, National
Autonomous University of Mexico, Mexico City 04510, Mexico

Abstract

Superalloys are a class of alloys renowned for their ability to retain their structural properties
under extreme and prolonged high-temperature conditions. Within this category, iron-based
superalloys have gained great importance, originating from the development of stainless steels
by combining iron with strengthening elements such as nickel, chromium, carbon, and others.
Their versatile and complex composition has made them widely applicable across various
industries and their complexity enables the design of a wide range of superalloys, allowing for
tailored properties to meet specific requirements. This versatility and outstanding characteristics
have increased their demand, reaching very advanced levels of sophistication.

Keywords

Iron Superalloys, Applications, Austenite, Gamma Phase, Thermal Treatment

Contents

1.   Introduction

Superalloys were developed in response to the increasing demands of the aerospace industry for structural stability under high-temperature conditions, leading to the creation of the first iron-based superalloys with tailored chemical compositions to enhance their properties. However, iron-based superalloys have received less research attention compared to nickel- and cobalt-based superalloys. Despite this, they continue to play a crucial role in aerospace applications [1–3]. These alloys are used in a wide range of components, most notably in gas turbines [4], as well as in other areas such as engines, metal processing equipment, medical devices, military equipment, and more [5–8].

Iron-based superalloys are regarded as the earliest category of superalloys, developed from stainless steel by incorporating various alloying elements into the composition. These iron-based composition offers a more cost-effective alternative in comparison to cobalt- or nickel-based superalloys and demonstrate high reliability at both room temperature and elevated temperatures exceeding 650°C [9,10]. These superalloys are also highly resistant to creep, wear, and oxidation corrosion, which are key characteristics of their design, where the wear resistance increases with higher carbon content[8,11], while the degree of oxidation corrosion decreases with the amount of chromium [12]. Other alloying elements typically used come from Group VIII A, which contribute a wide range of new properties [13,14].

Initially, iron-based superalloys were manufactured by adding titanium as an alloying element to high-chromium stainless steels to harden the metal. Additionally, iron-based superalloys were commonly combined with a significant amount of nickel and later, with continued research, elements like molybdenum and niobium were added to the composition for improved hardening [11,15,16]. Therefore, iron-based superalloys have been developed from stainless steels with high nickel and chromium content and carbon content, forming austenitic structures (face-centered cubic, FCC). This structure is complemented by precipitation formation and solid-solution strengthening in most cases, allowing for better control of their mechanical properties [12,17].

Iron-based superalloys can be grouped into different categories based on the strengthening mechanisms: i) those reinforced by martensitic transformation, ii) austenites strengthened by alloying elements or through hot-cold working processes, and iii) austenitic alloys reinforced by hardening. These alloys contain at least 25% nickel, which is essential for stabilizing the FCC structure formed by iron and nickel [8,18–20]. Based on their crystalline structure, various types of alloys have been classified as follows [21]:

Austenitic Alloys: These are the most widely produced iron-based alloys, recognized for maintaining their austenitic structure over a wide temperature range, from cryogenic levels up to the alloy's melting point [19].

Martensitic Alloys: Although these alloys offer lower corrosion resistance compared to ferritic and austenitic types, they are valued for their magnetic properties, exceptional strength, toughness, and machinability. They can also be hardened through heat treatments [22].

Precipitation-Hardened Martensitic Alloys: Capable of achieving higher hardness levels than standard martensitic alloys, these alloys also offer corrosion resistance comparable to austenitic alloys, thanks to precipitation hardening[23].

Ferritic Alloys: While ferritic alloys provide good corrosion resistance, they are less durable than austenitic varieties. These alloys typically contain minimal nickel, and some may include lead. However, the presence of chromium can lead to metal degradation, especially due to the formation of intermetallic phases during welding [24].

Duplex Alloys: These alloys exhibit a combined microstructure of austenite and ferrite in approximately equal proportions, offering improved strength compared to austenitic alloys and providing enhanced properties of corrosion, and cracking [25,26].

In some cases, only austenitic alloys are considered true superalloys. These alloys exhibit non-magnetic properties and remain ductile even at low temperatures[27]. The other types of alloys are often classified as high strength alloys and are usually used under high temperature conditions, while superalloys obtained by martensitic transformation are generally used at lower temperatures than austenitic alloys, allowing an effective operating temperature of 537 °C up to 760 °C [28]. These superalloys exhibit excellent resistance to oxidation and sulfidation due to their chromium content, which ranges from 5% to 36%[29–31], where chromium forms a coating layer of chromium oxide that serves as a protective surface against corrosion in extreme environments [32–34]. Therefore, the chromium content must have a minimum limit that ensures a uniform coating of the alloy's surface with chromium oxide.

In addition, strength and hardening properties are achieved through post-treatment processes and the inclusion of alloying elements with smaller atomic radii, such as carbon [35–40]. This class of superalloys is characterized by a compact FCC, which is further strengthened by solid-solution harden and precipitate-forming treatments. A summary of key iron-based superalloys is provided below with chemical composition (Table 1):

Table 1. Based-iron superalloys and chemical composition [41–44]

Alloy	Fe	Ni	Cr	C	Ti	Al	Co	Mo	Nb	W	Mn	Si	Cu	La	B	V
19-9DL	66.8	9.0	-	0.3	0.3	-	-	1.3	0.4	1.3	1.1	0.6	-	-	-	-
REX-78	60	18	14	0.01	0.6	-	-	4	-	-	-	-	-	-	0.015	-
W-545	55.8	26	13.5	0.08	2.85	0.2		1.5	-	-	-	-	-	-	0.05	-
Discaloy	55	26	14	0.06	1.7	0.25	-	3	-	-	-	-	-	-	-	-
V57	48.6	27	14.8	0.08	3	0.25		1.25	-	-	-	-	-	-	0.01	0.5
Incoloy 801	46.3	32	20.5	0.05	1.13	-	-	-	-	-	-	-	-	-	-	-
Incoloy 800HT	46	33	21	0.06-0.1	0.25-0.6	0.25-0.6	-	-	-	-	0.8	0.5	0.4	-	-	-
Incoloy 800H	45.8	33	21	0.05-0.1	0.15-0.6	0.15-0.6	-	-	-	-	-	-	-	-	-	-
Incoloy 800	45.7	32.5	21	0.05	0.38	0.38	-	-	-	-	-	-	-	-	-	-
Incoloy 802	44.8	33	21	0.35	-	-	-	-	-	-	-	-	-	-	-	-
Incoloy 909	42	38	-	0.01	1.5	0.03	13	-	4.7	-	-	0.4	-	-	-	-
Incoloy 907	42	38	-	0.01	1.5	0.03	13	-	4.7	-	-	0.15	-	-	-	-
Incoloy 903	41	38	0.1	0.04	1.4	0.7	15	0.1	3	-	-	-	-	-	-	-
Pyromet CTX-1	39	37.7	0.1	0.03	1.7	1	16	0.1	3	-	-	-	-	-	-	-
Inconel 706	37	42	16	0.01	1.8	0.2	-	-	2.9	-	-	0.05	-	-	-	-
Incoloy 901	36.2	43	12.5	0.1	2.7	-	-	6	-	-	-	-	-	-	-	-
Thermo-span	34.5	25	5.5	0.004	0.85	0.45	29	-	4.8	-	-	0.35	-	-	-	-
Incoloy 925	29	44	20.5	0.01	2.1	0.2	-	2.8	-	-	-	-	1.8	-	-	-
D-979	26.4	45	15	0.05	3	1	-	4	-	4	0.75	0.75	-	-	0.01	-
Inconel 783	25.5	29	3	-	0.1	5.4	34	-	3	-	-	-	-	-	-	-
Alloy 718	18.5	53	19	0.08	0.9	0.5	-	3	5.1	-	-	-	0.15	-	-	-
Hastelloy X	15.8	49	22	0.15	-	2	1.5	9	-	-	0.6	-	-	-	-	-

2. Composition and structural characteristics

The normal room-temperature structures of elemental metals vary and this can exist in allotropic forms, for example, iron exhibits a body-centered cubic (BCC) lattice [45,46], nickel has a FCC structure, known as the gamma (γ) phase [47,48], while cobalt can exist in allotropic forms such as hexagonal close-packed (HCP) and FCC [49,50]. This γ phase forms the foundational structure of many superalloys due to its excellent ductility and the presence of various secondary phases that contribute to its strength [51].

Iron exhibits several stable allotropic forms at different temperatures. In addition to its room-temperature BCC form, iron can exist in γ (FCC) and δ (BCC) configurations depending on the temperature [52,53]. Notably, iron has a Curie temperature of 768°C, above which it becomes paramagnetic while still maintaining its BCC structure, referred to as the β phase, which has a structure similar to the α phase with different magnetic properties [51,54], and transforms at atmospheric pressure into an FCC (γ) structure at temperatures above 911°C, remaining stable up to 1392°C (Figure 1) [55,56], and above this temperature, iron reverts to a BCC (δ) structure and remains stable until its melting point at 1538°C [57,58]. Similar behavior is observed for cobalt transforming completely into FCC structures at high temperatures and both iron and cobalt require other elements as alloying agents to maintain an FCC-type structure at room temperature. In the case of iron-based superalloys, the stabilization of the FCC-type structure at room temperature requires the addition of alloying agents such as nickel, which also acts as a reinforcementenhancing the alloy's mechanical properties [59,60].

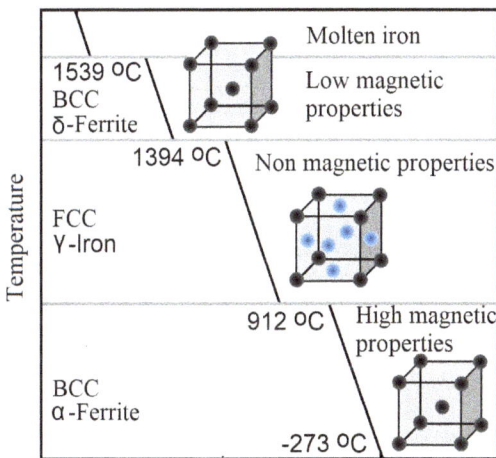

Figure 1. Iron phases at ranged temperature

The composition of superalloys plays a crucial role in determining their properties. Initially, these superalloys were primarily composed of an iron or nickel base with a significant amount of chromium. Over time, the addition of small quantities of other elements led to the formation of new secondary phases, further enhancing their performance. Other refractory alloying metals, including rhenium, tungsten, hafnium, molybdenum, and tantalum, etc., were also introduced (Table 2) [61–63]. Additionally, elements like silicon, boron, and zirconium were incorporated to provide significant additional reinforcement through solid solution effects and carbide formation [64].

In these superalloys, molybdenum is commonly used to enhance solid solution strengthening due to its larger atomic size compared to iron. Tungsten is also employed, albeit to a lesser extent, to achieve similar effects [65–68]. However, the presence of nickel in iron-based superalloys is essential, with its content reaching up to 53%. This high nickel concentration not only stabilizes the FCC(γ phase) but also promotes the formation of ordered precipitates during heat treatments, significantly strengthening the alloy [69]. These compositional factors are responsible for several key properties of superalloys, including:

- Exceptional mechanical strength at elevated temperatures, even near the melting point [11].

- High strength and ductility at cryogenic temperatures [70].

- Outstanding resistance to oxidation [71].

- Excellent resistance to corrosion and erosion under extreme environment of temperature[72].

On the other hand, carbon has always been present as a strengthening element in alloys, forming carbides that improve grain boundaries [73]. However, in the most recent developments of single-crystal alloys, this element may not be necessary [74], evolving with the addition of increasing amounts of various elements, each contributing specific mechanical, chemical, or structural benefits. This has resulted in the creation of highly complex austenitic superalloys containing a wide array of elements. One such element used in iron-based superalloys is phosphorus, which improves creep rupture resistance [75], likely by modifying grain boundaries to enhance performance at elevated temperatures.

The microstructural characteristics of iron-based superalloys exhibit similar properties to austenitic stainless steels [27]. Therefore, iron-based superalloys may have production and processing routes similar to those of austenitic stainless steels. These production techniques contributed to the improved properties of the earliest iron superalloys. Over time, the addition of elements such as aluminum, titanium, and niobium introduced new γ' and γ'' phases, which are known for their creep resistance [76], precipitating these alloys as an additional γ' phase. Unlike nickel-based superalloys, iron-based superalloys require a greater amount of titanium to achieve precipitation strengthening [77]. Additionally, niobiumis incorporated into certain superalloys, like Inconel 783, to form the γ'' phase, which enhances strength below 650°C [78,79]. However, the γ'' phase of the alloy at temperatures above 650 °C is transformed into the δ phase, which deteriorates the mechanical properties [80,81].

In this context, the stability of iron structures depends on temperature and chemical composition. Therefore, the presence of alloying elements and impurities can lead to the formation of various microstructures, each with distinct properties. Some elements will form precipitates and secondary phases, while others dissolve into the iron matrix. Nevertheless, regardless of their composition, these materials can be appropriately modified through heat treatments and thermomechanical processing.

Table 2. Commonly used alloying elements and their effects[5,15,30,82].

Element	Effect	Percentage (w)
Ni	Stabilizes the γ phase (FCC) and forms a hardening precipitate	9-53
Cr	Substantially enhances oxidation and corrosion resistance at elevated temperatures, promotes carbide formation, and improves solution hardening	5-36
Al	Enhances the precipitation hardening, oxidation resistance and corrosion at elevated temperatures, and also acts as γ' phase former	0-6
Ti	Enhances precipitation hardening, increases oxidation and corrosion resistance at elevated temperatures, and also serves as a γ' phase former and carbides	0-3
Co	Affect the amount of precipitate and improve the strength at high temperatures	0-20
Nb	Facilitates carbide formation, improves precipitation hardening and strengthens the alloy through solution hardening.	0-5
Ta	Promotes carbide formation, improves precipitation hardening, acts as a γ' phase former, improves oxidation resistance, and contributes to solution hardening	0-12
Mo	Promotes the formation of carbides, increases strength at high temperatures, and contributes to solution hardening	0-12
W	Promotes carbide formation, improves high temperature strength and contributes to solution hardening	0-12
Si	Enhances oxidation and corrosion resistance at elevated temperatures.	0-18

3. Alloyingphases

The composition of the alloy during precipitation defines the phases that form the microstructure, which subsequently influences the types of grains present in the superalloy [27]. The microstructure, composition, and phase constitution are key determinants of the physical and chemical properties of superalloys [21]. These alloys primarily exhibit an FCC-type matrix or γ phase, which is suitable for solid solution strengthening and also demonstrates ductile properties [83,84]. Additionally, they may exhibit various secondary phases, which serves as a coherent cubic reinforcement hardened through aging in the austenitic γ phase[85,86]. In the case of the γ' phase, it is formed by adding alloying elements like nickel, aluminum, or titanium, with careful control of the aluminum content being crucial, as excess aluminum can stabilize the initial iron phase (BCC), thereby hindering the formation of the desired FCC phase [77,87].

Other phases in iron-based superalloys have been identified, such as γ", which is characterized by a structure of two stacked cubes [88,89]. This phase can be undesirable in some cases, as it decomposes at temperatures above 650 °C into the δ phase, which is thermodynamically stable but leads to a concomitant deterioration in strength [90]. Other important secondary phases that influence the properties of these alloys include eta (η) and delta (δ) phases [76]. To minimize the formation of undesirable phases, rapid solidification is a commonly employed technique in alloy processing.

3.1   Gamma (γ) phase

The γ phase is characterized by an FCC structure and is the main phase present in superalloys, being recognized for its excellent ductility and toughness. Although its compact structure results in lower diffusion rates compared to the BCC phase. Despite this limitation, the γ phase is essential due to its superior properties and the stability it provides to superalloys under high-temperature conditions [87,91,92].

3.2   Gamma prime (γ′) phase

Iron-nickel alloys exhibit an intermetallic γ′ phase, which acts as a critical secondary phase in superalloys, significantly contributing to their high mechanical strength by increasing the precipitated phase. The γ′ phase forms through the addition of adequate amounts of aluminum and titanium, and its formation is favored at temperatures up to 650°C, where it precipitates as $Ni_3(Al, Ti)$ with a FCC structure [93,94]. However, exposure of this phase to temperatures above 650 °C can transform it into a detrimental η-type phase, characterized by a hexagonal close-packed structure (HCP), which reduces the alloy strength [76,95].

Other elements (Nb, Ta, and Cr) can also promote the homogeneous precipitation of the γ′ phase due to their close chemical compatibility with the matrix-precipitate lattice [96,97]. This compatibility results in ductility that is coherent with the matrix, enhancing strength without compromising fracture toughness, improving strength without compromising fracture toughness, and providing stability at high temperatures, up to 1100°C [11,73]. However, the oxidation and corrosion resistance of these brittle phases set limits on their potential applications.

This type of phase typically involves iron-based compositions with a nickel content between 25% and 35%, as seen in alloys like V57 [98]. Additionally, titanium is often incorporated to enhance the formation of the γ′ phase during aging treatments, which improves the lattice parameter of both γ and γ′ phase. This results in a coherent interface, essential for achieving effective strengthening [70,99,100]. However, in alloys with high titanium content, exposure to temperatures above 650°C may lead to the formation of a hexagonal $η-Ni_3Ti$ phase, which negatively impacts ductility and fracture toughness[76]. Nevertheless, this phase can be used to refine the microstructure of the superalloy through a fine-grain forging, leading to a more homogeneous distribution of the hardening phase and improved mechanical properties[94,101].

3.3   Gamma double prime (γ″) phase

Another important secondary phase in iron-nickel superalloys is the γ″ phase, characterized by a chemical composition of Ni_3Nb and a body-centered tetragonal (BCT) structure. This phase serves as a primary strengthening agent, with its formation facilitated by an appropriate concentration of niobium [88], as is the case with Inconel 706 [86] and Alloy 718 [70,102]. The formation of the γ″ phase is also influenced by the concentration of aluminum: lower aluminum concentrations favor the formation of the γ'' phase, while higher concentrations lead to the formation of the γ' phase[103].

The γ'' phase, with its BCT structure, is coherent with the γ phase[104]. However, this phase becomes unstable at temperatures exceeding 650°C, leading to the formation of other phases that can adversely affect the alloy's properties, such as the orthorhombic δ-Ni_3Nb phase, which forms from the BCT γ''-Ni_3Nb. These transformations can reduce the alloy's strength within this temperature range [22].

In summary, the γ' and γ'' phases are predominant in iron-nickel superalloys and act as primary reinforcing precipitates. These phases facilitate the formation of superdislocations, which influence the material behavior. Therefore, key secondary phases to control superalloy properties include carbides, and other undesirable phases such as η (eta) and δ (delta).

3.4   Carbides (M_xC_y)

Metal carbides (MC) are formed by adding carbon to elements such as iron, nickel, titanium, tantalum, hafnium, and niobium, with carbon concentrations up to 0.35% [22,98,105], providing a good oxidation resistance. In general, alloys reinforced by carbides, nitrides, and carbonitrides have good stability at temperatures up to 800°C due to the higher thermal stability of the secondary phases [106]. However, these alloys tend to have lower tensile strength, as their strengthening mechanism is less effective compared to that of precipitates, which have more ordered structures. Additionally, during heat treatment, carbides may decompose into lower carbides ($M_{23}C_6$ and M_6C), which predominantly form at grain boundaries, thereby enhancing creep resistance at elevated temperatures [107]. Therefore, carbides are one of the three primary phases in superalloys, alongside the γ phase and the secondary γ' and γ'' phases, which are crucial for strengthening. While carbides may offer partial strengthening, either directly or indirectly, their primary role is in improving hardening resistance and stabilizing grain boundaries against excessive shear [108,109].

4.   Processing of superalloys

Iron-based superalloys were initially cold forged, but constant progress in alloy development significantly increased the processing temperature of the materials, where the introduction of vacuum melting techniques has allowed precise control of the chemical composition of superalloys, minimizing contamination and improving their properties [11]. This progress has facilitated the development of various processing techniques that have driven technological innovations in the field. In this context, the processing method will be determined based on the final required properties of the material, making them suitable for specific applications. Some of the methods employed techniques include casting and forging, directional solidification, powder metallurgy, among others, often followed by heat treatments and coatings. These processes enable the material to meet the rigorous demands of industries that rely on advanced technologies [40,110,111].

4.1   Casting and forging of superalloys

These processing methods are capable of producing both polycrystalline and monocrystalline superalloys, where the polycrystalline superalloys offer greater fracture resistance, while single-crystal superalloys are notable for their superior creep resistance [112–114].

4.2   Investment casting of superalloys

This processing method, also known as lost-wax casting, involves encasing a wax pattern in a ceramic mold, where molten metal is subsequently poured into the mould, solidifying into the shape of the original wax pattern. This method is ideal for manufacturing complex components, such as turbine blades. The resulting superalloy is polycrystalline, with crystal grains nucleating and growing in multiple directions within the matrix, lacking a specific grain orientation[11,77].

4.3   Directional solidification of superalloys

Directional solidification utilizes a thermal gradient to promote grain nucleation and growth at a lower temperature surface. This process results in elongated grains oriented along the thermal gradient, which significantly improve creep resistance when stressed in the grain growth direction[110,115,116].

4.4   Single-crystal superalloys

This method is a process called crystal seeding, where a seed crystal acts as a template to guide the growth of larger crystals. This technique is time-intensive and typically requires further machining to refine the final product [117,118].

4.5   Powder metallurgy of superalloys

This is a contemporary processing method in which metals are pulverized into fine powders and then shaped by a heating process that reaches a temperature close to the melting point, as opposed to traditional casting methods where metals are melted. This method is widely used in the production of superalloys due to its efficiency, which reduces the need for extensive machining. In addition, powder metallurgy allows for mechanical alloying, where reinforcing particles can be incorporated into the superalloy through fracturing and welding cycles [111,119,120].

4.6   Sintering and hot isostatic pressing of superalloys

These techniques are used to form porous materials from metallic grains, which are then subsequently densified. Sintering involves bonding adjacent particles on the surface at temperatures below their melting point, creating a strong, weld-like connection. In contrast, hot isostatic pressing involves subjecting the sintered material to high isostatic pressure at an elevated temperature in all directions, resulting in a material with significantly higher density [121,122].

4.7   Additive manufacturing of superalloys

Selective laser melting (SLM) is an additive manufacturing process that is based on creating complex shapes from a 3D file. This process employs the use of a laser capable of sintering layers of metal powder, creating the material layer by layer in a similar way to 3D printing. Although it is effective in generating various highly complex materials, this process can generate a porous structure, often requiring additional processes to improve the density of the material and minimize porosity [123–125].

5.   Thermaltreatments

Heat treatments are crucial to modify the hardness and residual stress properties of metals, particularly after the deformation processing of superalloys. Common heat treatments include annealing, quenching, solution treatment, precipitation hardening, and full annealing. These processes are essential for enhancing strength, hardness, creep resistance, and ductility in the final material [11,110,126].

5.1   Annealing

Annealing is a treatment applied to alter the physical properties of metals by relieving stresses induced during shaping or high-temperature processes such as casting, rolling, forging, or cold forming [127,128]. One of its primary objectives is to eliminate internal stresses from solidification and deformation, particularly those associated with forging. This treatment reduces hardness and refines the grain structure, which improves ductility and enhances microstructural stability through complete recrystallization. The process also homogenizes coarse and deformed grains and is dependent on both temperature and duration [92,129,130].

The annealing temperature is critical in determining the final properties, where low temperatures maintain a fine grain size but may not achieve full recrystallization or some dissolve undesirable phases, such as η and δ. Meanwhile, elevated annealing temperature allows for complete recrystallization and removal of unwanted phases, resulting in significant grain growth. Therefore, the optimal annealing temperature is chosen based on the desired balance of properties, such as toughness, tensile rupture resistance, or creep resistance [107,131,132].

5.2   Solution heat-treatment

This treatment is an initial process aimed at softening the material and uniformly distributing internal stresses, resulting in a homogeneous structure, and this consists of heating the alloy to an elevated temperature to dissolve the elements into a solid solution, followed by rapid cooling (quenching) that traps the elements in place. This heat treatment creates optimal conditions for subsequent thermal processes, particularly in materials that have undergone plastic deformation, where structural changes can enhance creep resistance in deformed areas. Thus, solution heat treatment can eliminate localized stresses, creating a more uniform creep resistance throughout the structure [133–135].

This process consists of heating the superalloy to an elevated temperature that allows obtaining a monophase region, followed by rapid cooling achieving the most homogeneous solid solution possible, which can dissolve the secondary phases, providing additional solute for precipitation hardening. This process is similar to tempering, which is why it is also known as solution hardening, and is especially suitable for alloys that present significant changes in solubility with temperature. Consequently, the objective of the treatment is to achieve a more uniform alloy, eliminating the residual stresses generated during forming or welding, resulting in a relatively soft and ductile material[136,137].

5.3   Precipitation heat-treatment

Precipitation hardening is a heat treatment used to increase creep resistance by forming anomalies in the material's microstructure, providing excellent high-temperature strength [138,139]. This technique leverages the temperature-dependent solubility of solids to form uniformly distributed particles of various sizes, which prevent the formation of defects within the crystal lattice, resulting in hardening of the material. The principle of precipitation strengthening involves allowing reinforcement elements to precipitate in a phase that restricts the crystalline displacement of the superalloy. The precipitation phases are divided into γ' and γ'', which act as reinforcements at different temperatures. Unlike conventional quenching, this process requires maintaining the alloy at elevated temperatures for extended periods to allow for precipitation, a stage referred to as aging [140,141].

In contrast to solid solution strengthening, which relies on rapid cooling to form a single-phase solid solution, precipitation heat treatment incorporates particles often as impurities, to further increase material strength. This results in a higher yield strength for many ductile materials, achieving a desirable combination of strength and hardness tailored for specific applications.

6.   Applications of superalloys

Superalloys are extensively utilized in high-performance applications due to their remarkable properties, particularly their structural stability and corrosion resistance at temperatures near their melting point. These exceptional characteristics make superalloys essential for applications in extreme high-temperature environments [11,70].

Medical Sector: Superalloys are primarily used in systems that require high performance or specific applications, such as in medical devices, including hospital equipment, implants, and surgical instruments. In addition to their aforementioned corrosion resistance and excellent surface finish, superalloys also offer hygienic properties that significantly enhance safety in hospital applications [133,142].

Automotive Sector: The automotive industry uses superalloys in the manufacture of turbochargers and similar engine components as well as structural components, which require high structural stability and resistance to the extreme conditions to which they are subjected[111,143].

Power Generation Sector: This sector uses superalloys in heat exchanger systems, boilers, furnaces and steam and gas turbines due to their capacity to operate at high temperatures without presenting plastic deformations or fractures [103,144].

Oil and Gas Sector: In this industry, superalloys are primarily used to manufacture casings and drilling mandrels, where their excellent performance under high pressure and severe corrosion conditions is vital, particularly in deep-sea environments [145,146].

Chemical Sector: Superalloys are used in critical machine components for chemical processing, such as valves and instrumentation systems, where their resistance to corrosion and harsh environments is essential [11,70,103].

Electronics Sector: Superalloys find application in optical and electrical supports for telescopes and lasers due to dimensional stability, minimizing errors from thermal expansion. They are also utilized in components of laptops, phones, and other electronic devices as energy storage, where durability and performance are required [147,148].

Military Sector: In the military sector, superalloys are highly valued in the manufacture of gas turbine engines, aircraft, and naval vessels. Additionally, superalloys offer the advantages of

Superalloys: Fundamentals and Applications
Materials Research Foundations 178 (2025) 1-24

Materials Research Forum LLC
https://doi.org/10.21741/9781644903698-1

being lightweight and resistant to environmental degradation two critical properties for high-performance military equipment [149].

Aerospace Sector: The aerospace industry is one of the largest consumers of superalloys, particularly in the development of jet engines. Their favorable strength-to-lightweight ratio also boosts engine thrust performance, making them ideal for use in turbine components, such as turbine blades, which must endure extreme heat and significant centripetal forces [150–152].

Conclusion

In summary, iron-based superalloys are characterized by a chemical composition that allows them to have different properties. Nickel is the main component of these alloys after iron, which helps to stabilize the austenitic structure. In addition, the desired properties for a given application are achieved by subjecting the alloys to heat treatments. The excellent properties of hardness, creep resistance, and corrosion resistance that distinguish these materials, allow their implementation in various everyday sectors as well as in cutting-edge technologies.

Acknowledgment

The authors would like to thank the following institutions for their support: the University of Guadalajara for PROSNI 2022 (E.M.), the Dirección General de Asuntos del Personal Académico at the National Autonomous University of Mexico for Grant IN204223 (E.B.), and CONAHCyT for the postdoctoral fellowship (CVU 407270) awarded to G.G.F.R.

References

[1] E.O. Ezugwu, Key improvements in the machining of difficult-to-cut aerospace superalloys, Int. J. Mach. Tools Manuf. 45 (2005) 1353-1367. https://doi.org/10.1016/j.ijmachtools.2005.02.003

[2] E. Akca, A. Gürsel, A review on superalloys and IN718 nickel-based INCONEL superalloy, PEN. 3 (2015). https://doi.org/10.21533/pen.v3i1.43

[3] G. Prashar, K. Thakur,S. Singh, P. Singh, V.K. Srivastava, Superalloys for high-temperature applications: an overview, 14th Int. Conf. Mater. Process. Charact, Hyderabad, India, (2024) pp. 020022. https://doi.org/10.1063/5.0192482

[4] D. Chellaganesh, M.A. Khan, J.T.W. Jappes, Hot corrosion behaviour of nickel-iron-based superalloy in gas turbine application, Int. J. Ambient Energy. 41 (2020) 901-905. https://doi.org/10.1080/01430750.2018.1492446

[5] M. Cemal, S. Cevik, Y. Uzunonat, F. Diltemiz, ALLVAC 718 PlusTM superalloy for aircraft engine applications, in: R. Agarwal (Eds.), Recent Advances in Aircraft Technology; InTech, 2012, ISBN 978-953-51-0150-5. https://doi.org/10.5772/38433

[6] P. Sharma, D. Chakradhar, S. Narendranath. Analysis and optimization of WEDM performance characteristics of INCONEL 706 for aerospace application, Silicon. 10 (2018), 921-930doi. https://doi.org/10.1007/s12633-017-9549-6

[7] F. Garay, G. Suárez, K. De La Hoz, E. Urraya, P. Loaiza, L.J. Cruz, A. Ángel, M.D. Ortells, Computational analysis of the IN718 material microstructure (nickel-based INCONEL

superalloy), in: J.C. Figueroa-García, F.S. Garay-Rairán, G.J. Hernández-Pérez, Y. Díaz-Gutierrez (Eds.), Potential Applications in the Military Industry, Applied Computer Sciences in Engineering; Communications in Computer and Information Science; Springer International Publishing: Cham, 2020, pp. 350-359. ISBN 978-3-030-61833-9. https://doi.org/10.1007/978-3-030-61834-6_30

[8] N.M. Dawood, A.M. Salim, A review on characterization, classifications, and applications of super alloys, J. Uni. Babylon. Eng. Sci. 29 (2021).

[9] C.J. Tsai, T.K. Yeh, M.Y. Wang, High temperature oxidation behavior of nickel and iron based superalloys in helium containing trace impurities, Corros. Sci. Tech. 18 (2018), 8-15.

[10] M. Durand-Charre, Heat resisting steels and iron-containing superalloys in: Microstructure of Steels and Cast Irons; Engineering Materials and Processes, Springer Berlin Heidelberg, Berlin, Heidelberg, 2004, pp. 331-346. ISBN 978-3-642-05897-4. https://doi.org/10.1007/978-3-662-08729-9_20

[11] I.G. Akande, O.O. Oluwole, O.S.I. Fayomi, O.A. Odunlami, Overview of mechanical, microstructural, oxidation properties and high-temperature applications of superalloys, Mater.Today Proc. 43 (2021), 2222-2231. https://doi.org/10.1016/j.matpr.2020.12.523

[12] Y. Xu, J. Lu, W. Li, Z. Yang, Y. Gu, Chromium-dependent effect on oxidation behavior of Ni-Fe-based superalloy for ultra-supercritical steam turbine applications: influence of temperature and pure steam, Corrosion. 76 (2020) 941-953. https://doi.org/10.5006/3534

[13] K. Shinozaki, Welding and joining Fe and Ni-base superalloys, Weld. Int. 15 (2001) 593-610. https://doi.org/10.1080/09507110109549411

[14] C.T.A. Sims, Contemporary view of nickel-base superalloys, JOM. 18 (1966), 1119-1130. https://doi.org/10.1007/BF03378505

[15] G.D. Smith, S.J. Patel, The role of niobium in wrought precipitation-hardened nickel-base alloys, in: Proceedings of the Superalloys 718, 625, 706 and Various Derivative, TMS, 2005, pp. 135-154. https://doi.org/10.7449/2005/Superalloys_2005_135_154

[16] N.S. Stoloff, Wrought and P/M superalloys, in: Properties and Selection: Irons, Steels, and High-Performance Alloys, ASM Handbook Committee, ASM International, 1990, pp. 950-980. ISBN 978-1-62708-161-0. https://doi.org/10.31399/asm.hb.v01.a0001049

[17] K. Ma, T. Blackburn, J.P. Magnussen, M. Kerbstadt, P.A. Ferreirós, T. Pinomaa, C. Hofer, D.G. Hopkinson, S.J. Day, P.A.J. Bagot, et al. Chromium-based bcc-superalloys strengthened by iron supplements, Acta. Mater. 257 (2023) 119183. https://doi.org/10.1016/j.actamat.2023.119183

[18] Z. Li, Y. Zhang, K. Dong, Z. Zhang, Research progress of Fe-based superelastic alloys, Crystals. 12 (2022) 602. https://doi.org/10.3390/cryst12050602

[19] A.F. Padilha, P.R. Rios, Decomposition of austenite in austenitic stainless steels, ISIJ Int. 42 (2002) 325-327. https://doi.org/10.2355/isijinternational.42.325

[20] E.P. George, R.L. Kennedy, D.P. Pope, Review of trace element effects on high temperature fracture of Fe- and Ni-base alloys, Phys. Stat. Sol. A. 167 (1988) 313-333. https://doi.org/10.1002/(SICI)1521-396X(199806)167:2<313::AID-PSSA313>3.0.CO;2-5

[21] M. Durand-Charre, The Microstructure of Superalloys, M. Durand-Charre, (Ed.), First (ed.), Routledge, 2017, ISBN 978-0-203-73638-8. https://doi.org/10.1201/9780203736388

[22] H.A. Kishawy, A. Hosseini, Superalloys, in: Machining Difficult-to-Cut Materials, Materials Forming, Machining and Tribology; Springer International Publishing: Cham, 2019, pp. 97-137 ISBN 978-3-319-95965-8. https://doi.org/10.1007/978-3-319-95966-5_4

[23] V.F. Zackay, E.R. Parker, J.W. Morris, G. Thomas, The application of materials science to the design of engineering alloys, A review, Mater. Sci. Eng. 16 (1974) 201-221. https://doi.org/10.1016/0025-5416(74)90158-X

[24] C. Cabet, F. Dalle, E. Gaganidze, J. Henry, H. Tanigawa, Ferritic-martensitic steels for fission and fusion applications, J. Nucl. Mater. 523 (2019) 510-537. https://doi.org/10.1016/j.jnucmat.2019.05.058

[25] A.K. Maurya, C. Pandey, R. Chhibber, Dissimilar welding of duplex stainless steel with Ni alloys: A review, Int. J. Press. Vessels Pip. 192 (2021) 104439. https://doi.org/10.1016/j.ijpvp.2021.104439

[26] Y.H. Mozumder, K.A. Babu, R. Saha, V.S. Sarma, S. Mandal, Dynamic microstructural evolution and recrystallization mechanism during hot deformation of intermetallic hardened duplex lightweight steel, Mater. Sci. Eng. A. 788 (2020) 139613. https://doi.org/10.1016/j.msea.2020.139613

[27] K.J. Ducki, Analysis of the precipitation and growth processes of the intermetallic phases in an Fe-Ni superalloy, in: Superalloys, M. Aliofkhazraei, (Ed.), InTech, 2015 ISBN 978953-51-2212-8. https://doi.org/10.5772/61159

[28] C.I. Garcia, K. Cho, K. Redkin, A.J. Deardo, S. Tan, M. Somani, L.P. Karjalainen, Influence of critical carbide dissolution temperature during intercritical annealing on hardenability of austenite and mechanical properties of DP-980 steels, ISIJ Int. 51 (2011) 969-974. https://doi.org/10.2355/isijinternational.51.969

[29] J.H. DeVan, P.F. Tortorelli, The oxidation-sulfidation behavior of iron alloys containing 16 40 at% aluminum, Corros. Sci. 35 (1993) 1065-1071. https://doi.org/10.1016/0010-938X(93)90325-B

[30] A. Wiengmoon, J.T.H. Pearce, T. Chairuangsri, Relationship between microstructure, hardness and corrosion resistance in 20wt.%cr, 27wt.%cr and 36wt.%cr high chromium cast irons, Mater. Chem. Phys. 125 (2011) 739-748. https://doi.org/10.1016/j.matchemphys.2010.09.064

[31] Zumelzu, E.; Goyos, I.; Cabezas, C.; Opitz, O.; Parada, A. Wear and Corrosion Behaviour of High-Chromium (14-30% Cr) Cast Iron Alloys, J. Mater. Process. Technol. 128 (2002) 250-255. https://doi.org/10.1016/S0924-0136(02)00458-2

[32] M.F. Pillis, L.V. Ramanathan, Effect of alloying additions and peroxidation on high temperature sulphidation resistance of iron-chromium alloys, Surf. Eng. 22 (2006), 129 137. https://doi.org/10.1179/174329406X98412

[33] M.F. Pillis, L.V. Ramanathan, Effect of pre-oxidation on high temperature sulfidation behavior of fecr and fecral alloys, Mat. Res. 7 (2004) 97-102. https://doi.org/10.1590/S1516-14392004000100014

Superalloys: Fundamentals and Applications   Materials Research Forum LLC
Materials Research Foundations 178 (2025) 1-24   https://doi.org/10.21741/9781644903698-1

[34] M. Danielewski, K. Natesan, oxidation-sulfidation behavior of iron-chromium-nickel alloys, Oxid. Met. 122(1978) 27-245. https://doi.org/10.1007/BF00616098

[35] C. Wang, Y. Guo, J. Guo, L. Zhou, Microstructural stability and mechanical properties of a boron modified Ni-Fe based superalloy for steam boiler applications, Mater. Sci. Eng. A. 639 (2015) 380-388. https://doi.org/10.1016/j.msea.2015.05.026

[36] X. Xiao, H. Zhao, C. Wang, Y. Guo, J. Guo, L. Zhou, Effects of b and p on microstructure and mechanical properties of GH984 alloy, Acta. Metall. Sin. 49, (2013) 421. https://doi.org/10.3724/SP.J.1037.2013.00002

[37] Liu, P.; Zhang, R.; Yuan, Y.; Cui, C.; Liang, F.; Liu, X.; Gu, Y.; Zhou, Y.; Sun, X. Effects of Nitrogen Content on Microstructures and Tensile Properties of a New Ni-Fe Based Wrought Superalloy, Mater. Sci. Eng. A. 801 (2021) 140436.. https://doi.org/10.1016/j.msea.2020.140436

[38] A. Mitchell, Nitrogen in superalloys, High Temp. Mater. Process. 24 (2005) 101-110. https://doi.org/10.1515/HTMP.2005.24.2.101

[39] U. Ali, M.S. Qurashi, P.O. Lartey, I. Ali, P.K. Liaw, J.W. Qiao, Carbon and titanium effect on tensile behavior of aged a286 nickel-iron based superalloy, KEM. 963 (2023) 3-9. https://doi.org/10.4028/p-i2bzUj

[40] R. Frisk, N.A.I. Andersson, B. Rogberg, Cast structure in alloy a286, an iron-nickel based superalloy, Metals. 9 (2019) 711. https://doi.org/10.3390/met9060711

[41] J.D. WhITTENBERGER, A Review of: "Superalloys II", CT. Sims, N.S. Stoloff, W.C. Hagel (Eds.), A Wiley-Interscience Publication, John Wiley & Sons, New York, Hardcover, 1987, 615.

[42] Properties and Selection: Irons, Steels, and High-Performance Alloys, ASM Handbook Committee, ASM International, 1990, ISBN 978-1-62708-161-0.

[43] M.J. Donachie, S.J. Donachie, Superalloys: A technical guide, Second ed., ASM International, 2002, ISBN 978-1-62708-267-9. https://doi.org/10.31399/asm.tb.stg2.9781627082679

[44] B. Geddes, H. Leon, X. Huang, Superalloys: Alloying and performance; ASM International, 2010, ISBN 978-1-62708-313-3. https://doi.org/10.31399/asm.tb.sap.9781627083133

[45] W.A. Bassett, E. Huang, Mechanism of the body-centered cubic-hexagonal close packed phase transition in iron, Science. 238 (1987) 780-783. https://doi.org/10.1126/science.238.4828.780

[46] W.A. Bassett, M.S. Weathers, Stability of the body-centered cubic phase of iron: A thermodynamic analysis, J. Geophys. Res. 95 (1990) 21709-21711. https://doi.org/10.1029/JB095iB13p21709

[47] J.B. Singh, Phases in Alloy 625. in: Alloy 625, Materials Horizons: From Nature to Nanomaterials, Springer Nature Singapore, Singapore, 2022, pp. 29-65 ISBN 978-981 19156-1-1. https://doi.org/10.1007/978-981-19-1562-8_2

[48] P. Li, P. Zhang, F. Li, W. Jiang, Z. Cao, Pure face-centered-cubic (FCC) and hexagonal close-packed (hcp) nickel phases obtained in air atmosphere sol-gel process and fcc nickel

phase obtained in N2 protected sol-gel process, J. Sol-Gel. Sci. Technol. 68 (2013) 261 269.
https://doi.org/10.1007/s10971-013-3162-y

[49] D. Mukherji, P. Strunz, S. Piegert, R. Gilles, M. Hofmann, M. Hölzel, J. Rösler, The
hexagonal close-packed (HCP) ⇆ face-centered cubic (FCC) transition in Co-Re-based
experimental alloys investigated by neutron scattering, Metall. Mater. Trans. A. 2012, 43
(2012) 1834-1844. https://doi.org/10.1007/s11661-011-1058-4

[50] J.Y. Huang, Y.K. Wu, H.Q. Ye, K. Lu, Allotropic transformation of cobalt induced by ball
milling, Nanostructured Materials. 6 (1995) 723-7263. https://doi.org/10.1016/0965-
9773(95)00160-3

[51] T.M. Smith, B.D. Esser, N. Antolin, A. Carlsson, R.E.A. Williams, A. Wessman, T. Hanlon,
H.L. Fraser, W. Windl, D.W. McComb, et al. Phase Transformation strengthening of high-
temperature superalloys, Nat. Commun. 7 (2016) 13434.
https://doi.org/10.1038/ncomms13434

[52] T. Lee, M.I. Baskes, S.M. Valone, J.D. Doll, Atomistic modeling of thermodynamic
equilibrium and polymorphism of iron, J. Phys. Condens. Matter. 24 (2012) 225404.
https://doi.org/10.1088/0953-8984/24/22/225404

[53] D.E. Laughlin, The β iron controversy revisited, J. Phase Equilib. Diffus. 39 (2018) 274
279. https://doi.org/10.1007/s11669-018-0638-z

[54] H. Stuart, N. Ridley, Lattice parameters and curie-point anomalies of iron-cobalt alloys, J.
Phys. D: Appl. Phys. 2 (1969) 485-491. https://doi.org/10.1088/0022-3727/2/4/302

[55] M. Abuin, Z. Turgut, N. Aronhime, V. Keylin, A. Leary, V. DeGeorge, J. Horwath, S.L.
Semiatin, D.E. Laughlin, M.E. McHenry, Determination of pressure effects on the α → γ
phase transition and size of Fe in Nd-Fe-B spring exchange magnets, Metall. Mater. Trans. A.
46 (2015) 5002-5010. https://doi.org/10.1007/s11661-015-3120-0

[56] D. Rafaja, C. Ullrich, M. Motylenko, S. Martin, Microstructure aspects of the deformation
mechanisms in metastable austenitic steels. in: Austenitic TRIP/TWIP Steels and Steel-
Zirconia Composites, H. Biermann, C.G. Aneziris, (Eds.), Springer Series in Materials
Science, Springer International Publishing: Cham, 2020, pp. 325-377, ISBN 978-3-030-
42602-6. https://doi.org/10.1007/978-3-030-42603-3_11

[57] D. Golberg, M. Mitome, Ch. Müller, C. Tang, A. Leonhardt, Y. Bando, Atomic structures of
iron-based single-crystalline nanowires crystallized inside multi-walled carbon nanotubes as
revealed by analytical electron microscopy, Acta. Mater. 54 (2006) 2567 2576.
https://doi.org/10.1016/j.actamat.2006.01.040

[58] R. Schmid-Fetzer, Phase diagrams: The beginning of wisdom, J. Phase Equilib. Diffus. 35
(2014) 735-760. https://doi.org/10.1007/s11669-014-0343-5

[59] S.C. Deevi, V.K. Sikka, Nickel and iron aluminides: An overview on properties, processing,
and applications, Intermetallics. 4 (1996) 357-375. https://doi.org/10.1016/0966-
9795(95)00056-9

[60] C. Stallybrass, A. Schneider, G. Sauthoff, The strengthening effect of (Ni,Fe)Al precipitates
on the mechanical properties at high temperatures of ferritic Fe-Al-Ni-Cr alloys,
Intermetallics. 13 (2005) 1263-1268. https://doi.org/10.1016/j.intermet.2004.07.048

[61] W. Knabl, G. Leichtfried, R. Stickler, Refractory metals and refractory metal alloys, in: Springer Handbook of Materials Data, H. Warlimont, W. Martienssen (Eds.), Springer Handbooks, Springer International Publishing: Cham, 2018, pp. 307-337, ISBN 978-3-319-69741-3. https://doi.org/10.1007/978-3-319-69743-7_13

[62] T.G. Nieh, J. Wadsworth, Recent advances and developments in refractory alloys, MRS Proc. 322 (1993) 315. https://doi.org/10.1557/PROC-322-315

[63] X.J. Hua, P. Hu, H.R. Xing, J.Y. Han, S.W. Ge, S.L. Li, C.J. He, K.S. Wang, C.J. Cui, Development and property tuning of refractory high-entropy alloys: A review, Acta. Metall. Sin. (Engl. Lett.) 35 (2022) 1231-1265. https://doi.org/10.1007/s40195-022-01382-x

[64] N. Liu, X. Cao, T. Zhao, Z.W. Zhang, Progress of zirconium alloying in iron-based alloys and steels, Mater. Sci. Technol. 37 (2021) 830-851. https://doi.org/10.1080/02670836.2021.1958488

[65] W.D. Klopp, A review of chromium, molybdenum, and tungsten alloys, J. Less Common Met. 42 (1975) 261-278. https://doi.org/10.1016/0022-5088(75)90046-6

[66] Y.J. Hu, M.R. Fellinger, B.G. Butler, Y. Wang, K.A. Darling, L.J. Kecskes, D.R. Trinkle, Z.K. Liu, Solute-induced solid-solution softening and hardening in bcc tungsten, Acta. Mater. 141(2017) 304-316. https://doi.org/10.1016/j.actamat.2017.09.019

[67] A. Roy, P. Sreeramagiri, T. Babuska, B. Krick, P.K. Ray, G. Balasubramanian, Lattice distortion as an estimator of solid solution strengthening in high-entropy alloys, Mater. Charact. 172 (2021)1108777. https://doi.org/10.1016/j.matchar.2021.110877

[68] Y. Şahin, Recent progress in processing of tungsten heavy alloys, J. Powder. Technol. 2014 (2014) 1-22. https://doi.org/10.1155/2014/764306

[69] Z. Wu, H. Bei, Microstructures and mechanical properties of compositionally complex Co-free FeNiMnCr18 FCC solid solution alloy, Mater. Sci. Eng. A. 640 (2015) 217-224. https://doi.org/10.1016/j.msea.2015.05.097

[70] L.A. Lee, Hydrogen embrittlement of nickel, cobalt and iron-based superalloys. in: R.P Gangloff, B.P. Somerday (Eds.), Gaseous Hydrogen Embrittlement of Materials in Energy Technologies, Elsevier, 2012, pp. 624-667, ISBN 978-1-84569-677-1. https://doi.org/10.1533/9780857093899.3.624

[71] N.B. Maledi, J.H. Potgieter, M. Stephton, L.A. Cornish, L. Chown, R. Süss, R. Hot corrosion behaviour of pt-alloys for application in the next generation of gas turbines, International Platinum Conference 'Platinum Surges Ahead, 2006.

[72] T.S. Sidhu, S. Prakash, R.D. Agrawal, Hot corrosion behaviour of HVOF-sprayed NiCrBSi coatings on Ni- and Fe-based superalloys in Na2SO4-60% V2O5 environment at 900°C, Acta. Mater. 54 (2006) 773-784. https://doi.org/10.1016/j.actamat.2005.10.009

[73] J. Yan, Y. Gu, H. Li, F. Yang, Y. Yuan, P. Zhang, X. Zhao, Y. Dang, Z. Yang, Impact of aging temperature on the performance of a nickel-iron-based superalloy, Metall. Mater. Trans. A. 49 (2018) 1561-1570. https://doi.org/10.1007/s11661-018-4514-6

[74] M. Vollmer, T. Arold, M.J. Kriegel, V. Klemm, S. Degener, J. Freudenberger, T. Niendorf, Promoting abnormal grain growth in fe-based shape memory alloys through compositional adjustments, Nat. Commun. 10 (2019) 2337. https://doi.org/10.1038/s41467-019-10308-8

[75] S. Zhang, A. Zhang, C. Xue, D. Jia, W. Zhang, W. Wang, X. Xin, W. Sun, Influence of phosphorus addition on the stress rupture properties of direct aged IN706 superalloy, Crystals. 10 (2020) 641. https://doi.org/10.3390/cryst10080641

[76] M. Seifollahi, S. Kheirandish, S.H. Razavi, S.M. Abbasi, P. Sahrapour, Effect of ^|^eta; Phase on mechanical properties of the iron-based superalloy using shear punch testing. ISIJ Int. 53 (2013) 311-316. https://doi.org/10.2355/isijinternational.53.311

[77] M. Palm, F. Stein, Iron-based intermetallics, in: High-Performance Ferrous Alloys, R. Rana (Ed.), Springer International Publishing: Cham, 2021, pp. 423-458, ISBN 978-3030-53824-8. https://doi.org/10.1007/978-3-030-53825-5_10

[78] Q. Zhao, J. Wang, J. Liu, Y. Wu, X. Du, Life extension heat treatment of IN 783 bolts. Mater. Test. 62 (2020) 49-54. https://doi.org/10.3139/120.111452

[79] K.A. Heck, J.S. Smith, R. Smith, INCONEL® alloy 783: An oxidation-resistant, low expansion superalloy for gas turbine applications, J. Eng. Gas Turbines. Power. 120 (1998) 363-369. https://doi.org/10.1115/1.2818131

[80] K. Hou, M. Ou, M. Wang, H. Li, Y. Ma, K. Liu, K. Precipitation of η phase and its effects on stress rupture properties of K4750 alloy, Mater. Sci. Eng. A. 763, (2019) 138137. https://doi.org/10.1016/j.msea.2019.138137

[81] M. Anderson, A.L. Thielin, F. Bridier, P. Bocher, J. Savoie, δ phase precipitation in Inconel 718 and associated mechanical properties, Mater. Sci. Eng. A. 679 (2017) 48-55. https://doi.org/10.1016/j.msea.2016.09.114

[82] S. Kamal, R. Jayaganthan, S. Prakash, High temperature cyclic oxidation and hot corrosion behaviours of superalloys at 900°C, Bull. Mater. Sci. 33 (2010) 299-306. https://doi.org/10.1007/s12034-010-0046-4

[83] J. Huang, J. Yan, J. Lei, D. Yi, X. Zhu, J. Zhou, Y. He, P. Li, L. Huang, D. Liu, Tensile properties and deformation characteristics of nickel-iron-based superalloys for steam turbine rotors, Emergent Mater. 7 (2024) 3011-3028. https://doi.org/10.1007/s42247-024-00758-2

[84] S.C. Krishna, N.K. Gangwar, A.K. Jha, B. Pant, P.V. Venkitakrishnan, On the direct aging of iron based superalloy hot rolled plates, Mater. Sci. Eng. A. 648 (2015) 274-279. https://doi.org/10.1016/j.msea.2015.09.073

[85] O. Sifi, M.E.A. Djeghlal, Y. Mebdoua, S. Djeraf, F. Hadj-Larbi, The effect of the solution and aging treatments on the microstructures and microhardness of nickel-based superalloy, Appl. Phys. A. 126 (2020) 345. https://doi.org/10.1007/s00339-020-03517-2

[86] S. Zhang, L. Zeng, D. Zhao, T. Si, Y. Xu, Y. Wu, Z. Guo, Z. Fu, W. Wang, W. Sun, Comparison study of microstructure and mechanical properties of standard and direct aging heat treated superalloy Inconel 706, Mater. Sci. Eng. A. 839 (2022) 142836. https://doi.org/10.1016/j.msea.2022.142836

[87] E.A. Basuki, D.H. Prajitno, F. Muhammad, Alloys developed for high temperature applications, Proc. Int. Process. Metall. Conf. (2017). https://doi.org/10.1063/1.4974409

[88] J. Wang, X. Qin, S. Cheng, X. Guan, Y. Wu, L. Zhou, The microstructure and mechanical performance optimization of a new fe-ni-based superalloy for gen iv nuclear reactor: The critical role of nb alloying strategy, Mater. Charact. 205 (2023) 113240. https://doi.org/10.1016/j.matchar.2023.113240

[89] R. Unnikrishnan, A. Carruthers, S. Cao, S.R. Rogers, T.W. Kwok, R. Thomas, D. Dye, D. Bowden, J.A. Francis, M. Preuss, et al. Development of novel carbon-free cobalt free iron-based hardfacing alloys with a hard π-ferrosilicide phase, Materialia. 35 (2024) 102107. https://doi.org/10.1016/j.mtla.2024.102107

[90] J. Xu, Alloy Design and Characterization of Γ′ Strengthened Nickel-Based Superalloys for Additive Manufacturing, Linköping Studies in Science and Technology, Licentiate Thesis, Linköping University Electronic Press: Linköping, 2021, ISBN 978 917929-726-8. https://doi.org/10.3384/lic.diva-173042

[91] A. Domashenkov, A. Plotnikova, I. Movchan, P. Bertrand, N. Peillon, B. Desplanques, S. Saunier, C. Desrayaud, Microstructure and physical properties of a Ni/Fe-based superalloy processed by selective laser melting, Addit. Manuf. 15 (2017) 66-77. https://doi.org/10.1016/j.addma.2017.03.008

[92] Y. Xu, W. Li, X. Yang, Y. Gu, Y. evolution of grain structure, γ′ precipitate and hardness in friction welding and post weld heat treatment of a new Ni-Fe based superalloy, Mater. Sci. Eng. A. 788 (2020) 139596. https://doi.org/10.1016/j.msea.2020.139596

[93] D.V.V. Satyanarayana, G. Malakondaiah, Effect of prior oxidation on the creep behaviour of nickel and iron based superalloys, Trans. Indian. Inst. Met. 62 (2009)223 228. https://doi.org/10.1007/s12666-009-0034-9

[94] S. Qurashi, Y. Zhao, C. Dong, L. Wang, Y. Li, Transformation of η-phase into γ′-phase, γ-matrix and its effect on the high temperature hardness and compression of Fe-Ni based A286 heat-resistant superalloy, SSRN Electron. J. 2021. https://doi.org/10.2139/ssrn.3995181

[95] S. Cheng, J. Wang, Y. Wu, X. Qin, L. Zhou, Microstructure, thermal stability and tensile properties of a Ni-Fe-Cr based superalloy with different Fe contents, Intermetallics. 2023, 153 (2023) 1077855. https://doi.org/10.1016/j.intermet.2022.107785

[96] K. Du, Y. Zhang, Z. Zhang, T. Huang, G. Zhao, L. Liu, X. Wang, L. Sun, Elimination of room-temperature brittleness of Fe-Ni-Co-Al-Nb-V alloys by modulating the distribution of nb through the addition of V, Mater. Sci. Eng. A. 855 (2022) 143848. https://doi.org/10.1016/j.msea.2022.143848

[97] U. Wendt, Engineering Materials and Their Properties, in: K.H. Grote, H. Hefazi (Eds.), Springer Handbook of Mechanical Engineering, Springer Handbooks; Springer International Publishing: Cham, 2021, pp. 233-292, ISBN 978-3-030-47034-0. https://doi.org/10.1007/978-3-030-47035-7_8

[98] S. Gialanella, A. Malandruccolo, Superalloys, in: C.P. Bergmann (Ed.), Aerospace Alloys, Topics in Mining, Metallurgy and Materials Engineering, Springer International Publishing:

Cham, 2020, pp. 267-386 ISBN 978-3-030-24439-2. https://doi.org/10.1007/978-3-030-24440-8_6

[99] R. Soleimani Gilakjani, S.H. Razavi, M. Seifollahi, The Effect of niobium addition on the microstructure and tensile properties of iron-nickel base A286 superalloy, IJMSE. 18 (2021).

[100] M. Seifollahi, S.H. Razavi, Sh. Kheirandish, S.M. Abbasi, The role of η phase on the strength of A286 superalloy with different Ti/Al ratios, Phys. Metals Metallogr. 121 (2020) 284-290. https://doi.org/10.1134/S0031918X20030059

[101] R. Zhang, C. Zhang, Z. Wang, J. Liu, Evolution of recrystallization texture in A286 iron based superalloy thin plates rolled via various routes, Metals. 13 (2023) 1527. https://doi.org/10.3390/met13091527

[102] X. Sauvage, S. Mukhtarov, Microstructure evolution of a multiphase superalloy processed by severe plastic deformation, IOP Conf. Ser. Mater. Sci. Eng. 63 (2014) 012173. https://doi.org/10.1088/1757-899X/63/1/012173

[103] D.K. Ganji, G. Rajyalakshmi, Influence of alloying compositions on the properties of nickel based superalloys: A review, in: H. Kumar, P. Jain (Eds.), Recent Advances in Mechanical Engineering. Lecture Notes in Mechanical Engineering. Springer, Singapore, 2020.8-981-15-1071-7_44.

[104] S.J. Hong, W.P. Chen, T.W. Wang, A Diffraction Study of the γ'' Jb phase in INCONEL 718 superalloy, Metall. Mater. Trans. A. 32 (2001) 1887-1901. https://doi.org/10.1007/s11661-001-0002-4

[105] G.R. Thellaputta, Machinability of nickel based superalloys: A review, Mater. Today. 4 (2017) 3712-3721. https://doi.org/10.1016/j.matpr.2017.02.266

[106] M. Carsí, J. Llaneza, O.A. Ruano, Microstructure and stability conditions for hot deformation of a modified iron-based superalloy, Mater. Sci. Technol. 35 (2019) 2217-2224. https://doi.org/10.1080/02670836.2019.1667672

[107] P. Berthod, Microstructures and Metallographic Characterization of Superalloys, J. Mater. Sci. Technol. Res. 6 (2019). https://doi.org/10.31875/2410-4701.2019.06.6

[108] F. Badkoobeh, H. Mostaan, M. Rafiei, H.R. Bakhsheshi-Rad, F. Berto, F. Microstructural characteristics and strengthening mechanisms of ferritic-martensitic dual-phase steels: A review, Metals. 12 (2022) 101. https://doi.org/10.3390/met12010101

[109] A. Bahadur, Enhancement of high temperature strength and room temperature ductility of iron aluminides by alloying, Mater. Sci. Technol. 19 (2003) 1627-1634. https://doi.org/10.1179/026708303225008266

[110] H.K.D.H. Bhadeshia, Recrystallisation of practical mechanically alloyed iron-base and nickel-base superalloys. Mater. Sci. Eng. A. 223 (1997) 64-77. https://doi.org/10.1016/S0921-5093(96)10507-4

[111] T. Zhao-qiang, Z. Qing, G. Xue-yi, Z. Wei-jiang, Z. Cheng-shang, L. Yong, New development of powder metallurgy in automotive industry, J. Cent. South Univ. 27 (2020) 1611-1623. https://doi.org/10.1007/s11771-020-4394-y

[112] E. Akca, A. Gursel, A Review on superalloys and IN718 nickel-based INCONEL superalloy, Period. Eng. Net. Sci. 3 (2015). https://doi.org/10.21533/pen.v3i1.43

[113] D.C. Pratt, Industrial casting of superalloys, Mater. Sci. Technol. 2 (1896) 426-435. https://doi.org/10.1179/mst.1986.2.5.426

[114] G. Sjöberg, Casting Superalloys for Structural Applications, in: E.A. Ott, J.R. Groh, A. Banik, I. Dempster, T.P. Gabb, R. Helmink, X. Liu, A. Mitchell, G.P. Sjöberg, A. Wusatowska-Sarnek (Eds.), 7th International Symposium on Superalloys 718 and Derivatives, TMS, 2010. https://doi.org/10.7449/2010/Superalloys_2010_117_130

[115] H. Fu, X. Geng, High rate directional solidification and its application in single crystal superalloys, Sci. Technol. Adv. Mater. 2 (2001) 197-204. https://doi.org/10.1016/S1468-6996(01)00049-3

[116] J. Zhang, R.F. Singer, Hot tearing of nickel-based superalloys during directional solidification, Acta. Materialia. 50 (2002) 1869-1879. https://doi.org/10.1016/S1359-6454(02)00042-3

[117] Z.G. Gao, Numerical analysis of microstructure anomalies during laser welding nickel based single-crystal superalloy part III: Amelioration of solidification behavior, Mater. Sci. Forum. 1041 (2021) 47-56. https://doi.org/10.4028/www.scientific.net/MSF.1041.47

[118] W. Xia, X. Zhao, L. Yue, Z. Zhang, A review of composition evolution in Ni-based single crystal superalloys, J. Mater. Sci. Technol. 44 (2020) 76-95. https://doi.org/10.1016/j.jmst.2020.01.026

[119] J.E. Smugeresky, Characterization of a rapidly solidified iron-based superalloy, Metall. Trans. 13 (1982) 1535-1546. https://doi.org/10.1007/BF02644793

[120] J.M. Torralba, M. Campos, Toward high performance in powder metallurgy, Revista De Metalurgica. 50 (2014) e017. https://doi.org/10.3989/revmetalm.017

[121] Sintered Structural Components. https://serena.co.in/contact-us.html/2023 (accessed 10 October 2024)

[122] L. Ma, Z. Zhang, Effect of pressure and temperature on densification in electric field assisted sintering of Inconel 718 superalloy, Materials. 14 (2021) 2546. https://doi.org/10.3390/ma14102546

[123] M.G. Ozden, N.A. Morley, Laser additive manufacturing of Fe-based magnetic amorphous alloys, Magnetochemistry. 7 (2021) 2312-7481. https://doi.org/10.3390/magnetochemistry7020020

[124] J. Shi, Development of metal matrix composites by laser-assisted additive manufacturing technologies: A review, J. Mater. Sci. 55 (2020) 9883-9917. https://doi.org/10.1007/s10853-020-04730-3

[125] A. Kulkarni, Additive manufacturing of nickel based superalloy 2018, Phy. App ph. (2018)

[126] E.M. Lehockey, G. Palumbo, P. Lin, Improving the weldability and service performance of nickel-and iron-based superalloys by grain boundary engineering, Metall. Mater. Trans. A. 29 (1998) 3069-3079. https://doi.org/10.1007/s11661-998-0214-y

[127] J.H. Westbrook, R.L. Fleischer, Intermetallic Compounds: Principles and practice, (Third vol.), Progress, Wiley: Chichester, 2002, ISBN 978-0-471-49315-0. https://doi.org/10.1002/0470845856

[128] V.K. Sikka, Howmet Research Corporation, Whitehall, MI, USA.

[129] J. Ge, Post-process treatments for additive-manufactured metallic structures: A comprehensive review, J. Mater. Eng. Perform. 32 (2023) 7073-70122. https://doi.org/10.1007/s11665-023-08051-9

[130] P.K. Liaw, M.E. Fine, G. Ghosh, M.D. Asta, C.T. Liu, Computational and experimental design of fe-based superalloys for elevated-temperature applications, Technical Report. (2012). https://doi.org/10.2172/1047697

[131] S.A. Kareem, Hot deformation behaviour, constitutive model description, and processing map analysis of superalloys: An overview of nascent developments, J. Mater. Res. Technol. 26 (2023) 8624-8669. https://doi.org/10.1016/j.jmrt.2023.09.180

[132] F. Theska, On conventional versus direct ageing of alloy 718, Acta. Mater. 156 (2018) 116-124. https://doi.org/10.1016/j.actamat.2018.06.034

[133] M.L. Escudero, M.C. Garcia-Alonso, A. Gutierrez, M.F. Lopez, Surface analysis of a heat-treated, Al-containing, iron-based superalloy, J. Mater. Res. 13 (2011) 3411-3416. https://doi.org/10.1557/JMR.1998.0464

[134] S.P Coryell, K.O. Findley, M.C. Mataya, E. Brown, Evolution of microstructure and texture during hot compression of a Ni-Fe-Cr superalloy, Metall. Mater. Trans. A. 43 (2012) 633-649 https://doi.org/10.1007/s11661-011-0889-3

[135] S.K. Mukhtarov, A.G. Ermachenko, A.G. Influence of severe plastic deformation and heat treatment on microstructure and mechanical properties of a nickel-iron based superalloy, Rev. Adv. Mater. Sci. 31 (2012) 151-156.

[136] G.E. Fuchs, Solution heat treatment response of a third generation single crystal Ni-base superalloy, Mater. Sci. Eng. A. 300 (2001) 52-60 https://doi.org/10.1016/S0921-5093(00)01776-7

[137] Y. Xiaodai, L. Jiarong, S. Zhenxue, W.D. Xiaoguang, Designing of the homogenization solution heat treatment for advanced single crystal superalloys, Rare Met. Mater. Eng. 46 (2017) 1530-1535 https://doi.org/10.1016/S1875-5372(17)30159-5

[138] J. Ridhwan, E. Hamzah, E. Selamat, Z. Zulfattah, M.H.M. Hafidzal, Effect of aging treatment on the microstructures and hardness of Fe-Ni-Cr superalloy, Int. J. Automot. Mech. Eng. 8 (2013) 1430-1441 https://doi.org/10.15282/ijame.8.2013.30.0118

[139] S.K. Sahu, D.K. Mishra, A. Behera, R.P. Dalai, An overview on the effect of heat treatment and cooling rates on Ni-based superalloys, Mater. Today Proc. 47 (2021) 3309-3312 https://doi.org/10.1016/j.matpr.2021.07.146

[140] C. Papadaki, W. Li, A.M. Korsunsky, On the dependence of γ' precipitate size in a nickel-based superalloy on the cooling rate from super-solvus temperature heat treatment, Materials. 11 (2018) 1528 https://doi.org/10.3390/ma11091528

[141] M. Ni, S. Liu, C. Chen, R. Li, X. Zhang, K. Zhou, Effect of heat treatment on the microstructural evolution of a precipitation-hardened superalloy produced by selective laser melting, Mater. Sci. Eng. A. 748 (2019) 275-285 https://doi.org/10.1016/j.msea.2019.01.109

[142] N.M. Dawood, A.M. Salim, A review on characterization, classifications, and applications of super alloys, J. Univ. Babylon. Eng. Sci. 29 (2021).

[143] Z. Tan, Q. Zhang, X. Guo, W. Zhao, C. Zhou, Y. Liu, New development of powder metallurgy in automotive industry, J. Cent. South Univ. 27 (2020) 1611-1623 https://doi.org/10.1007/s11771-020-4394-y

[144] S.B. Mishra, K. Chandra, S. Prakash, Erosion-corrosion behaviour of nickel and iron based superalloys in boiler environment. Oxid. Met. 83 (2015) 101-117. https://doi.org/10.1007/s11085-014-9509-0

[145] R. Schafrik, R. Sprague, Superalloy technology - A perspective on critical innovations for turbine engines, KEM. 380 (2008) 113-134 https://doi.org/10.4028/www.scientific.net/KEM.380.113

[146] P.C. Gasson, The superalloys: Fundamentals and applications R. C. Reed Cambridge University Press, The Edinburgh Building, Shaftesbury Road, Cambridge, CB2 2RU, UK, 2006. 372pp. Illustrated. £80. ISBN 0-521-85904-2. Aeronaut. J. 112 (2008) 291 291 https://doi.org/10.1017/S0001924000087509

[147] K. Matsugi, Y. Murata, M. Morinaga, N. Yukawa, An electronic approach to alloy design and its application to Ni-based single-crystal superalloys, Mater. Sci. Eng. A. 172 (1993) 101-110 https://doi.org/10.1016/0921-5093(93)90430-M

[148] M. Prem Kumar, N. Arivazhagan, C. Chiranjeevi, Y. Raja Sekhar, N. Babu, M. Manikandan, Effect of molten binary salt on Inconel 600 and Hastelloy C-276 superalloys for thermal energy storage systems: A corrosion study, J. Mater. Eng. Perform. 33 (2023) 9070-9083. https://doi.org/10.1007/s11665-023-08641-7

[149] J.K. Tien, V.C. Nardone, The U.S. superalloys industry - Status and outlook, JOM. 36 (1984) 52-57 https://doi.org/10.1007/BF03338563

[150] H. Yurtkuran, An evaluation on machinability characteristics of titanium and nickel based superalloys used in aerospace industry. İmalatTeknolojileriveUygulamaları. 2 (2021) 10-29 https://doi.org/10.52795/mateca.940261

[151] A. Kale, N. Khanna, A review on cryogenic machining of super alloys used in aerospace industry, Procedia Manuf. 7 (2017) 191-197 https://doi.org/10.1016/j.promfg.2016.12.047

[152] A. Kracke, Superalloys, the most successful alloy system of modern times-past, present, and future, in: 7th International Symposium on Superalloys 718 and Derivatives, TMS. (2010). https://doi.org/10.7449/2010/Superalloys_2010_13_50

Superalloys: Fundamentals and Applications
Materials Research Foundations 178 (2025) 25-40

Materials Research Forum LLC
https://doi.org/10.21741/9781644903698-2

Chapter 2

Microstructure, Processing, Characteristics, and Engineering Application Specifically of Iron-based Superalloys

Uzma Hira[1]*, Yousra Akram[1]

[1]School of Physical Sciences, University of the Punjab, Lahore, Pakistan

*Uzma.sps@pu.edu.pk

Abstract

The struggle for finding materials with greater thermal stability, high strength, creep resistance, resistance to oxidizing environments, and corrosion, provides the basic impetus behind the creation of superalloys. Due to their possession of all these qualities, superalloys have brought innovations in jet and marine turbine engine applications because they possess extraordinary stability even at elevated temperatures in contrast to most of the materials used in the past. High fractions of the melting point can be used in superalloys, and this attribute has made them suitable for use in high-temperature applications. Superalloys possess special protective coatings responsible for their anti-corrosive characteristics and mechanical properties which remain stable at high temperatures, making it possible to manufacture their products under high stress. This chapter will further elaborate on the microstructure, processing, characteristics, and engineering application specifically of iron-based superalloys. It also shows the prospects and difficulties that may arise in superalloys machining.

Keywords

Iron-Based Superalloys, Microstructure, Thermal Stability, Protective Coatings, Strength

Contents

1. Introduction

Superalloys are generally strong, multi-component materials that have various resistance properties including high temperature stability, surface stability, and mechanical stress. Superalloys exhibit incomparable properties such as enhanced resistance against corrosion, good tensile strength, high structural stability, and the ability to withstand harsh environments of high temperature and pressure. The versatility of superalloys to factor its melting point is one of the fundamental reasons why it is widely used in many applications. Superalloys are manufactured predominantly by chemical and process development and have a crystal structure that is typical of face-centered cubic (FCC) austenitic. They are generally laminated in a single crystalline form. While grain boundaries may provide some measure of strength at low temperatures, however, they decrease the amount of creep resistance. Materials to be used at high temperatures must be stable for long periods for their efficient performance. They should possess excellent strength, creep resistance, increased homogeneity, and good resistance to corrosion and oxidation. The possession of these diverse properties by superalloys has made them a fascinating material for engineering components to be used in hot zones of jets, turbine blades, engines, combustion chambers, etc. Most advanced machines and engines use superalloys to enhance their efficiency and workability [1]. Fig. 1 shows some major types of superalloys with their specifications.

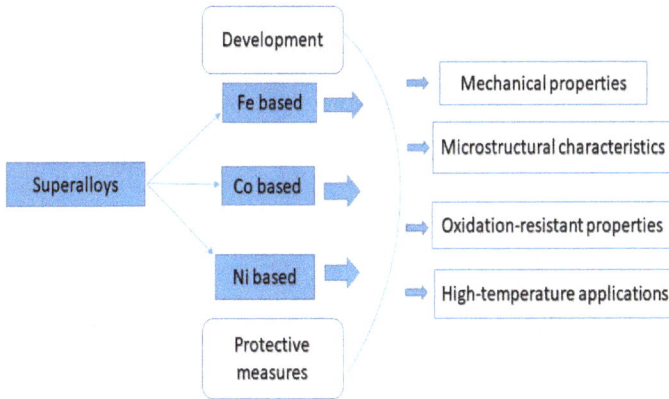

Figure 1 Some major types of superalloys.

2.   History-From where the name originated?

Originally, an alloy was a mixture of only two or three customary elements such as iron or nickel. The inclusion of some other elements like aluminum, chromium, carbon, and titanium to the basic matrix of iron, chromium, or nickel produces a new class of high temperature, erosion, corrosion, and creep resistance, known as superalloys [2]. These high-temperature materials were coined as superalloys for the very first time in the 1940s. In addition to the stability at high temperatures, superalloys can also retain their strength and rigidity. The term super before alloys was used herein as a result of their extraordinary characteristics despite being made of alloys. It is for the use of steam turbines that the early development of superalloys began in the early 20th century. Different forms of these superalloys were employed in heating elements and resistance wires. Turbine blades of jet engines needed a material that could cope with high temperatures, this need of the time led to the creation of superalloys during the Second World War. The first superalloy that was developed was based on the nickel-chromium system because that was the best known at that time, however, after the Second World War, further research and development of these alloys intensified because of the steady growth of the aerospace industry and increased demand for more efficient and reliable jet engines [3]. The work has been accomplished in the enhancement of quality, performance, and purity of the superalloys by the adoption of advanced manufacturing and processing techniques. Smart improvements have, however, also been developed to improve them in terms of their strength and their ability to be resistant to corrosive atmosphere, by the incorporation of new components like cobalt and molybdenum. [4]

Superalloys: Fundamentals and Applications   Materials Research Forum LLC
Materials Research Foundations 178 (2025) 25-40   https://doi.org/10.21741/9781644903698-2

3.   Iron (Fe) Based Superalloys (FBSs)

Like various other kinds of superalloys, iron-based superalloys are also a kind of superalloys. They are named so because of the predominant iron content in their composition. Some of the strains present in these stainless steel alloys include W, Ti, Si, Nb, Al, C, Cr, Ti, Y, Co, Si, and B. [5] While adopting some alloys of steel it is possible to obtain an oxidation and creep resistance comparable to that offered by nickel based supperalloys. Iron-based superalloys like nickel-based superalloys contain, the matrix phase of austenite iron (FCC) phase. Iron is an allotropic element and it can possess austenite structure in a certain temperature range only. Austenite character in iron-based superalloys has a significant contribution to their high performance and unique attributes [7]. The presence of a minor percentage of nickel element has a crucial role in maintaining austenite structure in the superalloys [8].

3.1   Significance of iron-based superalloys

These iron-based superalloys have become of tremendous importance based on their superior and unmatchable mechanical characteristics. There is increased use of those Fe-based superalloys that have been strengthened by precipitation hardening mechanisms in the manufacturing of those structures that require higher resistance against creep and oxidation for seamless operations [9]. Fe-based superalloys are gaining increased importance in arresting further degradation of the surface of metals of several structures that have been used in high-temperature environments by providing their replacement [10]. It is due to the material chemistry of Fe-based superalloys that provide the necessary mechanical strength to minimize eroded and corroded materials and to process at high temperatures [11]. The Fig.2 depicts the versatility and significance of iron-based superalloys.

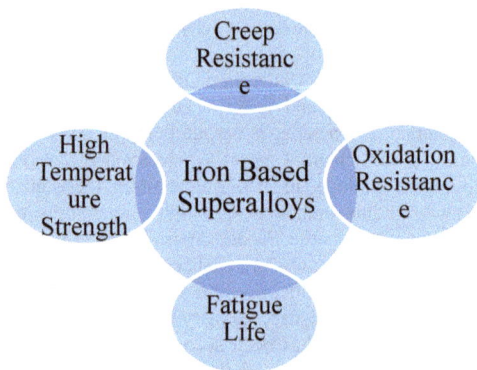

Figure 2 Some major attributes of iron-based superalloys.

3.2   Temperature range at which iron based superalloys operate

They can operate optimally at moderated temperatures, an essential feature for any superalloy. They can serve for a long time if they are used at a mild temperature of about 600-850°C [12]. Those iron-based superalloys that are precipitation-strengthened tend to have a lot of strength in this range of temperatures and these alloys can last for long without failure [13].

3.3   Composition and microstructure

3.3.1   Composition

Iron-based superalloys typically consist of a significant amount of iron (Fe) along with other elements like cobalt (Co), chromium (Cr), and nickel (Ni). They also contain minor amounts of other elements like titanium (Ti), molybdenum (Mo), aluminum, and niobium (Nb) etc. [14]. Fig. 3 shows the percentage of different elements in iron-based superalloys.

Composition

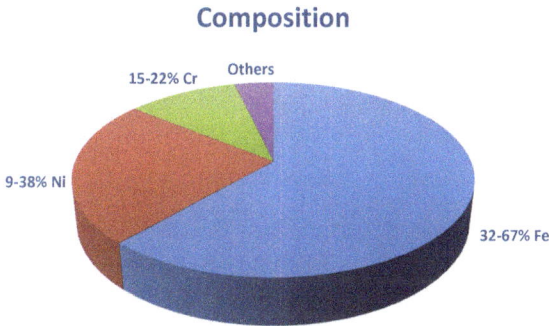

15-22% Cr   Others

9-38% Ni

32-67% Fe

Figure 3 Composition of iron-based superalloys.

3.3.1.1 Role of alloying elements

Aluminum

The first important alloying element is aluminum. Aluminum favors a good resistance to oxidation especially at high temperatures. A layer of aluminum oxide Al_2O_3 forms by the partial oxidation process. As aluminum oxide is a stoichiometric compound without interstitial spaces in its crystal structure, therefore, it prevents the movement of species responsible for the oxidation process and hence protects against corrosion. [15]. Aluminum is also responsible for weight reduction and maintaining austenite character in iron-based superalloys.

Role of Chromium

Chromium performs the same function as aluminum but with slightly less effectiveness since chromium can get re-oxidized to form chromium trioxide CrO_3 as soon as temperature exceeds 1000°C [16]. However, in the prevention of sulfidation or corrosion because of molten salts, chromium is many times more effective.

Superalloys: Fundamentals and Applications   Materials Research Forum LLC
Materials Research Foundations 178 (2025) 25-40   https://doi.org/10.21741/9781644903698-2

Role of rare earth elements

Other alloying elements such as lanthanides also doped to enhance their oxidation resistance ability at elevated temperatures by some more degrees. These doping elements include elements with greater atomic mass/number such as molybdenum, tungsten, tantalum, and rhenium which play an important role in strengthening alloys in their solid solutions. The presence of these big atoms locally imposes a stress field which can hinder the movement of dislocations [17].

3.3.2 Microstructure of iron-based superalloys

It is found that the matrix phase that exists in iron-based superalloys, is austenite iron, also known as face-centered cubic (FCC), just like it exists in nickel-based superalloys. The minor alloying elements found in these superalloys include W, Si, Nb, C, Al, Y, Ti, Co, and B [18]. The addition of Al at a low weight ratio in these superalloys results in enhanced oxidation resistance. The FCC matrix phase provides more thermal strength as compared to the BCC matrix phase, therefore, the concentration of Al must be kept low to enhance the stability of the FCC matrix phase as a higher concentration of Al encourages the stability of the BCC matrix phase. Hence FCC primary phase matrix is essential for their stability and efficient performance [19].

3.4 Types of iron-based superalloys

- Martensitic Superalloys
- Austenitic Superalloys

3.4.1 Martensitic superalloys

This type of superalloy is usually hardened and integrated by a martensitic type of transformation (a process by which a material changes its crystal structure without undergoing diffusion at a greater rate) [20].

Examples:

Pyromet 720: Provides high strength and resistance against oxidation, creep, and corrosion [21].

Ferrium S53: Provides high-temperature strength and oxidation resistance in the aerospace industry [22].

3.4.2 Austenitic superalloys

Austenitic superalloys are classified into two sub-classes depending upon their reinforcing mechanism. One class of austenitic superalloys is hardened by hot or cold working processes (typically forging at elevated temperatures of 1050-1200 °C, examined by finishing process at 650-900 °C. The second class of austenitic superalloys is hardened by precipitation hardening [23].

Incoloy 800: Because of the addition of nickel and chromium in their composition, they show extraordinary strength and resistance to oxidative reactions and carburization at elevated temperatures [24]. (See Table 1)

Superalloys: Fundamentals and Applications
Materials Research Foundations 178 (2025) 25-40

Materials Research Forum LLC
https://doi.org/10.21741/9781644903698-2

Table 1 Some common types of iron-based superalloys.

Type of Iron-based Superalloy	Properties	Application
Austenitic Stainless	Outstanding resistance to corrosion, good high-temperature strength, and good workability.	Power plant equipment like gas turbines, heat exchangers, furnaces, etc.
Ferritic Steel	With better resistance to oxidation and stress corrosion cracking, the cost is less than austenitic.	Automotive exhausts, heat exchange equipment, gas turbines.
Martensitic Steel	High strength, hardness, and moderate corrosion resistance.	Turbines, blades, and structural parts of the engine.
Iron-Nickel Superalloys (e.g., Incoloy, 800)	Strong at elevated temperatures, relatively good high-temperature oxidation, and rust resistant.	Aeronautical and automotive engines, marine gas turbines, and nuclear reactors.

3.5 Phases in Fe-based superalloys

Fe-based superalloys are made up of base elements and some amounts of doping elements. These elements are dispersed in their crystal lattice in the form of different phases to gain as much stability as possible. Following are the phases that exist in most of the superalloys [25].

3.5.1 Gamma phase

The gamma phase forms the basic matrix phase. This a primary phase of having an FCC crystal system. It maintains the structural stability of the precipitates and imparts ductility and strength to them [26].

3.5.2 Gamma prime γ′ phase

In iron-based superalloys, the Gamma prime (γ') precipitation strengthening mechanism is achieved by the right augment of Ni, Nb, Ti, Al, and so on. It contributes to the high strength of superalloys by inhibiting the dislocations within an ordered FCC structure. The gamma prime phase doesn't lower the fracture toughness of the superalloys and provides them significant strength due to its property of being ductile. Aluminum and titanium play an important role in imparting this property in them [27].

3.5.3 Carbides

Elements like chromium, tantalum, and tungsten, undergo carbide formation in the presence of carbon. These can form different types of carbides, such as MC, $M_{23}C_6$, and, M_6C in iron-based superalloys, Specific types of carbides may be precipitated in the base metal during some heating process (when undergoes casting), and such precipitates spread in the base metal to impede the movement of displacements and thus play a role in enhancing creep resistant properties. Carbides found in carbon-containing alloys occupy interstitial spaces as soon as they solidify. These solidified and packed carbides reduce the rate of creep, in addition to enhancing the interstitial cohesiveness. Commonly, these carbides are based on elements like titanium, tungsten, molybdenum, and vanadium. Examples of some carbides that can be present include M_6C carbide types which may be $Mo_3(Ni, Co)_3C$. They provide strengthening through grain boundary pinning, preventing grain boundary sliding for improved creep resistance [28].

3.6   Processing frameworks

- There are several methods for the processing of superalloys, they include

- Powder Metallurgy (PM)

- Vacuum Induction Melting (VIM)

- Secondary Melting

- Spray Casting

3.6.1   Powder metallurgy

Superalloys which are supposed to be used in heavy applications, are mostly processed by powder metallurgy. This processing technique is not much economical but it provides highly homogenized superalloys. The powder metallurgy processing of superalloys starts with atomization using molten metal derived from a vacuum induction melting process. The atomized powdered product is then subjected to an extrusion process for its proper fusion. The powder is then poured into a clean steel container and is evacuated to provide a way for outgassing. After completely evacuating the powder it is compressed using hot isotactic pressing or closed die forging. Isothermal forging is an ideal route for the production of powder metallurgy superalloys [29]. Fig. 4 displays an understanding of the steps involved in the powder metallurgy process.

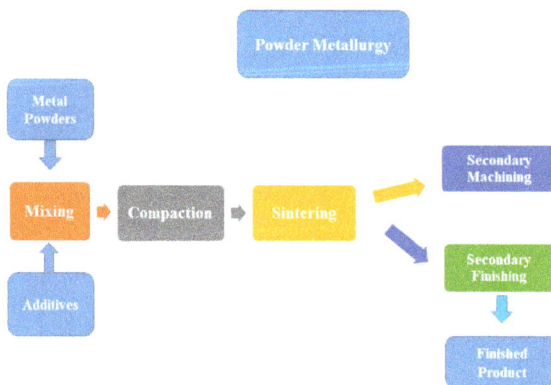

Figure 4 Steps involved in powder metallurgy process.

Superalloys: Fundamentals and Applications   Materials Research Forum LLC
Materials Research Foundations 178 (2025) 25-40   https://doi.org/10.21741/9781644903698-2

3.6.2 Vacuum induction melting

Vacuum induction melting is a form of primary melting processing technique before jumping to the secondary melting process. It is used for the refinement of metals and alloys under inert/vacuum conditions. It allows precise control over the concentration of alloying elements in superalloys. The first step involved in VIM is the charging of raw alloying elements into an induction furnace. The furnace allows the heating or melting of conductive material with the help of eddy current under vacuum conditions. The vacuum conditions prevent contamination from impure gases like oxygen, nitrogen, or hydrogen. Vacuum conditions help in achieving uniform distribution of the doping elements in the molten phase, and the molten product is then poured into molds. Molten metal solidifies in the molds forming structures with minimal defects and high strength [30].

3.6.3 Secondary melting

After vacuum induction melting (VIM), a secondary melting process is performed to improve the chemical homogeneity in alloys. It also ensures the removal of ceramic inclusions, resulting from the VIM [31]. The Fig. 5 describes the secondary melting process in a stepwise manner.

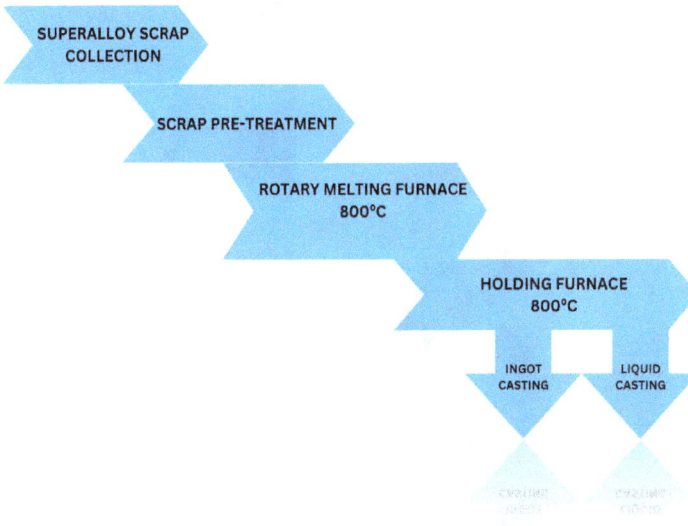

Figure 5 Schematic diagram of steps involved in secondary melting process.

3.6.4 Spray casting

Starting with VIM, the product obtained is transferred to a hopper and is quantified by appropriate methods. The quantified flow of alloy is then atomized into fine droplets through an interaction with argon gas that causes the collision. A pre-heated carbon steel mandrel that circles beneath the spray is used to deposit the material when making billets. The spray-casted alloy can be subjected to deformation processing. With the mechanism of ring rolling, deformation processing modifies and improves the microstructure, and cross-section of roughly made superalloys. Deformation-processed superalloys are passed through major heat treatments which involve an annealing process. The annealing process helps in improving the mechanical and strengthening properties of superalloys. The superalloys must be surface coated to protect them from oxidation and corrosion. Three important types of surface coatings include TBCs, overlay, and diffusion (aluminide). These coatings are proven to provide excellent protection of superalloys against thermal and oxidative corrosion [32]. See Fig 6 for an understanding of the spray casting method of processing superalloys.

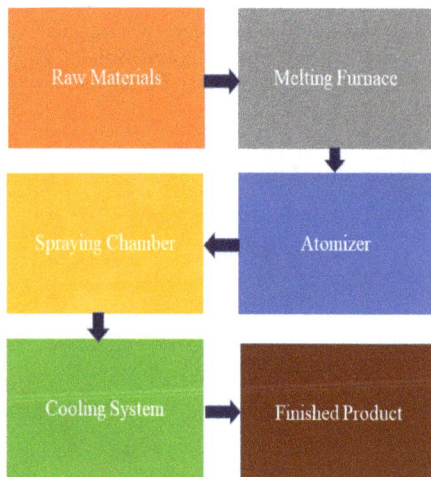

Figure 6 Spray casting process.

3.7 Physical properties

Iron-based superalloys are a group of high-performance materials containing iron as the base element along with nickel, chromium, and other alloying elements. These are intended to possess adequate strength and durability and act as a barrier between the alloy and oxidizing species [33]. Table 2 illustrates some of their physical attributes with standard specifications.

Materials Research Forum LLC
https://doi.org/10.21741/9781644903698-2

Table 2 Some physical attributes of iron-based superalloys.

Physical Attributes	Typical value/range	Description
Hardness (Vickers)	200-400 HV	Giving them a heat treatment can enhance their hardness.
Density	7.5-8.3 g/cm^3	Depends on the alloy composition.
Thermal Conductivity	10-20 W/m.K	Their thermal conductivity is usually lower than pure iron: and varies with alloying composition.
Specific Heat Capacity	450-550 J/kg.K	Comparable with other superalloys
Young's Modulus	200-225 GPa	Contributes to the high stiffness.
Yield Strength	650-1250 MPa	Provides good resistance against deformation

3.8 Mechanical attributes

Iron-based superalloys are designed in such a way that they can exhibit exceptional strengthening and mechanical attributes even at elevated temperatures. Some of their important mechanical attributes are listed below.

High Temperature Endurance

Iron-based superalloys do not lose their strength at high temperatures and do not undergo degradation at elevated temperatures, therefore they are perfect materials for applications where the conditions of extreme temperature exist. The existence of stable microstructure, and precipitation hardening strength are the key factors responsible for such attributes in superalloys.

Corrosion Durability

The presence of chromium content and other alloying elements in minor quantities in iron-based superalloys provides a protective layer against oxidation and corrosion. It helps in enhancing the shelf life of the superalloys, and materials can remain intact for a longer time even at harsh conditions of temperature and oxidative environments.

Creep Resistance

The slow deformation of materials under constant working load over an extended period especially at elevated temperatures is referred to as creep. Iron-based superalloys offer great resistance against creep. Their creep resistance property enables them to maintain their structural integrity over an extended period [34].

Wear Resistance

The carbon content in iron-based superalloys is responsible for its property of wear resistance. The more carbon content, the more will be the wear resistance property. This property is beneficial in applications where materials have to undergo rubbing, and sliding at high temperatures [35]. The alloys 613, 612, and 611 are believed to possess the maximum value of wear resistance.

Mechanical Strength

These materials can withstand sudden shocks without undergoing fracturing. This property makes them suitable for applications where they have to face varying temperatures [36].

Precipitation Strengthening

The precipitation-strengthening elements in the superalloys are allowed to form a precipitate phase to attain the resistance against displacement of the alloy crystal. The precipitation phases are further subdivided into γ' phase and γ" phase. These two phases act differently to improve the strengthening properties of superalloys. Those iron-based superalloys that are strengthened by precipitation strengthening mechanisms exhibit excellent mechanical properties below 800°C. The number of precipitated phases in iron-based superalloys is less as compared to nickel-based superalloys, it reflects their high ductility than nickel-based superalloys [37].

Hardness

The carbide formation in the crystal structure of iron-based superalloys imparts their hardness and toughness. The carbide precipitates present in the matrix of iron-based superalloys prevent the dislocation of the alloying elements within the structure hence resulting in increased hardness. The hardness of the superalloys can be changed by changing the heating conditions and time parameters. Finer grain results in more hardness by increasing more build-up of dislocation at the grain edges.

Tensile Strength

The tensile strength also increases by increasing the carbide precipitate content in the matrix of superalloys. Increased carbide precipitates block the dislocations within the crystal structure and result in great tensile strength [38].

4.   Applications of iron-based superalloys

Superalloys have gained huge importance because of their extensive use in high-temperature applications. They are also being used in heat-exchanging tubes, for turbines in the aerospace industry. Some of the applications of superalloys in different industries are listed below.

Air Crafts and turbines for land

Iron-based superalloys are well known for applications that require high creep resistance and good temperature resistance. Iron-based superalloys possess a lower coefficient of thermal expansion which makes them a valuable choice for applications where the main design challenge is thermal stability. Therefore, they are used as the manufacturing materials of disks, bolts, shaft cases, turbine blades, and combustors in aircraft and turbine engines [39].

Exhaust Valves of Diesel Engines

Iron-based super alloys due to their property of hardness have been widely used as hard-faced heads of exhaust valves in heavy-duty engines. Hard-facing alloys are usually expensive, but iron-based superalloys are excellent examples of cheap and easy-to-fabricate materials to be used in heavy-duty engines [40].

Metallurgy

FERRALIUM 225 is an example of an iron-based superalloy, and being an economical material finds adverse uses in metallurgy, gas and oil industries, petrochemicals, and paper-making industries. FERRALIUM 225 has replaced many previously good stainless materials which are not adequate anymore in many applications. Resistance to corrosive environments, high ductility, and excellent fatigue life make them perfect for marine applications [41].

Gas and Oil Industry

Iron-based superalloys are extensively used in the oil and gas field for many applications. The components to be used in oil processing equipment have to face corrosive and harsh environments, therefore they must be made up of materials that can endure high temperatures, and superalloys are one of those materials. NITRONIC 50, a perfect example of iron-based superalloys, provides a combination of properties like corrosion resistance and integrity and proves to be a good choice for use in oil and gas processing industries [42].

5.   Conundrums and future endeavors

The production of superalloys was initially driven by the aerospace industry, but now the demand for superalloys is increasing extensively for all-purpose applications. One of the main challenges in the production of superalloys is the costly procedures involved in the production of complex and heavy parts. The goal is to design and develop cost-effective materials like iron-based superalloys on a large scale and make them suitable for each kind of application. Another necessary and important aspect associated with future endeavors of iron-based superalloys is to enhance their operating temperature range. Nano-particles of superalloys would also prove to be an innovative material replacing typical stainless materials [43].

Conclusion

Iron-based super alloys are superb materials with great strength, high ductility, good corrosion resistivity, and oxidation resistance. They are made up of the predominant components of iron along with other strengthening alloying elements like nickel, chromium, molybdenum, aluminum, carbon, etc. Through a proper grain refinement and precipitation strengthening mechanism, iron-based superalloys can be engineered with the desired properties required for specific applications.

References

[1] W. Betteridge, S.W.K. Shaw, Development of superalloys, Mater. Sci. Technol. 3 (1987) 682-694. https://doi.org/10.1179/mst.1987.3.9.682

[2] A. Kracke, A. Allvac, Superalloys, the most successful alloy system of modern times-past, present and future, Proceedings of the 7th International Symposium on Superalloy 718 (2010) 13-50. https://doi.org/10.7449/2010/Superalloys_2010_13_50

[3] S.C. Krishna, N.K. Gangwar, A.K. Jha, B. Pant, P.V. Venkitakrishnan, On the direct aging of iron-based superalloy hot rolled plates, Mater. Sci. Eng. A. 648 (2015) 274-279. https://doi.org/10.1016/j.msea.2015.09.073

[4] K.Y. Shin, J.W. Lee, J.M. Han, K.W. Lee, B.O. Kong, H.U. Hong, K.Y. Shin, J.W. Lee, J.M. Han, K.W. Lee, B.O. Kong, H.U. Hong, Transition of creep damage region in dissimilar welds between Inconel 740H Ni-based superalloy and P92 ferritic/martensitic steel, Mater. Charact. 139 (2018) 144-152. https://doi.org/10.1016/j.matchar.2018.02.039

[5] A. Fukunaga, Differences between internal and external hydrogen effects on slow strain rate tensile test of iron-based superalloy A286, Int. J. Hydrog. Energy. 47 (2022) 2723-2734. https://doi.org/10.1016/j.ijhydene.2021.10.178

[6] E.M. Lehockey, G. Palumbo, P. Lin, Improving the weldability and service performance of nickel-and iron-based superalloys by grain boundary engineering, Metall. Mater. Trans. 29 (1998) 3069-3079. https://doi.org/10.1007/s11661-998-0214-y

[7] N.M. Dawood, A.M. Salim, A review on characterization, classifications, and applications of superalloys, J. Univ. Babylon Eng. Sci. (2021) 53-62

[8] I.G. Akande, O.O. Oluwole, O.S.I. Fayomi, O.A. Odunlami, Overview of mechanical, microstructural, oxidation properties and high-temperature applications of superalloys, Mater. Today Proc. 43 (2021) 2222-2231. https://doi.org/10.1016/j.matpr.2020.12.523

[9] M. Seifollahi, S. Kheirandish, S.H. Razavi, S.M. Abbasi, P. Sahrapour, Effect of η phase on mechanical properties of the iron-based superalloy using shear punch testing, ISIJ Int. 53 (2013) 311-316. https://doi.org/10.2355/isijinternational.53.311

[10] S.B. Mishra, K. Chandra, S. Prakash, Erosion-corrosion behaviour of nickel and iron based superalloys in boiler environment, Oxid. Met. 83 (2015) 101- 117. https://doi.org/10.1007/s11085-014-9509-0

[11] I.G. Wright, V.K. Sethi, A.J. Markworth, A generalized description of the simultaneous processes of scale growth by high-temperature oxidation and removal by erosive impact, Wear. 186 (1995) 230-237. https://doi.org/10.1016/0043-1648(95)07129-6

[12] Z.H. Zhong, Y.F. Gu, Y. Yuan, Z. Shi, A new wrought Ni-Fe-base superalloy for advanced ultra-supercritical power plant applications beyond 700°C, Mater. Lett. 109 (2013) 38-41. https://doi.org/10.1016/j.matlet.2013.07.060

[13] B.B. Zhang, F.K. Yan, M.J. Zhao, N.R. Tao, K. Lu, Combined strengthening from nano twins and nanoprecipitates in an iron-based superalloy, Acta Mater. 151 (2018) 310-320. https://doi.org/10.1016/j.actamat.2018.04.001

[14] K. Li, J. Liu, Z. Cai, C. Han, X. Li, Microstructure and creep properties of graded transition joints between nickel-based alloys and martensitic heat-resistant steels, Mater. Sci. Eng. A. 872 (2023) 144962. https://doi.org/10.1016/j.msea.2023.144962

[15] M. Abbasi, D. Kim, J. Shim, W. Jung, Effects of alloyed aluminum and titanium on the oxidation behavior of INCONEL 740 superalloy, J. Alloys Compd. 658 (2016) 210-221. https://doi.org/10.1016/j.jallcom.2015.10.198

[16] H. Wang, J. Yang, J. Meng, S. Ci, Y. Yang, N. Sheng, Y. Zhou, X. Sun, Effects of B content on microstructure and high-temperature stress rupture properties of a high chromium polycrystalline nickel-based superalloy, J. Alloys Compd. 860 (2021) 157929. https://doi.org/10.1016/j.jallcom.2020.157929

[17] A.V. Guimarães, R.M. Silveira, N. Jaffrezou, M.C. Mendes, D.S. Santos, L.H de Almeida, L.S. Araujo, Influence of yttrium alloying on improving the resistance to hydrogen embrittlement of superalloy 718, Int. J. Hydrog. Energy. 58 (2024) 479-484. https://doi.org/10.1016/j.ijhydene.2024.01.229

[18] O.G. Kwon, M.S. Han, Failure analysis of the exhaust valve stem from a Waukesha P9390 GSI gas engine, Eng. Fail. Anal. 11 (2004) 439-447. https://doi.org/10.1016/j.engfailanal.2003.05.015

[19] Y. Zhang, D. Zhan, X. Qi, Z. Jiang, Effect of solid solution temperature on the microstructure and properties of ultra-high-strength ferrium S53® steel, Mater. Sci. Eng. A 730 (2018) 41-49. https://doi.org/10.1016/j.msea.2018.05.099

[20] H. Torres, M. Varga, M.R. Ripoll, High-temperature hardness of steels and iron based alloys, Mater. Sci. Eng. A 671 (2016) 170-181. https://doi.org/10.1016/j.msea.2016.06.058

[21] M. Fulger, D. Ohai, M. Mihalache, M. Pantiru, V. Malinovschi, Oxidation behavior of Incoloy 800 under simulated supercritical water conditions, J. Nucl. Mater. 385 (2009) 288-293. https://doi.org/10.1016/j.jnucmat.2008.12.004

[22] Z.R. Khayat, T.A. Palmer, Impact of iron composition on the properties of an additively manufactured solid solution strengthened nickel base alloy, Mater. Sci. Eng. A. 718 (2018) 123-134. https://doi.org/10.1016/j.msea.2018.01.112

[23] K. Wieczerzak, M. Stygar, T. Brylewski, R. Chulist, P. Bała, J. Michler, Robert Chulist, Piotr Bała, Johann Michler, Kinetics and mechanisms of high-temperature oxidation in BCC and FCC high-alloy Fe-based alloys with high volume fraction of carbides, Mater. Des. 244 (2024) 113163. https://doi.org/10.1016/j.matdes.2024.113163

[24] Y. Huang, R. Zhang, Z. Zhou, P. Zhang, J. Yan, Y. Yuan, Y. Gu, C. Cui, Y. Zhou, X. Sun, Microstructure optimization for higher strength of a new Fe-Ni-based superalloy, Mater. Sci. Eng. A. 865 (2023) 144632. https://doi.org/10.1016/j.msea.2023.144632

[25] R. Shi, D.P. McAllister, N. Zhou, A.J. Detor, R. DiDomizio, M.J. Mills, Y. Wang, Growth behavior of γ'/γ" coprecipitates in Ni-Base superalloys, Acta Mater.164 (2019) 220-236. https://doi.org/10.1016/j.actamat.2018.10.028

[26] S. Antonov, J. Huo, Q. Feng, D. Isheim, D.N. Seidman, R.C. Helmink, E. Sun, S. Tin, Eugene Sun, Sammy Tin, σ and η Phase formation in advanced polycrystalline Ni base superalloys, Mater. Sci. Eng. A. 687 (2017) 232-240. https://doi.org/10.1016/j.msea.2017.01.064

[27] S. Zhou, M. Hu, C. Li, Q. Guo, L. Yu, H. Ding, Y. Liu, Microstructure- performance relationships in Ni-based superalloy with co-precipitation of γ' and γ" phases, Mater. Sci. Eng. A. 855 (2022) 143954. https://doi.org/10.1016/j.msea.2022.143954

[28] X. Song, Y. Wang, X. Zhao, J. Zhang, Y. Li, Y. Wang, Z. Chen, Analysis of carbide transformation in MC-M23C6 and its effect on mechanical properties of Ni-based superalloy, J. Alloys Compd. 911 (2022) 164959. https://doi.org/10.1016/j.jallcom.2022.164959

[29] C. Li, J. Teng, B. Yang, X. Ye, J. Liu, Y. Li, Correlation between microstructure and mechanical properties of novel Co-Ni-based powder metallurgy superalloy, Mater. Charact.181 (2021) 111480. https://doi.org/10.1016/j.matchar.2021.111480

[30] R.L. Kennedy, R.M. F. Jones, R.M. Davis, M.G. Benz, W.T. Carter, Superalloys made by conventional vacuum melting and a novel spray-forming process, Vacuum. 47 (1996) 819-824. https://doi.org/10.1016/0042-207X(96)00074-7

[31] H. Cui, Y. Tan, R. Bai, Y. Li, L. Zhao, X. Zhuang, Y. Wang, Z. Chen, P. Li, X. You, C. Cui, Effect of melt superheat treatment on solidification behavior and microstructure of new

Ni-Co-based superalloy, J. Mater. Res. Technol. 15 (2021) 4970-4980.
https://doi.org/10.1016/j.jmrt.2021.10.122

[32] P. Mathur, S. Annavarapu, D. Apelian, A. Lawley, Spray casting: an integral model for process understanding and control, Mater. Sci. Eng. A. 142 (1991) 261- 276.
https://doi.org/10.1016/0921-5093(91)90665-A

[33] Z. Peng, J. Zou, Y. Wang, L. Zhou, Y. Tang, Effects of solution temperatures on creep resistance in a powder metallurgy nickel-based superalloy, Mater. Today Commun. 28 (2021) 102573. https://doi.org/10.1016/j.mtcomm.2021.102573

[34] B. Ohl, L. Owen, H. Stone, D.C. Dunand, Microstructure and mechanical properties of L12-strengthened Co-Ni-Fe-based superalloys, Mater. Sci. Eng. A. 884 (2023) 145276.
https://doi.org/10.1016/j.msea.2023.145276

[35] A.J. Goodfellow, E.I.G. Nava, K.A. Christofidou, N.G. Jones, C.D. Boyer, T.L. Martin, P.A.J. Bagot, M.C. Hardy, H.J. Stone, The effect of phase chemistry on the extent of strengthening mechanisms in model Ni-Cr-Al-Ti-Mo based superalloys, Acta Mater. 153 (2018) 290-302. https://doi.org/10.1016/j.actamat.2018.04.064

[36] H. Pei, S. Wang, X. Gao, Z. Wen, J. Wang, X. Ai, Z. Yue, Thermomechanical fatigue behavior and failure mechanism of a nickel-based directional solidification column crystal superalloy, Eng. Fract. Mech. 292 (2023) 109674.
https://doi.org/10.1016/j.engfracmech.2023.109674

[37] B. Ohl, L. Owen, H. Stone, D.C. Dunand, Microstructure and mechanical properties of L12-strengthened Co-Ni-Fe-based superalloys, Mater. Sci. Eng. A. 884 (2023) 145276.
https://doi.org/10.1016/j.msea.2023.145276

[38] G. Gudivada, A.K. Pandey, Recent developments in nickel-based superalloys for gas turbine applications, J. Alloys Compd. 963 (2023) 171128.
https://doi.org/10.1016/j.jallcom.2023.171128

[39] A. Günen, M. Keddam, S. Alkan, A. Erdoğan, M. Çetin, Microstructural characterization, boriding kinetics and tribo-wear behavior of borided Fe-based A286 superalloy, Mater. Charact. 186 (2022) 111778. https://doi.org/10.1016/j.matchar.2022.111778

[40] M.I. Khan, M.A. Khan, A. Shakoor, A failure analysis of the exhaust valve from a heavy-duty natural gas engine, Eng. Fail. Anal. 85 (2018) 77-88.
https://doi.org/10.1016/j.engfailanal.2017.12.001

[41] C.R.F. Azevedo, H.B. Pereira, S. Wolynec, A.F. Padilha, An overview of the recurrent failures of duplex stainless steels, Eng. Fail. Anal. 97 (2019) 161-188.
https://doi.org/10.1016/j.engfailanal.2018.12.009

[42] B. Sutton, E. Herderick, R. Thodla, M. Ahlfors, A. Ramirez, Heat treatment of alloy 718 made by additive manufacturing for oil and gas applications, JOM. 7 (2019) 1134-1143.
https://doi.org/10.1007/s11837-018-03321-7

[43] Z. Huda, P. Edi, Materials selection in the design of structures and engines of supersonic aircraft: A review, Mater. Des. 46 (1980-2015) 46 (2013) 552-560.
https://doi.org/10.1016/j.matdes.2012.10.001

Superalloys: Fundamentals and Applications
Materials Research Foundations 178 (2025) 41-67

Materials Research Forum LLC
https://doi.org/10.21741/9781644903698-3

Chapter 3

Fundamentals and Applications of Nickel-based Superalloy

Uzma Hira[1], Iqra Ijaz[1]

[1]School of physical sciences, University of the Punjab, Lahore, Pakistan

uzma.sps@pu.edu.pk

Abstract

Nickel-based superalloys are essential for high-temperature applications due to their exceptional strength, corrosion, and oxidation resistance. Their microstructure, featuring a gamma matrix (γ), gamma-prime (γ') precipitates, carbides and borides provide stability at elevated temperatures. Synthesis techniques include casting, powder metallurgy, and additive manufacturing each having its own specification. There have been several generations of nickel-based superalloys, and each new generation came up with increased creep resistance and working temperature. Superalloys based on nickel were first developed with an emphasis on solid-solution strengthening; later generations added intricate alloying elements to improve precipitation hardening. Because of their remarkable strength and stability at very high temperatures, these superalloys find extensive application in industrial gas turbines, aircraft, and power generation. Future advancements focus on additive manufacturing, nano engineering, and new alloy compositions to further enhance performance in extreme environments.

Keywords

Nickel-based Superalloys, Gamma Matrix, Gamma-Prime, Microstructure, Additive Manufacturing. Creep Resistance

Contents

1.   Introduction

1.1   Definition and Overview

A superalloy is a type of metallic alloy that can be utilized at elevated temperatures, frequently surpassing 0.7 times of the absolute melting point [1]. The two main requirements for design of superalloys are resistance to oxidation and creep. High-performance alloys are another term that can be used to describe superalloys. It is important to distinguish "superalloys" from "alloys," as the latter have weaker performance characteristics. Thus, nickel-based superalloy is an emerging class of superalloys predominantly based on nickel, combined with iron, chromium, cobalt, and other metals to achieve desired properties. The preferred material for the hottest engine parts that must run at temperatures over 800 °C is a superalloy based on nickel also termed as Nimonic [2, 3]. The emergence of what we know as superalloys is looked at from the beginnings of high-temperature alloys in light of engineers is a desire for materials that can perform effectively under diverse temperature, stress, and environmental conditions. Superalloys are special because of their design characteristics, which include their mechanical strength, longevity, and resistance to a variety of things that could harm more traditional materials. Nickel based superalloys are acknowledged because of their material composition, which can be modified to yield specific features based on the intended application either in aerospace industry, chemical industry, nuclear reactors or power generation turbines [4].

1.2   Historical Development and Evolution

The history of nickel-based super alloys dates back to 1900s, when metallurgy was moving forward from the age of iron and copper in a drive to discover something much stronger and more corrosion resistant [5]. Observing the strength of rising heated air led humans to recognize the link between efficiency and high temperature. This insight led to the development of thermodynamic concepts, including the Brayton cycle, which is used in rotating engines. Intrigued by this concept, engineers were addressing their attention to something more reliable and efficient even at higher temperatures.

In 1903-1904 the first gas turbine was engineered and used to produce electricity in Europe. Under the supervision of Dr. Stanford A. Moss from Cornell University, New York a joint work with US army resulted in the invention of turbo supercharge engines that were used in airplanes during the World War I [6].

The mid-1900s were marked by significant changes and innovations brought about by World War II. By that time it was understood that not only strength and high temperatures efficient material, but also more reliable and economic friendly material was required [7]. So, after 1940s and World War II, superalloys have seen continuous advancement, growth and innovation.

World War II brought about a lot of rapid changes and innovations in the 1940s. This was soon supplemented by the needs for industrial "heavy duty" gas turbines, which required not just strength and high temperatures but also genuine economic feasibility and dependability.

The Huntington alloy products division of the international nickel company began developing a range of nickel-chromium-iron alloys in 1939. To optimize their high temperature strength, heat treatment producers involving solution treatment and precipitation hardening treatment were developed [8].

Subsequently, throughout the 1940s during World War II, the development of new alloy compositions and techniques led to significant advancements in superalloys. This was initially prompted by the military's use of jet engines and later by industry's requirement for industrial gas turbines. In order to increase the superalloys' strength and thermal stability, additional alloying elements including titanium and aluminum were added in the 1950s. During this time, famous alloys like Inconel 718 and Nimonic 80A were produced [9].

New processing methods including single crystal growth and directed solidification were developed in the 1970s and 1980s. Grain boundaries are weak spots in the material that were removed by these techniques, greatly enhancing the mechanical characteristics and creep resistance properties of superalloys. The possibility of high-entropy alloys, which combine several primary elements to produce improved qualities, has been investigated recently. These alloys mark a significant advancement in the creation of high-performing materials. The examples demonstrate how, in the 1950s and 1960s, alloy development practically took off, and how, in the 1970s and 1980s, to till now modern process development did the same [10].

1.3 Importance and Applications

As long as using superalloys is economical, they are employed in high-performance applications where their improved qualities may be utilized. Because of their remarkable qualities in harsh environmental conditions, such as good creep resistance, high-temperature strength, corrosion resistance, and fatigue performance, nickel-based superalloys are frequently used as essential materials in high-temperature structures in industrial gas turbines and aero engines [11]. Superalloys based on nickel feature remarkable microstructures that offer them superior high-temperature characteristics. Strength should decrease as temperature rises as is the case in almost all materials. Heat energy causes atoms to vibrate more quickly, increasing the likelihood that they may elude one another. But superalloys based on nickel are the exception [12].

Because of their anomalous yield strength, nickel-based superalloys are far stronger than most materials at a high percentage of their melting point. They perform so well at high temperatures because of this.

Superalloys based on nickel can withstand temperatures well over 1000 °C without losing strength or degrading. They are therefore invaluable in situations where resistance to thermal creep deformation and mechanical strength are both essential. The superior resistance of these superalloys to oxidation and corrosion is crucial for parts that are subjected to severe conditions,

Superalloys: Fundamentals and Applications   Materials Research Forum LLC
Materials Research Foundations 178 (2025) 41-67   https://doi.org/10.21741/9781644903698-3

including those found in gas turbines and jet engines. Aluminum and chromium are two elements that are used for supporting in the formation of protective oxide coatings. Because of their strong creep-rupture and tensile strengths, nickel-based superalloys can tolerate high mechanical stress without deforming [13]. Because of their remarkable fatigue resistance, they can withstand repeated loading and unloading without fracturing or experiencing other structural defects. Beyond aerospace, nickel-based superalloys find extensive use in gas turbines for power production, marine propulsion, chemical processing facilities, and even in medical equipment like implants and prosthetic limbs. Because of these structural and technical developments, nickel-based superalloys are now unique and extremely significant materials.

2. Fundamental Properties

These combined characteristic properties provide lifespan, safety, and efficiency for nickel-based superalloys in some of the most demanding engineering applications.

2.1 Mechanical Properties

The components that make up superalloys are essential in producing an exceptional blend with different mechanical characteristics [14]. At extreme temperatures, extraordinarily high mechanical properties, including ductility, hardness, tensile strength, creep and fatigue resistance are attained by carefully controlling the alloy composition, microstructure, and heat treatment methods, allowing nickel-based superalloys to operate effectively in demanding applications. Generally speaking, adding elements such as Al, Ti, Ta, Cr, W, Nb, C, B, Mo, V, Zr, Re, etc. could improve the mechanical strength and properties of nickel super alloys [15-17]. The mechanical properties of superalloys based on nickel are greatly influenced by their microstructure. The Hall-Petch relationship suggests the strength of metals and alloys improves with smaller grain size.

Fine grain microstructure typically limits tensile characteristics, while coarse grain microstructure enhances creek and fatigue characteristics at higher temperatures. Thus the dependency of mechanical strength of superalloys on their microstructure can be improved by grain boundary strengthening techniques. Among the techniques that alter the microstructure are carbide transformations, heat treatment, and additive fabrication [18]. Machine learning methods are used to predict the mechanical characteristics of materials such as strength, durability, and heat resistance. This aids scientists and engineers in developing superalloys with enhanced performance properties. They can use machine learning to swiftly find useful compositions and processing processes without relying on considerable trial and error, accelerating the development of these high-performing materials [19].

2.2 Thermal Properties

Superalloys based on nickel are well known for having exceptional thermal characteristics, which makes them perfect for high-temperature purposes like turbine engines. More thermal conductivity in the superalloys ensures better cooling and heat dissipation in the blade design. The change in thermal conductivity depends on the number of different types of materials that affect thermal conductivity based on nickel, and the variation in thermal conductivity is influenced by the chemical composition of superalloys [20]. Mechanical qualities of nickel based super alloys are improved by adding different chemical compositions; however, these chemical compositions cause point defects in the nickel lattice, resulting in lowering the alloy's thermal conductivity. In this regard, it is noticeable that the conductivity of nickel based super alloys remains relatively unchanged when ThO_2 is added to pure nickel. At room temperature, nickel, the constituent of nickel-based superalloys, has a thermal conductivity of roughly 106 $Wm^{-1}K^{-1}$. Similarly, thermal expansion is another important engineering factor while employing nickel based super alloys for higher temperature purposes [21].

Deformation and degradation due to oxidation at elevated temperature environment can be avoided by the creation of a tight, continuous surface scale that serves as a diffusion barrier and does not spall off during heat cycling. Elements like aluminum, chromium may create the protective oxides Cr_2O_3 and Al_2O_3, nickel-chromium alloys with high aluminum content, such 713C and B-1900, are often thought to have great resistance to oxidation [22].

2.3 Chemical Properties

The chemical makeup of alloying elements can significantly alter chemical properties like resistance to oxidation and corrosion [23]. This resistance is mostly attained by carefully choosing and combining the alloying components to create stable, shielding oxide layers on the surface of the alloy. Ni-based superalloy oxidation behavior is heavily influenced by alloy composition, operating temperature, and cyclic conditions. Fortunately, for high temperature applications, both Ni and Ni_3Al exhibit comparatively high oxidation resistance as compared to other alternative alloy systems. Elements like aluminum and chromium generate stable and protective oxide layers of alumina Al_2O_3 and chromia Cr_2O_3, which can lower the growth rate of oxide scale, thereby establishing a stable and long-lasting oxidation and corrosion resistance barrier across the surface [24].

Two heat-corrosion mechanisms that operate in the 600–950°C temperature range are high and low-temperature corrosion, also referred to as Type I and Type II. Type I corrosion, also known as "sulfidation," occurs in the temperature range of 800-950°C and is primarily caused by oxygen reacting with sulphur in the air or other impurities [25]. As a result, it can damage the surface of an alloy by penetrating deeply and destroying the nickel surface coating. Typically, a layer of chromia Cr_2O_3 formed with a high chromium content substantially decreases the reaction rate. The temperature range of 600–700°C is where Type II corrosion attack occurs most frequently. In this temperature range, nickel or cobalt in combination with a low melting eutectic phase can generate corrosion pits that interact with surface oxide materials that are eroding from the surface. Because this kind of erosion is susceptible to fuel's sulphur concentration, it can be lessened by adding more chromium and aluminum [26, 27].

3. Composition and Alloying Elements

3.1 Role of Nickel

The main component and backbone of nickel-based super alloys, nickel, comprising of more than 50% of alloy's composition has certain special and equisetic qualities. With a face-centered cubic crystal structure, it can withstand a wide range of temperatures and acts as a matrix to incorporate other alloying elements. Face centered crystal structure of nickel enables it to move thus provides better dislocation density [28]. Because of its exceptional strength and resilience to withstand deformation and retain strength at high temperatures, this solid solution matrix can endure applications that involve high temperatures making nickel-based superalloys a perfect candidate

Materials Research Forum LLC
https://doi.org/10.21741/9781644903698-3

for usage in jet engine and power plant turbine blades and other components. Furthermore, nickel has outstanding resistance to oxidation and corrosion, which is essential for durability. Its durability is further increased by the fact that it may create stable, protective oxide layers on its surface. Additionally, nickel possesses strong mechanical qualities, such as high tensile strength and toughness, which are necessary to survive the mechanical pressures that come with operation [29]. In general, nickel improves resistance to carburisation, oxidation, nitridation, and halogenation as well as strength at higher temperatures. In addition, it enhances stress corrosion and cracking at high temperatures and offers metallurgical stability.

Depending upon the operating conditions, nickel's ability to tune its properties with other alloying elements by incorporating them in solid matrix for optimum results makes it indispensable for advanced engineering applications.

3.2   Chromium and Oxidation Resistance

The major purposes of the addition of chromium, a week solid-solution strengthener, are oxidation, hot-corrosion resistance and sulphidation [30]. Depending upon the targeted application the percentage weightage of chromium added to the nickel based super alloy may vary from 10-30%. As chromium provides good balance of oxidation and corrosion resistance so depending upon its specific application amount of chromium can be vary to withstand environmental conditions and stress [27].

In high temperatures and oxidative conditions, chromium develops a persistent and protective layer of chromium oxide on the alloy's surface. This shield protects the underlying metal from corrosion and stops additional oxidation, which is crucial in high-temperature applications like industrial gas turbines and turbine engines. A suitable range for chromium is 5.0–9.5% weight of its weight correspondingly for binary systems, Cr × Al, Cr × Co, and Cr × Mo provide synergistic benefits. To protect against heat corrosion, alloys used in power-generating turbines have a higher percentage of Cr [31].

In addition of providing oxidation and thermal resistance, chromium contributes to mechanical strength and durability of alloy, helping it in carbides formation that enhance alloys creep resistance and thus overall improving its mechanical capability. The presence of chromium in the alloy matrix also helps to stabilize the austenitic structure, which is necessary for retaining ductility and toughness at high temperatures.

3.3   Strengthening Elements: Cobalt, Aluminum, Titanium

Cobalt, aluminum and titanium solid solution strengtheners are added to form sub-micron size cuboidal shaped gamma prime phase γ' in the face-centered cubic (fcc) gamma phase (γ) of nickel matrix to impart strength and durability to the matrix [32, 33]. All these elements are capable of forming Ni_3X type compounds where X stands for aluminum, cobalt or titanium each of which are stable at relatively high temperature thus imparting strength to nickel super alloys. Despite the fact that these compounds' structures seem to differ greatly from one another and from nickel's fcc structure, all of these structures have very similar atomic arrangement [34].

As explained in Table 1, aluminum is an alloying element that forms aluminum oxide at high temperatures and resists oxidation and also offers strength at elevated temperatures. It works with Ti to maintain age hardening in some alloys [35, 36].

Superalloys: Fundamentals and Applications   Materials Research Forum LLC
Materials Research Foundations 178 (2025) 41-67   https://doi.org/10.21741/9781644903698-3

Ti presence significantly enhanced solid solution strength in the later γ' phase. Although Ti is the most potent solid solution strengthener in the γ phase, its influence was minimal due to its low concentration in the matrix. Similarly, cobalt contributes to the mechanical stability and consistent performance of nickel-based super alloys under thermal and mechanical stress by giving them additional stability by strengthening the γ' phase [37].

Table 1. Properties imparted by the addition of various additive elements in nickle superalloy.

Additives of nickel based superalloy	Provided property
Chromium, Tungsten, Iron, Tantalum, Molybdenum,	Higher strength
Boron, Carbon, Zirconium	Creep resistance
Chromium, Tantalum, Aluminum	Oxidation resistance
Aluminum, Titanium	High temperature strength

3.4   Refractory Elements: Tungsten, Molybdenum, Tantalum, Rhenium

In accordance with Metals Handbook, published by The American Society for Metals [38] refractory metals are the metals having melting points above the melting point range of iron, cobalt, and nickel. This expanded definition would include tungsten, molybdenum, tantalum, niobium, chromium, vanadium, rhenium and a few lesser-known metals in the category of refractory elements [39]. A high hardness at ambient temperature and a melting point above 2000 °C are two characteristics they all have in common. They have a relatively high density and are chemically inert.

Turbine disc materials and other high temperature applications need to have a low density and low raw material prices in addition to having a high yield strength, microstructural stability, creep resistance, fatigue strength, and oxidation resistance because they typically run at maximum service temperatures of 750–950 °C. As a result, the incorporation of other metals with these exceptional qualities is inevitable. As the refractory elements belong to various periodic table groups, their physical characteristics differ greatly from one another. These elements increase an alloy's ability to withstand oxidation and spalling thereby increasing grain boundary strength as they are solid solution strengtheners. These refractory elements exhibit a balanced distribution between the γ and γ' phases, raising the overall volume fraction of the alloy [40]. While most W and Ta atoms are found in the Al site of the γ' in Ni_3Al phase, the majority of Mo and Re atoms are mostly distributed in the γ-Ni phase providing necessary strength against high temperature [25].

3.5   Minor Additives: Hafnium, Zirconium, Carbon, Boron

Even in minimal amounts, trace elements, also known as microalloying elements, can significantly affect the alloy's characteristics and performance [41]. These microalloying elements like hafnium, zirconium, carbon, boron is added to the nickel superalloys to form a submicron size cuboidal shape γ' precipitates in Ni_3Al phase in the gamma matrix. In addition to that, these minor additives also provide solid solution strengthening of γ' phase. Being a potent carbide maker, carbon aids in the creation of carbides such as MC, $M_{23}C_6$, and M_6C, which reinforce the alloy by obstructing the movement of dislocations. Thus, the addition of few additives even in smallest proportion of about 0.5wt-% can improves alloy's cast ability and weldability [42].

4. Microstructure

Nickel-based superalloys are primarily composed of a matrix phase γ, from which a fine, uniform dispersion of γ', one or more carbide phases, and typically at least one other minor phase, flakes [43]. Overall composition of the material is one of the most significant factors influencing the microstructure. Additionally, the γ-phase plays a crucial role in controlling the microstructure because, in almost all the alloys, the alloying elements are soluble in the nickel within a limited temperature range. Consequently, γ-phase needs consideration.

4.1 Gamma (γ) Phase

The primary matrix phase of solid solution in nickel based super alloys is the gamma phase comprising of nickel along with cobalt (Co), chromium (Cr), molybdenum (Mo), and tungsten (W) [44]. The γ-phase comprises of a solid solution having a face centered cubic lattice structure and an atom species distribution of alloying elements that is random. The role of the alloying elements in Ni-based alloys must be taken into account since the FCC matrix in these alloys acts as the solvent for a number of alloying additions. The elements Mo, W, and Cr are significant for their impact on carbide formation and stability along with their role in solid-solution strengthening; Cr is also required for corrosion resistance. It has been discovered that adding boron and zirconium prolongs the alloys' stress-rupture life as they fill in the spaces at the grain boundaries thus creating a stabilized structure. This phase is present in all nickel-based superalloys as the matrix contributing to structures mechanical strength, oxidation resistance and stability.

4.2 Gamma Prime (γ') Phase

The majority of the strength of nickel-based superalloys comes from an intermetallic homogeneous precipitation phase based on Ni_3AlTi [45]. Therefore, aluminum and titanium are the key elements in this phase that precipitates coherently with the FCC gamma matrix. Ni_3Al is an ordered system with Ni atoms on the cube faces and the Al atoms at the cube edges as shown in Fig. 1. Both the gamma and gamma prime phase have very slight difference in lattice constant (0.5%) thus retaining the FCC structure as well. The material's remarkable resistance to creep deformation and its strength at elevated temperatures can be attributed mostly to the γ'. Temperature and chemical composition affect the amount of γ' [46]. Although both titanium and aluminum provides strength but almost all commercial alloys contain both Al and Ti because titanium bearing alloys are more stable at higher temperatures and less prone to overaging than those containing solely Al. This phase often appears as evenly distributed cuboidal particles in the γ matrix.

4.3 Gamma Double Prime (γ'') Phase

Alloys that require increased strength at lower temperatures, such as turbine discs, can benefit from the use of another phase called γ''. This phase is found in nickel superalloys that include substantial additions of vanadium or niobium (Inconel 718) in these cases, the γ'' is composed of Ni_3Nb or

Ni$_3$V. Body centered tetragonal (BCT) crystal structure is present in γ'' phase. As this provides strength at reduced temperatures, it is usually unstable above 650 °C [47].

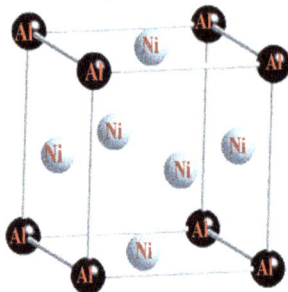

Figure 1. Crystal structure of gamma prime γ' phase.

4.4   Carbides and Borides

Carbon and carbide formation are regarded as a disadvantageous in most nonferrous alloys and even in some ferrous systems because they typically have a negative impact on mechanical strength [48]. Nevertheless, carbide production is necessary in Ni-based superalloys to stabilize the structure against high-temperature deformation. Metal carbides (MC) network like MC, M$_{23}$C$_6$, and M$_6$C are produced when carbon is provided 0.02 to 0.2% along with reactive elements like titanium, tantalum, hafnium, and niobium that represents M [49]. The primary processes leading to the creation of these carbides are given in equations 1 and 2 as follow:

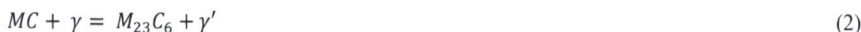

$$MC + \gamma = M_6C + \gamma' \tag{1}$$

$$MC + \gamma = M_{23}C_6 + \gamma' \tag{2}$$

Grain-boundary motion is inhibited by these networks of carbides, which increases creep and stress-rupture qualities. The kind and shape of the carbide present determines how it affects high-temperature characteristics [50].

A comparatively low density of boride particles is produced when boron separates to form grain boundaries.

5.   Manufacturing Processes

Casting techniques are as old to history as is the metallurgy itself. Superalloys are made through a number of crucial processes that are meant to provide materials with remarkable mechanical properties and high-temperature characteristics. Firstly, the exact formulation and melting of the alloy's component parts come and then casting into the required shapes using vacuum induction melting or investment casting is the next step. Followed by subsequent phases of heat treatment to optimize the microstructure and mechanical properties, as well as finishing procedures including machining and surface treatment. In order to meet the final requirements needed for demanding applications in the aerospace, power generation, and industrial sectors, these steps must be followed, the details of each step involved are given below:

5.1 Casting Techniques

5.1.1 Investment Casting

Investment casting, sometimes referred to as precision casting or lost-wax casting, is a common manufacturing technique used to prepare intricate and high-quality components of superalloys. Investment casting has been around since the ancient Egyptians utilized it in various forms between 4000 and 3000 B.C., but it wasn't until the second world war that its industrial significance and popularity increased because of the increased need for parts of aircraft engines and the airframe industry [51].

Being able to create extremely precise dimensional pattern, complex shapes and no parting line with a shallow depth feature makes this technique exceptional. It involves these basic steps:

5.1.1.1 Master Pattern Creation

A professional mould maker or sculptor uses wax, clay, wood, steel, or plastic to create a master pattern based on the specifications for the finished product. These are referred to as "master patterns" because they have two shrinkage allowances—one for wax and the other for casting material shrinkage. This pattern is created by injecting melted wax into metal die and allowing it to get hard [52]. Based on the size and material to be utilized this pattern creation can take some time and trials to get the required size and shape.

5.1.1.2 Shell Building and Stuccoing

The wax pattern is dipped into a slurry of finely crushed refractory material and emptied, producing a "prime coat," which is a homogeneous surface coating. To get a very smooth finish a very fine layer of homogenous particle size is applied. Refractory materials include fine silica, alumina or zirconia and other binders, such as water, ethyl silicate, and acids are also added. Once the prime coat is dried and hardened, the pattern is repeatedly coated with ceramic material in order to achieve strength and thickness. This process is called "stuccoing" which increase the thickness from 5-15mm [53].

5.1.1.3 Dewaxing

After the ceramic mould is fully dried, it is placed in a de-waxing autoclave upside down from which the wax is melted, recovered and recycled. This inverted position allows the removal of any residual wax and vaporize any subsequent residue leaving behind a hollow ceramic mould. Ceramic mould is heated (90- 175 °C) for 4-5 hours to drive off any water of crystallization. Most prime shell failures occur during dewaxing stage thus a crucial and critical step [54].

5.1.1.4 Burnout Preheating and Casting

The mould is burnout and preheated at a temperature of about 870 °C-1050 °C before pouring the melted molten alloy. Molten metal up to 3000 °C is then poured into the sintered ceramic mould and allowed to cool. Preheating enables the mould to keep molten metal for a longer period of time, improving its ability to fill all mould details and raising dimensional accuracy. To achieve high-quality casting and prevent oxidation, molten superalloy is poured into the warmed ceramic mould under vacuum or inert gas [55]. Although gravity pouring is the easiest, there are alternative ways to make sure the mould is completely filled. Centrifugal casting, tilt casting, vacuum casting, or positive air pressure casting may all help in mould filling when intricate, thin parts are involved.

5.1.1.5 Shell Knockout and Post Processes

On the solidification of the metal, the mould is broken and the metal casting is removed using methods such as mechanical chipping, hammering, vibrating, high-pressure water jetting, and media blasting. Finally, the casting is subjected to heat treatment, surface treatments and machining to finalize the component.

5.1.2 Directional Solidification

Directional solidification also known as progressive solidification is another casting technique aims to control the solidification process in order to get a columnar growth of casting material thus enhancing its strength and creep resistance [56]. From the mould preparation to pouring of molten metal it follows the same basic step as in investment casting. However, during the drying and post casting step the mould containing the molten alloy moves across a temperature gradient in a furnace. A water-cooled chill plate or other comparable cooling mechanism is used to cool the mold's base while keeping the top heated. As a result, a directed solidification front is produced, which advances upward and encourages the formation of a single crystal or aligned columnar grains.

Directional solidification is also one of the purification method that can be employed to purify the final solidified metal enriched with impurities. The preferred method for casting high temperature nickel-based superalloys used in aircraft turbine engines is directional solidification as grain boundaries perpendicular to the plane of stress are reduced through directional solidification [57]. Thus, directional solidification method is employed to prepare high performance super alloys components.

5.1.3 Single Crystal growth

As mentioned above, directional solidification is used for the formation of single crystal growth. On the basis of cooling techniques employed during directional solidification, Bridman method is employed industrially as it provides improved mechanical properties, reduce production cost and specifically design to prepare single crystal turbine blades. In order to generate single crystals with the correct crystallographic orientations, the approaches include heating the polycrystalline material to its melting point and then progressively cooling it from the end of the container that resides the seed crystal [58]. As the seed crystal provides a template, the atoms in the molten material can align themselves in a regular, uniform around the seed crystal. Compared to polycrystalline alloys, single crystal superalloys have better resistance to creep and thermal fatigue because they lack grain boundaries as the growth of single crystal is slow and require highly controlled movement of molten metal. The whole process is carried out in Bridgman furnace that contains basically three different temperature zones:

- The uppermost region where the temperature is leveled above the melting point.

- The lower zone, where temperature is lower than that of melting point.
- Serving as a buffer between the two is an adiabatic zone [59]

Schematic diagram of Bridman furnace is shown in Fig. 2 that clearly indicates these three temperature zones.

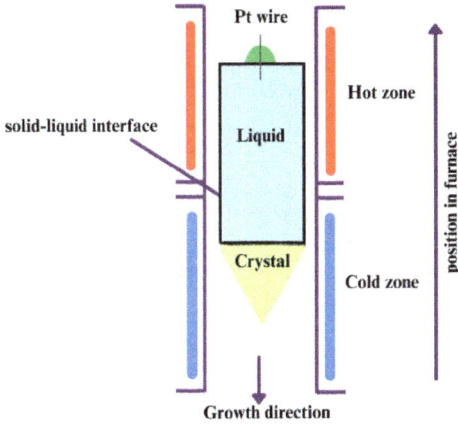

Figure 2. Schematic diagram of bridman furnace.

5.2 Powder Metallurgy

The production of powder metallurgy-based nickel alloys dates back to 1960s [60]. Until now, several developmental adjustments have been implemented. Superalloys like Waspalloy, which only contains a minimal quantity of strengthening elements (Al, Ti, and Nb) combined with Ni and Cr, can be produced using the traditional casting technique. However, a lot of additional elements may generate segregation during the casting process, which can have a substantial effect on the thermal properties of casting alloys. Furthermore, casting alloys frequently fracture during hot-working procedures. The powder metallurgy process has thus been developed. Superalloys manufactured via powder metallurgy have strong fatigue and yield strength, as well as a consistent composition and structure [61]. The powder metallurgy technique is commonly utilized to produce high-performance aero-engine components.

5.3 Additive Manufacturing

Additive manufacturing (AM) or additive layer manufacturing (ALM), the terms used in industrial production to describe 3D printing, a computer-controlled process that deposits materials, usually in layers, to generate three-dimensional objects [62]. These are made layer by layer, just like 3D printing, as opposed to traditional production, which frequently requires cutting or other techniques to eliminate extra material. Building molecules layer by layer requires the following basic steps:

- Make a digital 3D model with computer-aided design (CAD) software.

- Transform the model into a file format that the 3D printer can read, such as STL. Using slicing software, split the model into tiny horizontal layers.

- Fill the 3D printer with the selected material (such as metal, plastic, resin, etc.).

- In accordance with the sliced model, the printer deposits or fuses material layer by layer. Diverse additive manufacturing approaches employ various techniques (such as laser sintering, extrusion, or jetting etc to be discussed later) to construct the object.

- To get the desired qualities and finish after printing, the object might need to go through post-processing operations like cleaning, curing, heat treatment, or surface finishing [63].

Several unique AM processes exist, each with its own specific criteria. These include:

Binder jetting

Binder jetting is a 3D printing technique in which layers of powdered material are produced and a liquid binder is used to hold the layers together. The procedure uses a print head moving along the x, y, and z axes, similar to how a standard 3D printer operates, to produce the object layer by layer [64].

Directed energy deposition

A concentrated energy source, such as a laser, plasma arc, or electron beam, is used in the Directed Energy Deposition (DED) 3D printing technique to melt a material that is subsequently sprayed via a pointed nozzle [65]. Similarly, in directed energy deposition arc method (DED-arc), arc welding power source is used to build a 3-D shape that commonly utilizes wire as material source.

Material extrusion

To create a three-dimensional object, a spool of material typically thermoplastic polymer is continuously forced into a heated nozzle that moves horizontally while the bed moves vertically thus selectively build layer by layer. Generally speaking, In comparison to other additive manufacturing techniques, material extrusion is slower and less accurate.

Powder bed fusion

A laser or an electron beam is used in powder bed fusion (PBF) techniques to melt and fuse powdered materials together after which excess of the material is blasted away. Many additive manufacturing processes, including as electron beam melting (EBM), selective laser sintering (SLS), selective heat sintering (SHS), direct metal laser melting (DMLM), and direct metal laser sintering (DMLS), are incorporated in powder bed fusion [66].

Laminated object manufacturing

The quick prototyping method known as laminated object manufacture (LOM) was first created by Helisys Inc. Laminates of metal, plastic, or paper with adhesive coatings are used in LOM technology as a 3D printing material. These material sheets are adhered to one another layer by layer and then shaped using a computer-controlled laser or blade. It continues to be the quickest and cost-effective methods available for building 3D prototypes [67].

Superalloys: Fundamentals and Applications
Materials Research Foundations 178 (2025) 41-67

Materials Research Forum LLC
https://doi.org/10.21741/9781644903698-3

Vat polymerization

Layer by layer, the model is built using a vat of liquid photopolymer resin in a process known as vat polymerization. When necessary, an ultraviolet (UV) light is utilized to dry or stiffen the resin, and once each new layer is hardened, a platform is used to move the object being created lower.

Material jetting

Layer by layer, droplets of the build-up material typically photopolymers are deposited; they are subsequently cured by UV radiation. Material jetting works identically like a two-dimensional ink jet printer to produce objects [64]. Material is dropped onto the build platform via two methods: continuously and through Drop on Demand (DOD).

6. Advanced Superalloys

As time went on, new generations of superalloys evolved, each of which with substantial innovations and better properties than the one before it. These advancements were made by adding some new components to the mixture as well as modifying the composition of the alloying elements. Additionally, Ni-superalloys can be divided into six generations based on their elemental compositions; the sixth generation of Ni-based superalloys was created very recently, in the early 21st century [68]. To achieve these desired qualities, new super alloys need costly alloying metals like ruthenium and rhenium

6.1 First Generation Superalloys

In the 1950s, vacuum melting and casting were developed, leading to the development of alloys from wrought superalloys. The principal alloying elements found in first-generation Ni-superalloys are Cr, Co, Mo, Al, Ti, Ta, W, and occasionally Nb, V, Y, or Ca, which are added to improve certain properties of the alloys including creep resistance and oxidation resistance. Various elemental compositions were developed and improved for better creep resistance and oxidation resistance but it was soon realized that when the alloying composition exceeds a certain limit these alloying elements eventually cause the weakening of γ phase. René N4 synthesized by GE [69] and PWA1480 developed by Pratt and Whitney [70] were the first commercialized nickel based superalloys belonging to first generation.

6.2 Second-Generation Superalloys

Turbine engine exposure with the first-generation superalloys has led in additional alloy research with a fundamental plan target of increasing metal temperature by around 30°C.

To avoid the problems faced in first generation superalloys and keeping the γ' volume percentage above 60%, refractory alloying elements needed to be considerably increased. Thus the addition of small amount of Rhenium gave birth to second generation superalloys. Although rhenium is a relatively costly additive, it improves fatigue resistance and creep strength. Aside from creep resistance properties, other characteristics considered during development of second-generation superalloys includes microstructure stability, castability, oxidation resistance and thermal fatigue resistance properties. PWA1484 [70] was created with 20wt% total refractory materials, reduced Cr from 10wt% of PWA1480 [71] to 5wt%, and included 3wt.% Re. CMSX4, Rene5, and SC180 are other second-generation superalloys with 3wt% Re and comparable metal temperature capabilities.

6.3   Third-Generation Superalloys

Second-generation single crystal alloys have been successfully used in gas turbines, although there is a need for increased temperature stability. Encouraged by the 2nd generation's improved metal temperature capabilities of 30°C with the addition of 3wt% Re, pursued the formation of 3rd generation nickel based super alloy with about 6wt% Re and a minor increase in Mo. The creep qualities were seen to significantly improve with such combination [72]. However, the solid solution limit was reached at these high Re concentrations, and this was observed to cause the production of topologically closed-packed (TCP) phases at high temperatures that were detrimental to creep properties. These phases are typically rich in additive elements and form a complex crystal structure of closed packed layer of atoms. Tetragonal σ phase and orthorhombic P phase get aligned with the FCC matrix of nickel thus depleting its strengthening elements from its microstructure. After tailoring Cr, Co, and W level to 2, 3, and 5 wt%, respectively, Cannon Muskegon Corporation, USA, optimized their third- generation single crystal alloy CMSX10 using total refractory components of roughly 20 wt% with 6 wt% Re [73].

6.4   Fourth and Fifth- Generation Superalloys

Ruthenium was added to third-generation superalloys to create the fourth generation of Ni-superalloys. Ru was seen to both prevent the formation of the TCP phase and stabilize the microstructure, which is why this was achieved [74]. Additionally, it was seen to greatly improve the working temperatures and creep characteristics. These fourth generation superalloys didn't find to have any grain boundary strengthening elements. While the flaws in fourth generation were overcome in fifth generation superalloys where the amount of ruthenium has been increased to 6wt% and Mo content was also increased from 2.8-3.8%. This seem to stabilize the microstructure and improves the stability as well. TMS-162 and TMS-173 were the example of fifth generation superalloys. The lower oxidation resistance of TMS-162 and TMS-173 was then further improved in sixth generation superalloys [75, 76].

As technology improves and material demands become more complicated, the development of novel superalloys with improved characteristics is crucial. These new challenges need superalloys to satisfy higher performance standards in different severe situations, pushes limits of what these materials can offer.

6.5   Oxide Dispersion Strengthened (ODS) Superalloys

The uniform dispersion of fine nanoscale oxide particles (like Y_2O_3 yttrium oxide) inside a metallic matrix is the defining characteristic of oxide dispersion strengthening (ODS) alloys [77]. By adding other elements to the matrix, alloy oxide nanoparticles can be used to control their size and can be tuned towards more refined design. The aim of adding nanoscaled oxide particles to nickel-based superalloys is to enhance their mechanical characteristics, such as strength and resistance to creep, especially in high-temperature environments. These superalloys are frequently used in conditions where traditional superalloys may fail, such as gas turbines, nuclear reactors, and aircraft components. The presence of γ' forming agents titanium and aluminum in nickel based superalloys makes them useful in high temperature properties [78].

Superalloys: Fundamentals and Applications
Materials Research Foundations 178 (2025) 41-67

Materials Research Forum LLC
https://doi.org/10.21741/9781644903698-3

6.6 High-Entropy Alloys

Conventional commercial alloys can accomplish excellent creep strength, heat resistance, and strength to weight ratio, but they have drawbacks such as a high density and a reduced specific weight, which are detrimental to structural applications. In addition to that, the comparable temperature limits of conventional superalloys have been approached, attempts have been made to improve cost performance by lowering pricey components like rhenium [79]. To obtain greater high-temperature performance, alternative alloy designs are required because there is limited room for additional composition alterations. HEAs thus, are appealing materials for numerous purposes due to their unique features resulting from primary high entropy effect. These alloys were developed using equiatomic substitution, which involves substituting individual components with chemically equivalent amount [80]. To effectively utilize HEAs and all of its diversity, a thorough assessment of their mechanical properties at high temperatures is currently required.

7. Applications

7.1 Aerospace Industry

Modern jet engine blades must withstand what is arguably one of the most extreme environments any engineered substance could ever encounter, working at temperatures near the material's melting point, spinning hundreds of times per second, and supporting a load equal to a family car's weight. Nickel based superalloys have outstanding resistance to oxidation, corrosion and high temperature, which is necessary for long-term fitness in aerospace applications [81]. As the Fig. 3, explains about 70% of nickel based superalloy consumption occur in aerospace industry, it is usually used in the manufacturing of:

NICKLE SUPER ALLOY APPLICATIONS

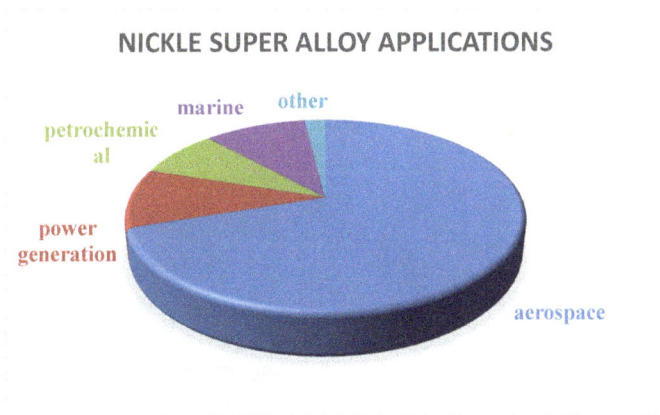

Figure 3. Consumption of nickel superalloy in various fields.

7.1.1 Turbine Blades

The capacity of nickel based superalloys to retain mechanical strength at temperatures higher than 1,000°C significantly enhances engine performance and efficiency. Their unique ability to withstand high temperatures, makes them ideal for use in the hot section of jet engines as engine parts especially turbine blades and discs.

The aerofoil segment of turbine blade often experiences temperatures ranging from 650 to 980°C and longitudinal loads up to about 20,000 psi (138 MPa). This demands high tensile strength, fatigue resistance and high temperature durability. The most often utilized material, Inconel 718, makes up half of an aeroplane engine. Ni-Fe-Cr alloy in industrial aircraft currently in use, nickel-based superalloys comprise 40–50% of the engine weight. These alloys are particularly utilized in the combustion and turbine parts, where operating temperatures exceed 1250 °C. Grain boundary structure can be carefully controlled through the complex casting process, used to make nozzle director blades and turbine wings [82]. Extensively used nickel-based alloys on commercial scale includes Inconel (587, 600, 625, 706, X750, etc.), Nimonic (75, 80A, 90, 263, PE 11, C-263, etc.), Rene (41, 95), Udimet (400, 500, 630, 700, etc.)

7.1.2 Combustor Liners

High compression systems, such as the combustion chambers of jet engines, combine high pressure air and fuel to burn at a steady pressure as indicated in Fig. 4.

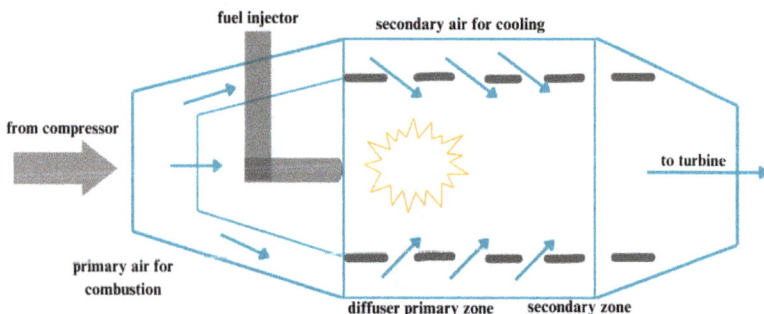

Figure 4. Schematic diagram of a combustor [84].

Because they are situated between the engine's turbine and compressor sections, these liners are made to endure exceptionally high pressures and temperatures. The main requirements are resistance to oxidation, as well as resistance to thermal fatigue and buckling [83]. To make manufacturing easier, the material must also have good formability and weldability. Because of these factors, this item is usually made of high strength, high temperature resistant materials, such as nickel-based superalloys, in sheet form.

7.2 Power Generation

A result of the growing demand for engines with improved efficiency leds to the development of superalloys that prevent harmful grain boundary effects, which weaken material at high temperatures. These superalloys make it possible for turbines to run at higher temperatures, which boosts productivity and lowers fuel consumption, two essential components of both conventional and renewable energy systems. Additionally, because of their resilience, turbine components have a longer lifespan, which lowers maintenance costs and plant downtime.

7.2.1 Gas Turbines

Advanced nickel based alloys designed for gas turbine blade applications have a reduced chromium content and strictly refrain from containing compositions that could result in the creation of topological close-packed (TCP) and other phases, which may adversely affect the performance of the component parts. A combustion chamber is required for gas turbines in order to burn fuel and generate the hot gases that power the turbine. The development of single-crystal superalloys that prevent harmful grain boundary effects, which weaken material at high temperatures, is a result of the growing demand for engines with improved efficiency [85]. A significant amount of work has been invested into creating alloys for IGT applications that will match the demands for increased strength, resistance to oxidation and corrosion, and high amounts of Cr, which are essential as the low level of Cr make them susceptible to corrosion.

7.2.2 Steam Turbines

The production of electrical power by fossil fuel-fueled power plants is responsible for about 40% of global CO_2 emissions. Thus, improving the efficiency of fossil fuel-fueled power plants is a very efficient way to lower greenhouse gas emissions and enhance energy conservation. The creation of a new type of steam turbine that can be used in cutting-edge coal-fired power plants 9with potential operating temperatures higher than 700 °C. Some of commercially available and utilized nickel based superalloys for turbine blades and engine parts applications are N754, IN X-750, Nimonic 115, Nimonic 75, Nimonic 80, Nimonic 90, Rene 41 etc [86].

7.3 Marine Applications

The increasing demand for energy has prompted oil and gas production efforts worldwide. During exploration and development, one has to cope with more hostile situations such as deep offshore wells, increased temperatures and pressures, sour wells, etc. Because of this, building materials that are very resistive to corrosion in the harsh environments of production and refining are required. The highly durable nickel-base superalloy grades 686 and 725 have shown to be great options for fasteners, springs, and other maritime applications. Similar to superalloy grade 686, the Ni-base superalloy grade C 276 also has high corrosion resistance to seawater, particularly under conditions of crevice corrosion [87]. Thus, they are utilized for the manufacturing of exhaust systems, turbines, and other parts that are subjected to seawater's corrosive properties as well as high temperatures. They are also utilized in shipbuilding for parts that need to be extremely durable and strong.

7.4   Chemical and Petrochemical Industries

Because of their resistance to corrosion and high pressure conditions, nickel-based superalloys are used in the oil and gas industry in down hole tools, wellhead components and pipelines. Because of their chemical resistance and thermal stability, which guarantee long-term reliability, these alloys are crucial for petrochemical amenities' reactors, heat exchangers, and high-temperature equipment. Within solid solution strengthened alloys, the two most significant alloys are Ni-based grade 625 (UNS N 06625) and Fe-Ni based superalloy grade 825 (UNS N 08825) [88]. The workhorse of the present oil and gas sector, grade 825 is used extensively in pipelines and flowlines. It is employed in sour gas wells under cold-worked conditions. Because of its high weldability and great behaviour in the majority of oil and gas conditions, grade 625 is now the preferred option for designers when it comes to welding applications.

7.5   Other Applications

In addition to their excellent usage in various fields, nickel based superalloys are also used for the manufacturing of hospital equipment, implants, and surgical. These superalloys are in use because they are corrosion-resistant, safe, and capable of producing a highly refined surface finish.

As the automotive industry grows, nickel alloys are being used in an increasing amount of car manufacture to enhance vehicle performance. Superalloys are utilized extensively in auto parts these days, including nozzles, exhaust valves, ignition prechambers, exhaust gas cleaner fasteners and turbochargers [11].

8.   Challenges and Future Directions

8.1   Cost Considerations

Because nickel, cobalt, and other alloying components are expensive raw materials, nickel-based superalloys are costly. Production expenses are further raised by the complicated manufacturing procedures like heat treatment and precise casting. Because of these reasons, it's critical to maximize resource utilization and look into affordable options without sacrificing effectiveness.

8.2   Manufacturing Challenges

Because of the extreme strength and toughness of nickle based superalloys, their production requires complex procedures such melting, casting, forging, and machining. Three major issues are ensuring consistent microstructure, preventing flaws, and maintaining tight tolerances. To solve these problems, cutting-edge methods like additive manufacturing are being researched [89].

8.3   Recycling and Sustainability

Because it is hard to separate the alloying elements and keep the material pure, recycling nickel-based superalloys is tough. Yet, the need for sustainability is propelling initiatives to create effective recycling techniques and lower industrial waste.

8.4   Future Research Trends

Future studies on nickel-based superalloys will concentrate on decreasing weight, strengthening corrosion resistance, and increasing performance at even greater temperatures. To get over the present restrictions, advanced manufacturing methods including powder metallurgy and additive manufacturing are being investigated [13]. Furthermore, efforts are being made to create superalloys that rely less on expensive or rare components. In this regard synthesis of nostructured superalloys with compositional design using machine learning are becoming increasing interest of engineers. The focus on environmental impact of these alloys and green manufacturing techniques for their production is also under consideration.

Conclusion

Superalloys based on nickel are essential parts of high-performance applications, especially in the chemical processing, aerospace, and power generation sectors. These alloys get great attention due to their exceptional resistance to corrosion and oxidation, even at elevated temperatures like 1200 °C, as well as their exceptional mechanical strength and resistance to thermal creep deformation. Strengthening the superalloy through a variety of procedures, including solid solution strengthening and precipitation hardening. The addition of elements like chromium, aluminum, and titanium improves the alloy's characteristics by producing stable oxide layers that shield against oxidation as the directly strengthen their microstructure and thus improves the grain boundary resistance. Even with their great performance, they still face difficulties with expensive costs and complicated manufacture. To satisfy the increasing needs for current technology, research and development efforts are still being made to push the limits of these materials in an effort to increase their sustainability, efficiency, and flexibility.

References

[1] D. Satyanarayana, N. Satyanarayana, E. Prasad Nickel-based superalloys, Aerosp, Mater. Mater. Technol. 1 (2017) 199-228. https://doi.org/10.1007/978-981-10-2134-3_9

[2] C.T Sims, A contemporary view of nickel-base superalloys, JOM. (1966) 1119-1130. https://doi.org/10.1007/BF03378505

[3] S. Tin, T.M. Pollock, Nickel-based superalloys, in: T.I-P. Shih, V. Yang (Eds.), Turbine Aerodynamics, Heat Transfer, Materials and Mechanics, American Institute of Aeronautics and Astronautics, (2014) 423-466. https://doi.org/10.2514/5.9781624102660.0423.0466

[4] J.J Little, A field dislocation mechanics approach to emergent properties in two-phase nickel-based superalloys, University of Birmingham. (2020).

[5] H.A. Kishawy, A. Hosseini, Superalloys, Machining difficult-to-cut materials: basic principles and challenges, (2019) 97-137. https://doi.org/10.1007/978-3-319-95966-5_4

[6] C.T Sims, A history of superalloy metallurgy for superalloy metallurgists, Superalloys, (1984) 399-419. https://doi.org/10.7449/1984/Superalloys_1984_399_419

[7] R.C Reed, The superalloys: fundamentals and applications. Cambridge university press (2008)

[8] A. Kracke, and A. Allvac, Superalloys, the most successful alloy system of modern times-past, present and future, Proc. Int. Symp. on Superalloy, (2010). https://doi.org/10.1002/9781118495223.ch2

[9] A.F Giamei, Development of single crystal superalloys, a brief history, AM&P Technical Articles, (2013) 26-30. https://doi.org/10.31399/asm.amp.2013-09.p026

[10] A. Nowotnik, Development of nickel based superalloys for advanced turbine engines, Mater. Sci. forum. Trans Tech Publ. (2014). https://doi.org/10.4028/www.scientific.net/MSF.783-786.2491

[11] G.R Thellaputta, P.S. Chandra, and C. Rao, Machinability of nickel based superalloys, a review, Mater. Today Proc. 4 (2017) 3712-3721. https://doi.org/10.1016/j.matpr.2017.02.266

[12] R. Darolia, Development of strong, oxidation and corrosion resistant nickel-based superalloys, critical review of challenges, progress and prospect, Inter. Mater. Rev. 6 (2019) 355-380. https://doi.org/10.1080/09506608.2018.1516713

[13] A. Thakur, and S. Gangopadhyay, State-of-the-art in surface integrity in machining of nickel-based super alloys, IJMTM 100 (2016) 25-54. https://doi.org/10.1016/j.ijmachtools.2015.10.001

[14] A. Strondl, M. Palm, J. Gnauk, and G. Frommeyer, Microstructure and mechanical properties of nickel based superalloy IN718 produced by rapid prototyping with electron beam melting (EBM), Mater. Sci. Technol. 5 (2011) 876-883. https://doi.org/10.1179/026708309X12468927349451

[15] H.T. Lee, and W.H. Hou, Development of fine-grained structure and the mechanical properties of nickel-based Superalloy 718, Mater. Sci. Engr. A 555 (2012) 13-20. https://doi.org/10.1016/j.msea.2012.06.027

[16] Z. GaoLe, High-temperature mechanical properties of nickel-based superalloys manufactured by additive manufacturing, Mater. Sci. Technol. 14 (2020) 1523-1533. https://doi.org/10.1080/02670836.2020.1799137

[17] D. Zhang, Dynamic mechanical behavior of nickel-based superalloy metal rubber Mater. Des. 56 (2014) 69-77. https://doi.org/10.1016/j.matdes.2013.10.088

[18] E. Alabort, Grain boundary properties of a nickel-based superalloy: characterisation and modelling. Acta Mater. 151 (2018) 377-394. https://doi.org/10.1016/j.actamat.2018.03.059

[19] Y.T. Tang, A.J. Wilkinson, and R.C. Reed, Grain boundary serration in nickel-based superalloy inconel 600 generation and effects on mechanical behavior Metall. Mater. Trans A, 49 (2018) 4324-4342. https://doi.org/10.1007/s11661-018-4671-7

[20] M. Zielińska, M. Yavorska, M. Poręba, and J. Sieniawski, Thermal properties of cast nickel based superalloys, Arch. Mater. Sci. Eng., 44 (2010) 35-38.

[21] M. Karunaratne, Modelling the coefficient of thermal expansion in Ni-based superalloys and bond coatings, J. Mater. Sci. 51 (2016) 4213-4226. https://doi.org/10.1007/s10853-015-9554-3

[22] H. Hamdi, H.R. Abedi, Thermal Stability of Ni-based Superalloys Fabricated Through Additive Manufacturing, A Review, JMR&T, (2024). https://doi.org/10.1016/j.jmrt.2024.04.161

[23] A. Tyagunov, O. Milder, and D. Tarasov, Application of artificial neural networks for prediction of nickel-based superalloys service properties based on the chemical composition, WSEAS Transactions on Environment and Development, 15 (2019) 113-119.

[24] J. Zuback, Impact of chemical composition on precipitate morphology in an additively manufactured nickel base superalloy, J. Alloys Compd. 798 (2019) 446-457. https://doi.org/10.1016/j.jallcom.2019.05.230

[25] D.K. Ganji, and G. Rajyalakshmi, Influence of alloying compositions on the properties of nickel-based superalloys, a review, Recent Advances in Mechanical Engineering, Select Proceedings of NCAME 2019, (2020) 537-555. https://doi.org/10.1007/978-981-15-1071-7_44

[26] B. Nithin, Effect of Cr addition on γ-γ' cobalt-based Co-Mo-Al-Ta class of superalloys, a combined experimental and computational study, J. Mater. Sci. 52 (2017) 11036-11047. https://doi.org/10.1007/s10853-017-1159-6

[27] S.J. Park, Effects of Cr, W, and Mo on the high temperature oxidation of Ni-based superalloys, MDPI Mater. 12 (2019) 2934. https://doi.org/10.3390/ma12182934

[28] A. Jena, and M. Chaturvedi, The role of alloying elements in the design of nickel-base superalloys, J. Mater. Sci. 19 (1984) 3121-3139. https://doi.org/10.1007/BF00549796

[29] A. Chakraborty, Role of alloy composition on micro-cracking mechanisms in additively manufactured Ni-based superalloys, Acta Mater., 255 (2023) 119089. https://doi.org/10.1016/j.actamat.2023.119089

[30] P. Kontis, The effect of chromium and cobalt segregation at dislocations on nickel-based superalloys, Scr. Mater. 145 (2018) 76-80. https://doi.org/10.1016/j.scriptamat.2017.10.005

[31] L. Zhang, C.T. Peng, J. Shi, and R. Lu, Surface alloying of chromium/tungsten/stannum on pure nickel and theoretical analysis of strengthening mechanism, Appl. Surf. Sci., 532 (2020) 147477. https://doi.org/10.1016/j.apsusc.2020.147477

[32] A. Bauer, S. Neumeier, F. Pyczak, M. Göken, Creep strength and microstructure of polycrystalline γ'-strengthened cobalt-base superalloys, Superalloys 12 (2012) 695-703. https://doi.org/10.1002/9781118516430.ch77

[33] Y.C. Lin, and C.Y. Wang, Alloying-element dependence of structural, elastic and electronic properties of nickel-based superalloys: Influence of γ' volume fraction, J. Alloys Comp. 838 (2020) 155141. https://doi.org/10.1016/j.jallcom.2020.155141

[34] H. Mallikarjuna, N. Richards, W. Caley, Effect of alloying elements and microstructure on the cyclic oxidation performance of three nickel-based superalloys, Mater. 4 (2018) 487-499. https://doi.org/10.1016/j.mtla.2018.11.004

[35] Y. Chiu, and A. Ngan, Effects of boron doping on the grain-growth kinetics and mechanical properties of γ/γ' nickel-aluminum alloys, Metall. Mater. Trans. A, 31 (2000) 3179-3186. https://doi.org/10.1007/s11661-000-0097-z

[36] Y. Chen, The strengthening effects and mechanisms of alloying elements on interfaces for multiphase Ni-based superalloys, A first-principles study, JMR&T, 23 (2023) 4802-4813. https://doi.org/10.1016/j.jmrt.2023.02.119

[37] Y. Murata, K. Suga, and N. Yukawa, Effect of transition elements on the properties of MC carbides in IN-100 nickel-based superalloy, J. Mater. Sci. 21 (1986) 3653-3660. https://doi.org/10.1007/BF00553814

[38] M. Bauccio, ASM metals reference book, ASM. Int. (1993).

[39] E. Caldwell, F. Fela, and G. Fuchs, The segregation of elements in high-refractory-content single-crystal nickel-based superalloys, JOM., 56 (2004) 44-48. https://doi.org/10.1007/s11837-004-0200-9

[40] Y. Ji, Effect of refractory elements M (= Re, W, Mo or Ta) on the diffusion properties of boron in nickel-based single crystal superalloys, Vacuum, 211 (2023) 111923. https://doi.org/10.1016/j.vacuum.2023.111923

[41] K. Povarova, Influence of rare-earth metals on the high-temperature strength of Ni 3 Al-based alloys, Russian Metall. (Metally), (2011) 47-54. https://doi.org/10.1134/S0036029511010137

[42] R.T. Holt, W. Wallace, Impurities and trace elements in nickel-base superalloys, Int. Met. Rev. 21 (1976) 1-24. https://doi.org/10.1179/095066076790136762

[43] Q. Zhang, Study of microstructure of nickel-based superalloys at high temperatures, Scr. Mater. 126 (2017) 55-57. https://doi.org/10.1016/j.scriptamat.2016.08.013

[44] S. Zhao, X. Xie, G.D. Smith, S.J. Patel, Microstructural stability and mechanical properties of a new nickel-based superalloy, Mater. Sci. Eng. A 355 (2003) 96-105. https://doi.org/10.1016/S0921-5093(03)00051-0

[45] S. Zhao, X. Xie, G.D. Smith, S.J. Patel, Gamma prime coarsening and age-hardening behaviors in a new nickel base superalloy, Mater. lett. 58 (2004) 1784-1787. https://doi.org/10.1016/j.matlet.2003.10.053

[46] A. Goodfellow, Gamma prime precipitate evolution during aging of a model nickel-based superalloy, Metall. Mater. Trans. A 49 (2018) 718-728. https://doi.org/10.1007/s11661-017-4336-y

[47] M.M. Barjesteh, S.M. Abbasi, K.Z. Madar, and K. Shirvani, The effect of heat treatment on characteristics of the gamma prime phase and hardness of the nickel-based superalloy Rene® 80, Mater. Chem. Phy. 227 (2019) 46-55. https://doi.org/10.1016/j.matchemphys.2019.01.038

[48] J. Singh, and K. Ravikanth, Roles of Refractory Solutes on the Stability of Carbide and Boride Phases in Nickel Superalloys, JPED. (2024) 1-25. https://doi.org/10.1007/s11669-024-01136-5

[49] L. Rakoczy, Analysis of γ' precipitates, carbides and nano-borides in heat-treated Ni-based superalloy using SEM, STEM-EDX, and HRSTEM, Mater. 13 (2020) 4452. https://doi.org/10.3390/ma13194452

[50] Z. Asghary, S. Abbasi, M. Seifollahi, and M. Morakabati, Boron effect on phase transformation of σ and M23C6 in nimonic 105 superalloy, Mater. Res. Express, 6 (2019) 116529. https://doi.org/10.1088/2053-1591/ab446f

Materials Research Forum LLC
https://doi.org/10.21741/9781644903698-3

[51] J.E. Kanyo, S. Schafföner, R.S. Uwanyuze, and K.S. Leary, An overview of ceramic molds for investment casting of nickel superalloys, J. Eur. Ceram. Soc., 40 (2020) 4955-4973. https://doi.org/10.1016/j.jeurceramsoc.2020.07.013

[52] Q. Song, High-temperature flexural strength of aluminosilicate ceramic shells for the investment casting of nickel-based superalloy, IJMC. 18 (2024) 962-974. https://doi.org/10.1007/s40962-023-01061-2

[53] S. Jones, and C. Yuan, Advances in shell moulding for investment casting, J. Mater. Process.Technol. 135 (2003) 258-265. https://doi.org/10.1016/S0924-0136(02)00907-X

[54] H. A. Mehrabi, K. Salonitis, and M. Jolly, Sustainable Investment Casting. in 14th World Conference in Investment Casting, (2016).

[55] S. Pattnaik, D.B. Karunakar, and P.K. Jha, Developments in investment casting process, A review, J. Mater. Process.Technol. 212 (2012) 2332-2348. https://doi.org/10.1016/j.jmatprotec.2012.06.003

[56] E. Rzyankina, Numerical and experimental investigation of directional solidification in vacuum investment casting of superalloys, Cape Peninsula University of Technology, (2013).

[57] D. Szeliga, K. Kubiak, M. Motyka, and J. Sieniawski, Directional solidification of Ni-based superalloy castings: thermal analysis, Vacuum, 131 (2016) 327-342. https://doi.org/10.1016/j.vacuum.2016.07.009

[58] H.N. Mathur, Nucleation of recrystallisation in castings of single crystal Ni-based superalloys, Acta Mater. (2017) 112-123. https://doi.org/10.1016/j.actamat.2017.02.058

[59] J. Zhang , Recent progress in research and development of nickel-based single crystal superalloys, Acta. Metall. Sin. 129 (2023) 1109-1124.

[60] R. Jiang , Y. Song, and P. Reed, Fatigue crack growth mechanisms in powder metallurgy Ni-based superalloys, A review, Int. J. Fatigue, 141 (2020) 105887. https://doi.org/10.1016/j.ijfatigue.2020.105887

[61] C.L. Jia, C.C. Ge, and Q.Z. Yan, Innovative technologies for powder metallurgy-based disk superalloys: Progress and proposal, Chin. Phys. B 25(2016) 026103. https://doi.org/10.1088/1674-1056/25/2/026103

[62] M.M. Attallah, R. Jennings, X. Wang, and L.N. Carter, Additive manufacturing of Ni-based superalloys, the outstanding issues, MRS Bull. 41 (2016) 758-764. https://doi.org/10.1557/mrs.2016.211

[63] B. Graybill, Additive manufacturing of nickel-based superalloys, Int. MSEC, ASME. (2018) https://doi.org/10.1115/MSEC2018-6666

[64] M. Li , Metal binder jetting additive manufacturing, a literature review, J. Manuf. Sci. Eng., 142 (2020) 090801. https://doi.org/10.1115/1.4047430

[65] R. Wang, Microstructure characteristics of a René N5 Ni-based single-crystal superalloy prepared by laser-directed energy deposition, Addit. Manuf. 61 (2023) 103363. https://doi.org/10.1016/j.addma.2022.103363

[66] M.P. Haines, V.V. Rielli, S. Primig, and N. Haghdadi, Powder bed fusion additive manufacturing of Ni-based superalloys: a review of the main microstructural constituents and characterization techniques, J. Mater. Sci. 57 (2022) 14135-14187. https://doi.org/10.1007/s10853-022-07501-4

[67] A. Hariharan, Misorientation-dependent solute enrichment at interfaces and its contribution to defect formation mechanisms during laser additive manufacturing of superalloys, Phy. Rev. Mater. 3 (2019) 123602. https://doi.org/10.1103/PhysRevMaterials.3.123602

[68] N. Das, Advances in nickel-based cast superalloys, T. I. metals, 63 (2010) 265-274. https://doi.org/10.1007/s12666-010-0036-7

[69] E.W. Ross, and K.S. O'Hara, Rene N4: a first generation single crystal turbine airfoil alloy with improved oxidation resistance, low angle boundary strength and superior long time rupture strength, Superalloys, (1996) 19-25. https://doi.org/10.7449/1996/Superalloys_1996_19_25

[70] A. Cetel, and D. Duhl, Second generation nickel-base single crystal superalloy, Superalloys, (1988) 235-244. https://doi.org/10.7449/1988/Superalloys_1988_235_244

[71] K. Harris, Development of the rhenium containing superalloys CMSX-4 & CM 186 LC for single crystal blade and directionally solidified vane applications in advanced turbine engines, Superalloys (1992) 297. https://doi.org/10.7449/1992/Superalloys_1992_297_306

[72] A. Singh, Mechanisms related to different generations of γ' precipitation during continuous cooling of a nickel base superalloy, Acta Mater, 61(2013) 280-293. https://doi.org/10.1016/j.actamat.2012.09.058

[73] T.M. Smith, Producing Next Generation Superalloys Through Advanced Characterization and Manufacturing Techniques. in Case Western Reserve University Seminar Series, (2020).

[74] S. Walston, Joint development of a fourth generation single crystal superalloy. 10th Int. Symp. on Superalloys, (2004). https://doi.org/10.7449/2004/Superalloys_2004_15_24

[75] N. Petrushin, E. Elyutin, E. Visik, and S. Golynets, Development of a single-crystal fifth-generation nickel superalloy, Russ. Metall. (Met.), 11 (2017) 936-947. https://doi.org/10.1134/S0036029517110118

[76] K. Kawagishi, Development of an oxidation-resistant high-strength sixth-generation single-crystal superalloy TMS-238, Superalloys, 9 (2012)189-195. https://doi.org/10.1002/9781118516430.ch21

[77] G. Wang, Process optimization and mechanical properties of oxide dispersion strengthened nickel-based superalloy by selective laser melting, Mater. Des. 188 (2020) 108418. https://doi.org/10.1016/j.matdes.2019.108418

[78] S. Pasebani, Oxide dispersion strengthened nickel based alloys via spark plasma sintering, Mater. Sci. Eng. A 630 (2015) 155-169. https://doi.org/10.1016/j.msea.2015.01.066

[79] S. Chikumba, and V.V. Rao. High entropy alloys, development and applications, ICLTET 2015, (2015).

[80] A. Yeh, Developing new type of high temperature alloys high entropy superalloys, Int. J. Metall. Mater. Eng. 1 (2015) 1-4. https://doi.org/10.15344/2455-2372/2015/107

[81] R. Smith, G. Lewi, and D. Yates, Development and application of nickel alloys in aerospace engineering, AEAT. 73 (2001) 138-147. https://doi.org/10.1108/00022660110694995

[82] M. Perrut, P. Caron, M. Thomas, A. Couret, High temperature materials for aerospace applications: Ni-based superalloys and γ-TiAl alloys, C. R. Phys. 19 (2018) 657-671. https://doi.org/10.1016/j.crhy.2018.10.002

[83] A. R. Jabalquinto, A. V. Tellez, P. Zambrano-Robledo, and B. B. Reyes, Feasibility of manufacturing combustion chambers for aeronautical use in Mexico, JART. 14 (2016) 167-172. https://doi.org/10.1016/j.jart.2016.05.003

[84] P.M. Sforza, Combustion Chambers for Air-Breathing Engines, in: Theory of Aerospace Propulsion, Butterworth-Heinemann, Boston. (2012) 127-159. https://doi.org/10.1016/B978-1-85617-912-6.00004-9

[85] A. Sato, Y.L. Chiu, R. Reed, Oxidation of nickel-based single-crystal superalloys for industrial gas turbine applications, Acta Mater. 59 (2011) 225-240. https://doi.org/10.1016/j.actamat.2010.09.027

[86] D. Klarstrom, L. Pike, V. Ishwar, Nickel-base alloy solutions for ultrasupercritical steam power plants, Procedia Eng. 55 (2013) 221-225. https://doi.org/10.1016/j.proeng.2013.03.246

[87] H. Alves, and M.S. Niederau, Successful applications of nickel alloys and high alloyed stainless steels in seawater service, (2008). https://doi.org/10.5006/C2008-08259

[88] M. N. Rao, Application of superalloys in petrochemical and marine sectors in India, Trans. Indian Inst. Metals 61 (2008) 87-91. https://doi.org/10.1007/s12666-008-0012-7

[89] M. Hardy, Solving recent challenges for wrought Ni-base superalloys, Metall. Mater. Trans. A. 51 (2020) 2626-2650. https://doi.org/10.1007/s11661-020-05773-6

Superalloys: Fundamentals and Applications
Materials Research Foundations 178 (2025) 68-80

Materials Research Forum LLC
https://doi.org/10.21741/9781644903698-4

Chapter 4

Cobalt based Superalloy

Uzma Hira[1*], Nimrah Maqsood[1]

[1]School of Physical Sciences, University of the Punjab, Lahore, Pakistan

*Uzma.sps@pu.edu.pk

Abstract

Superalloys indicates to a class of alloys which may retain their mechanical properties despite prolonged high-temperature environments. Superalloys were originally designed to get usein turbo-superchargers and aviation turbine engines. Cobalt based superalloys have a melting point that is greater than alloys based on nickel, making them intriguing materials. These novel superalloys have excellent high-temperature characteristics since they are enhanced by an ordered γ' phase (L12 phase). Researchers must use alloying methodologies to investigate high-dimensional composition and temperature space in order to generate them. The advancements of cobalt-based superalloys can benefit new technology due to their distinct microstructure features, uncovered in this work, which can also offer insights into the microstructural engineering of other produced superalloys.

Keywords: Creep Resistance, Power of Cobalt, Thermodynamic Stability, Oxidation

Contents

1.   Introduction: Unveiling the Power of Cobalt

Cobalt (Co) with the atomic number Z of 27is a byproduct of copper (Cu) and nickel (Ni) mining. As a rare magnetic element, Co has qualities comparable to iron (Fe) and nickel (Ni). There are its two valence states i.e. *Co (II)* and *Co (III)*. Co *(II)*is used more frequently in the chemical manufacturing sectors. It exists predominantly in nature as sulphides, arsenides and oxides [1]. The majority of cobalt output is in the metallic form, which is used to make superalloys of Co. The "tough metal" represents compounds that comprise WC (*81-94%*), as well as matrices made of Co (*5-20%*) and nickel [2].

Cobalt, as silvery grey metal, is an important component of hard magnets. These magnets have a high coercivity and can be permanently magnetized by using a magnetic field. Its diverse uses are based on these key features i.e. ferromagnetism, hardness, wear resistance, poor electrical and thermal conductivity, a high M.P, numerous valences, and strong blue colors when mixed with silica. It is also utilized as a cathode in rechargeable batteries and a superalloy in jet engine turbines [3, 4]. In 2011, global consumption of cobalt was about *75,000* metric tons and the top three cobalt-consuming countries were *China, Japan*, and the *United States*. Pneumonitis, fibrosis, and asthma are among the lung conditions that have been connected to occupational cobalt exposure. It has been demonstrated that cobalt alters genes and disrupts the cellular balance of iron, calcium, and reactive oxygen species [5].

2.   Evolution of Superalloys

Superalloys are alloys that can retain their mechanical qualities even after being exposed to extreme environment for a long time period [6]. These alloys are among the most intricate metallic materials created for use in commercial, industrial, and military settings. Cobalt superalloys are good to bear high temperature and harsh settings as they have exceptional durability and good oxidation resistance strength [7]. These superalloys are made of elements such as **cobalt, chromium, nickel, tungsten** and **aluminum**. Cobalt-based superalloys have evolved over time, their usage in more demanding applications has been made possible by these developments, and future research promises [8].

H. Brearley and E. Maure discovered austenitic stainless steels between 1910 and 1915. γ-Fe with FCC structure was chosen for its great thermal tolerance. This resulted in the invention of

Superalloys: Fundamentals and Applications   Materials Research Forum LLC
Materials Research Foundations 178 (2025) 68-80   https://doi.org/10.21741/9781644903698-4

superalloys, which are high-temperature alloys. The earliest known substance with superalloy properties included Austenitic steel with 34% nickel, 11% Chromium, and 0.3% C hardening. Before 2nd World War, the term "superalloy" was not widely used.

Imphy, a French business, submitted a claim in 1917 for an alloy for land-based gas turbines that had 60% nickel and 11% Cr and was toughened with C and W. *Pierre Chevenard*, a French engineer, then submitted another claim for Austenitic stainless steel made of 35% Ni, 12% Cr, and toughened with 0.5% C to combat grain boundary deteriorationin of steam turbine blades [9].

In the 1920s and 1930s, vacuum induction melting technique was used to create Ni-Fe-Cr alloys. In 1918, Nichrome was introduced by British patent, a Ni-Cr alloy that is regarded as the forerunner of upcoming superalloys i.e. Nimonics and Inconels. In 1929, Nichrome creep resistance was significantly increased by *Bedford, Pilling, and Merica*.

The launch of *Heinkel and Whittle* engines constituted a watershed moment in the development of superalloys, with gas turbine intake temperatures reaching 780 °C. Microstructure and process optimization were the focus of research from the 1930s to the 1950s. World War II accelerated superalloy research in the 1940s, due to the first deployment of turbine engines in fighter planes and the demand for heavy-duty, dependable gas turbines [10].

During World War II, various scientific investigations concentrated on producing high-temperature alloys, which resulted in the creation of novel alloys. Between the 1940s and 1950s, tremendous development led in the introduction of various new alloys, which were frequently cast for simplicity of manufacture [11].

3.   Cobalt based Superalloys

Superalloys are clearly "super" because they have superior high-temperature properties that allow engineers to attain dependability and cost-effectiveness. Others describe superalloys as materials that are ideally suited for applications requiring components to be subjected to temperatures higher than 650 °C for prolonged time period. They are taken as exceptional materials due to following physical characteristics;

- Competencetooperateat high temperatures

- Excellentresistanttomechanicaldegradationovera longtime period [12]

- They are resistanttocorrosion

The most significant type of superalloy is Cobalt based. Co has FCC (Face centered cubic) crystal structure at 25 degrees Celsius, over *417 °C*, it changes to HCP. Cobalt-based superalloys vary from Fe-based and Ni-based alloys in that they lack ordered coherent precipitates (γ' or γ''). The combination of carbides and solid solution hardeners significantly strengthens cobalt-based superalloys. Iron, nickel, and chromium are the primary alloying elements with cobalt; hence, the most frequent carbides i.e. Co_6C, Co_7C_3, and $Co_{23}C_6$ exists in Co-based superalloys. Cobalt-based superalloys cannot generate MC-type carbides because they lack Ta, Ti, Zr and Hf.

Co-based superalloys offer high heat damage, but their strength and resistance are lesser than Ni-base superalloys [13]. However, they can keep their strength at relatively high temperatures. As a result, their uses are comparable to, but less demanding than, those of Ni-based systems. Co-

Superalloys: Fundamentals and Applications   Materials Research Forum LLC
Materials Research Foundations 178 (2025) 68-80   https://doi.org/10.21741/9781644903698-4

based superalloys, like nickel-based system of superalloys, are primarily utilized in applications with temperatures ranges from 650 to 1150 °C (*1200 to 2100 °F*), although their applications are less prevalent. They are costly than Fe-based and Ni-based superalloys, which creates a further barrier to their widespread distribution [14]. Superalloys are further classified into three categories i.e. Iron based superalloy, Cobalt based superalloy and Nickle based superalloy with cast and wrought macrostructures as shown in the Fig. 1.

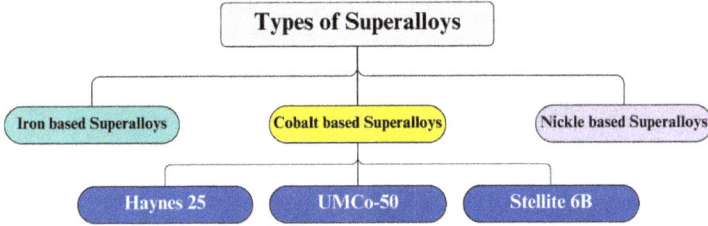

Figure 1. Types of superalloys.

3.1   Types of Cobalt based Superalloys

Superalloys based on cobalt may be divided into many categories according to their intended uses, microstructure, and composition [15]. There are abundant examples of Co-based superalloys as shown in Fig. 2. The basic examples are *Haynes25, UMCo-50, and Stellite6B.*

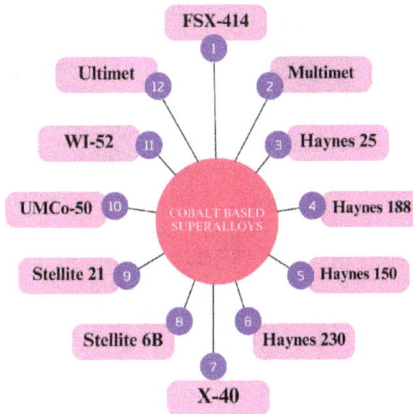

Figure 3. Other types of cobalt based superalloys.

Haynes 25, a famous Co-base wrought superalloy, created by machining, cold working and welding. It offers high thermal stability and galling resistance, as well as oxidation resistance up to 980 °C. It also resists oceanic conditions, i.e. acidic, basic and bodily fluids.

UMCo-50, which contains 21% Fe. It is not as strong as Haynes 25; thus, it is best suited for small level applications.

Stellite 6B, 3rd example of a Co-base superalloy, which is resistant to oxidation owing to its chromium component. The development of complex carbides results in a high hot hardness. Stellites are most commonly employed in applications that need high temperature strength and wear resistance [16].

There exist various types of cobalt based superalloys with their chemical compositions. Some of them are mentioned in Table 1.

Table 1. Examples of cobalt based superalloys with their chemical compositioninsights.

Superalloys	Chemical composition insights
V-36	20% Ni, 25% Cr, Co(bal), 4% Mo, 2.3% Nb, 2.4% Fe, 1% Mn, 0.32% C
S-816	20% Ni, 20% Cr, Co(bal), 4% Mo, 4% W, 4% Nb, 3% Fe, 1.2 % Mn, 0.4% C
Haynes 150	28% Cr, 50.5% Co, 4% Mo, 0.75% Si, 0.02% P, 0.002% S 20% Fe
Stellite 6B	3% Ni, 30% Cr, 52% Co, 1.5% Mo, 4.5% W, 2% Si, 3% Fe, 2% Mn, 1.1% C
MP159	25.5% Ni, 19% Cr, 35.7% Co, 7% Mo, 4% W, 0.6% Nb, 0.2% Al, 9% Fe, 3% Ti
MP35N	35% Ni, 20% Cr, 35% Co, 10% Mo
L-605	10% Ni, 20% Cr, 52.9% Co, 15% W, 0.05%C

4. Defining Characteristics and Significance

4.1 Creep Resistance

Creep resistance is a material's capability to tolerate gradual, permanent deformation under sustained tension over time, especially at high temperatures. Creep damage is a serious problem in gas turbine parts, discs, bolts and blades, due to high temperatures and stress. Creep occurs under three conditions: time, temperature, and stress [17]. Co-based superalloys are highly needed to overcome creep resistance while other superalloys show their greater endurance below 900 °C and above that their durability decreases.

Superalloys: Fundamentals and Applications Materials Research Forum LLC
Materials Research Foundations 178 (2025) 68-80 https://doi.org/10.21741/9781644903698-4

4.2 Corrosion Resistance

Corrosion is an intricate natural process in superalloys. Corrosion resistance is the capacity of a metal to withstand damage induced by oxidation or other chemical processes, frequently under extreme environmental circumstances. This feature is critical for materials used in building, manufacturing, and other industries where long-term durability is required [18]. Cobalt based superalloys containing a larger percentage of chromium are better at preventing corrosion at high temperatures to generate stable oxide layers.

4.3 Fatigue Resistance

Fatigue resistance states as material's tendency to tolerate cyclic loads without failing over time. This feature is critical for materials used in applications that experience repeated stress cycles, such as bridges, airplanes, and automobile components. The cyclical stresses may be machine-driven, current (thermal) or a combination of both. Thus, cobalt based superalloys have a high melting point (over 1300 °C) that makes them perfect for guiding vanes in aviation engines [19]. This prevents defects like burning or distortion caused by the inclusion of refractory metal.

5. Compositional Insights

To optimize its qualities for rigorous operating circumstances, several alloying elements are added to their unique composition, which is primarily based on cobalt [20]. Table 2. Shows the compositional insights of Co-based superalloys.

5.1 Microstructural Phases

Cobalt's crystal structure is face-centered cubic (FCC), and it serves as the basis for its superalloy. It is made up of an austenitic matrix known as the gamma (γ) phase, as well as other secondary phases i.e. carbides, gamma prime phase and TCP. At room temperature, the structure of cobalt is hexagonal close-packed (HCP), which changes to FCC at high temperatures or when coupled with other elements such as iron [22]. At all temperatures, Nickel sustains only one crystal structure i.e. an FCC. Superalloys made of iron and cobalt are stabilized by adding alloying metals, notably nickel, to increase their characteristics. These alloys are predominantly made up of Fe, Ni, and Co as well as other elements i.e. Cr, W, Mo, Ta, Nb, Ti, and Al, which are classified as Group VIII B elements of periodic table.

They may also be researched at the atomic level by examining crystals or *microstructures* under a microscope. The intricate microstructure of cobalt-based superalloys improves their strength, durability, and performance at high temperatures [23].

5.1.1 Gamma Phase (γ)

Cobalt based superalloys' principal phase is the gamma (γ) phase, which has an FCC structure. It is an advantageous phase because of its ductility and hardness. An FCC structure possess lesser diffusivity than the BCC structure because of its tight packing. This is the key to better properties of superalloy at elevated temperatures [24].

5.1.2   Gamma Prime phase (γ')

Gamma prime is a secondary phase in superalloy microstructure that forms when *Al* and *Ti* are combined to precipitate the *Gamma prime phase* with a*FCC* structure. It is ductile and coherent with the matrix [25], enhancing strength. This phase is critical for their creep resistance.

5.1.3   Carbide Phase

Metal carbides are carbon-based compounds containing reactive elements i.e. Ti, Ta, Hf, and Nb. They precipitate at high temperatures and disintegrate during heat treatment, yielding lesser carbides such as M23C6 and M6C. Carbides, which are found in all three superalloy classes, increase rupture strength at high temperatures and can give partial strengthening effects, either directly boosting superalloy strength by dispersion hardening or indirectly by alleviating grain boundaries [26].

6.   Optimization

Designing materials with various desirable qualities is difficult, particularly in complicated systems. Machine learning is commonly used to optimize multi-component Co-base superalloys, with a focus on microstructural stability, γ' solvus temperature, volume fraction, density, processing window, freezing range, and oxidation resistance. The γ' solvus temperature is important for advanced single crystal materials as it defines their maximum temperature capacity. Other desirable attributes include processing window, freezing range, density, and microstructural stability, all of which *EGO* takes into account while optimizing.

6.1   Efficient Global Optimization

The efficient global optimization (EGO) approach is a strong optimization technique used to improve material design. The EGO technique uses an adaptive iterative loop. In each iteration, the technique employs the Expected Improvement (EI) criteria; this adaptive approach aids in gradually improving the model's accuracy and identifying optimum solutions. EI is calculated by the following formula;

$$EI = \sigma \left[\phi (z) + z \, \Phi (z) \right]$$

Where;

$$Z = \frac{T\gamma' \, solvus - \mu*}{\sigma}$$

- $T\gamma'$ solvus is the predicted γ' solvus temperature.
- $\mu*$ is the maximum γ' solvus temperature in the current dataset.
- σ is the uncertainty of the prediction.
- $\phi (z)$ is the standard normal density
- $\Phi(z)$ cumulative distribution function

Superalloys: Fundamentals and Applications
Materials Research Foundations 178 (2025) 68-80

Materials Research Forum LLC
https://doi.org/10.21741/9781644903698-4

7.   Stability and Oxidation

A novel Co-base superalloy (Co22) with a 22-weight percent Co content is presented in this chapter. It is derived from UMCo50. Due to its cheaper cost, this alloy is increasingly frequently utilized in industrial production since it saves money on materials while maintaining high temperature oxidation resistance and structural stability [27]. Early ideas of metal oxidation emerged, and studies based on the oxidation of cobalt superalloys were extensively publicized.

Oxidation tests were carried out on commercial Co-base superalloys L-605 and HA-188 by *C. A. Barrett et al.* They discovered that Cr_2O_3 occurs after 100 hours at 1150°C to stop more oxidation, although gasification reduces protection [28].

7.1 Thermodynamic Stability at Elevated Temperatures

Strong thermodynamic stability is exhibited by cobalt-based superalloys at high temperatures, which is essential to their performance in demanding applications. The following are the main facets of their stability [29].

7.1.1   Factors Contributing to Thermodynamic Stability

Phase Stability

The Strongness and stability are achieved by maintaining stable phases at high temperatures in cobalt-based superalloys, which are frequently based on the γ and γ' phases [30].

Alloying Elements

Tantalum, tungsten, and chromium are all alloying elements that aid to stabilize phases at high temperatures, improve oxidation resistance, and reinforce processes like as precipitation and solid solution hardening.

Phase Transformation Temperatures

During alloy design, critical temperatures for phase transitions, such the γ to γ' transformation, are carefully maintained to maintain the desired microstructure and characteristics over the operational temperature ranges [31].

8.   Applications

Additive Manufacturing (3D Printing): Cobalt-based alloys may be used to create complex geometries and customized components using advanced additive manufacturing processes, expanding their applications in the aerospace and medical industries [32].

Structural field: Through alloy design, surface treatments, and coatings, cobalt-based alloys' resistance to corrosion and oxidation, increasing their suitability for severe environments, these materials are now widely used in gas turbine and jet engines that function at tremendous temperature [33].

Biomedical field: The advancement of cobalt-based alloys for biomedical uses, such as enhanced dental materials and biodegradable implants, is the focus of the healthcare industry [34, 35].

Industrial field: Superalloys have evolved as a result of the development of gas turbines and jet engines that run at high temperatures, making them more efficient and less polluting. However, these high temperatures pose problems, such as significant corrosion and material deterioration [36]. Their primary use is in aircraft and heat exchangers etc. Over *70%* of superalloy utilization occurs in the aerospace industry [37, 41]. The main consumers of Cobalt based superalloys are shown in Fig.3.

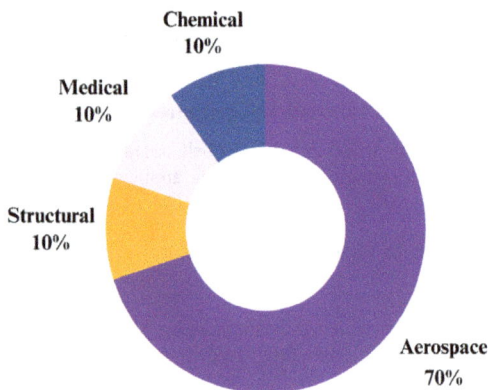

Figure 3. Main consumers of Cobalt based superalloys.

Conclusion

Co-base superalloys having low density and high specific yield strength, fueled by the inclusion of Nb and Ta, at increased temperatures present a current class of superalloys and open up new avenues for high-temperature alloy development. Superalloys based on cobalt are essential for applications because of their remarkable resistance to oxidation and corrosion, thermal stability, and mechanical strength. Because of their resistance to thermal stress, they are perfect for turbine blades and other components that are frequently exposed to temperature variations. Because of their microstructural stability, these alloys also continue to be dependable and durable throughout time. Technological developments in alloying and manufacturing processes keep improving their functionality and industrial uses.

References

[1] L.F. Călinoiu, R. Odochean, G. Martău, L. Mitrea, S.A. Nemes, B. Ștefănescu, D.C. Vodnar, In situ fortification of cereal by-products with vitamin B12: An eco-sustainable approach for food fortification, Food Chem. 460 (2024) 140766. https://doi.org/10.1016/j.foodchem.2024.140766

[2] S.E. Smith, J.K. Loosli, Cobalt and vitamin 12 in ruminant nutrition: a review. JDS. 40 (1957) 1215-1227. https://doi.org/10.3168/jds.S0022-0302(57)94618-0

Materials Research Forum LLC
https://doi.org/10.21741/9781644903698-4

[3] C. Boedler, Metal passivity as mechanism of metal carcinogenesis: Chromium, nickel, iron, copper, cobalt, platinum, molybdenum, Toxicol. Environ. Chem. 89 (2007) 15-70. https://doi.org/10.1080/02772240601008513

[4] M. Kovochich, A. Monnot, D. G. Kougias, S. L. More, J. T. Wilsey, Q. Qiu, L.E.L. Perkins, P. Hasgall, M. Taneja, E. E. Reverdy, J. Sague, S. Marcello, K. Connor, J. Scutti, W.V. Christian, P. Coplan, L.B. Katz, M. Vreeke, M. Calistri-Yeh, B. Faiola, K. Unice, G. Eichenbaum, Carcinogenic hazard assessment of cobalt-containing alloys in medical devices: Review of in vivo studies, RTP 122 (2021) 104910. https://doi.org/10.1016/j.yrtph.2021.104910

[5] T. Froeliger, A. Després, L. Toualbi, D. Locq, M. Veron, G. Martin, R. Dendievel, Interplay between solidification microsegregation and complex precipitation in a γ/γ' cobalt-based superalloy elaborated by directed energy deposition, Mater. Charact. 194 (2022) 112376. https://doi.org/10.1016/j.matchar.2022.112376

[6] N. Jegadeeswaran, K.U. Bhat, M.R. Ramesh, Improving hot corrosion resistance of cobalt based superalloy (Superco-605) using HVOF sprayed oxide alloy powder coating, INDIAN I METALS 68 (2015) 309-316. https://doi.org/10.1007/s12666-015-0605-x

[7] S.K. Sahay, B. Goswami, Recent developments in Co-base alloys, SSP 150 (2009) 197-219. https://doi.org/10.4028/www.scientific.net/SSP.150.197

[8] K. Harris, J. B. Wahl, Developments in superalloy castability and new applications for advanced superalloys, JMST 25 (2009) 147-153. https://doi.org/10.1179/174328408X355442

[9] W. Betteridge, S.W.K. Shaw, Development of superalloys, JMST 3 (1987) 682-694. https://doi.org/10.1179/026708387790329847

[10] D.L. Klarstrom, Wrought cobalt-base superalloys. JMEP 2 (1993) 523-530. https://doi.org/10.1007/BF02661736

[11] R. Casas, F. Gálvez, M. Campos, Microstructural development of powder metallurgy cobalt-based superalloys processed by field assisted sintering techniques (FAST), Mater. Sci. Eng. A 724 (2018) 461-468. https://doi.org/10.1016/j.msea.2018.04.004

[12] D.C. Pratt, Industrial casting of superalloys, JMST 2 (1986) 426-435. https://doi.org/10.1179/mst.1986.2.5.426

[13] J. Stringer, High-temperature corrosion of superalloys, JMST 3 (1987) 482-493. https://doi.org/10.1080/02670836.1987.11782259

[14] G. Chen, F. Liu, F. Chen, Y. Tan, Y. Cai, W. Shi, X. Ji, S. Xiang, Ultrahigh strength-ductility synergy via heterogeneous grain structure and multi-scale L12-γ' precipitates in a cobalt-based superalloy GH159, Mater. Sci. Eng. A 904 (2024) 146687. https://doi.org/10.1016/j.msea.2024.146687

[15] N. Baler, P. Pandey, K. Chattopadhyay, G. Phanikumar, Influence of thermomechanical processing parameters on microstructural evolution of a gamma-prime strengthened cobalt based superalloy during high temperature deformation, Mater. Sci. Eng. A 791 (2020) 139498. https://doi.org/10.1016/j.msea.2020.139498

Superalloys: Fundamentals and Applications
Materials Research Foundations 178 (2025) 68-80

Materials Research Forum LLC
https://doi.org/10.21741/9781644903698-4

[16] V. Sklenička, M. Kvapilová, P. Král, J. Dvořák, M. Svoboda, B. Podhorná, J. Zýka, K. Hrbáček, A. Joch, Degradation processes in high-temperature creep of cast cobalt-based superalloys, Mater. Charact. 144 (2018) 479-489. https://doi.org/10.1016/j.matchar.2018.08.006

[17] A. Korashy, H. Attia, V. Thomson, S. Oskooei, Fretting wear behavior of cobalt - Based superalloys at high temperature - A comparative study, Tribol. Int. 145 (2020) 106155. https://doi.org/10.1016/j.triboint.2019.106155

[18] W. Gui, H. Zhang, M. Yang, T. Jin, X. Sun, Q. Zheng, Influence of type and morphology of carbides on stress-rupture behavior of a cast cobalt-base superalloy, J. Alloys Compd. 728 (2017) 145-151. https://doi.org/10.1016/j.jallcom.2017.08.287

[19] S. Atamert, J. Stekly, Microstructure, wear resistance, and stability of cobalt based and alternative iron based hardfacing alloys, Surf. Eng. 9 (1993) 231-240. https://doi.org/10.1179/sur.1993.9.3.231

[20] Y.J. Ren, T. Dai, X.H. Guo, J. Shen, Y.L. Lv, J. Chen, Y. Niu, Scaling behavior of four Co-20Ni-xCr-yAl (x= 8, 15 wt.%; y= 3, 5 wt.%) alloys exposed to 1 atm O2 at 1000° C and 1100° C, Corros. Sci. 191 (2021) 109719. https://doi.org/10.1016/j.corsci.2021.109719

[21] E. Grundy, Other applications of superalloys, JMST 3 (1987) 782-790. https://doi.org/10.1179/mst.1987.3.9.782

[22] Kumar, V. Anil, R.K. Gupta, SVS N. Murty, A.D. Prasad, Hot workability and microstructure control in Co20Cr15W10Ni cobalt-based superalloy, J. Alloys Compd. 676 (2016) 527-541. https://doi.org/10.1016/j.jallcom.2016.03.186

[23] H.R. Abedi, O.A. Ojo, X. Cao, Effect of cooling rate on precipitation behavior of gamma prime in a newly developed co-based superalloy, Jom 72 (2020) 4054-4059. https://doi.org/10.1007/s11837-020-04241-1

[24] W. Gui, X. Zhang, H. Zhang, X. Sun, Q. Zheng, Melting of primary carbides in a cobalt-base superalloy, J. Alloys Compd. 787 (2019) 152-157. https://doi.org/10.1016/j.jallcom.2019.02.041

[25] S.H.M. Anijdan, A. Bahrami, A new method in prediction of TCP phases formation in superalloys, Mater. Sci. Eng. A 396 (2005) 138-142. https://doi.org/10.1016/j.msea.2005.01.012

[26] A.M. Jokisaari, S.S. Naghavi, C. Wolverton, P.W. Voorhees, O.G. Heinonen, Predicting the morphologies of γ' precipitates in cobalt-based superalloys, Acta Mater. 141 (2017) 273-284. https://doi.org/10.1016/j.actamat.2017.09.003

[27] S. Pu, J. Zhang, Y.F. Shen, L.H. Lou, Recrystallization in a directionally solidified cobalt-base superalloy, Mater. Sci. Eng. A. 480 (2008) 428-433. https://doi.org/10.1016/j.msea.2007.07.028

[28] M.K. Kandula, M.P. Singh, E. Neelamegan, A. Paul, C. Kamanio, High-throughput pseudo-binary diffusion couple approach for alloy design in cobalt-based superalloys, Mater. Charact. 210 (2024) 113842. https://doi.org/10.1016/j.matchar.2024.113842

[29] S. Xi, J. Yu, L. Bao, J. Li, Q. Tao, Z. Li, R. Shi, C. Wang, X. Liu, Predicting atomic structure and mechanical properties in quinary L12-strengthened cobalt-based superalloys using machine learning-driven first-principles calculations, Mater. Today Commun. 38 (2024) 107774. https://doi.org/10.1016/j.mtcomm.2023.107774

[30] X. Qi, Q. Hou, M. Chen, S. Zhu, M. Li, C. Zhou, Microstructure and hot corrosion behavior of Al-Si-Hf coating on new γ′-strengthened cobalt-based superalloy, Surf. Coat. 405 (2021) 126519. https://doi.org/10.1016/j.surfcoat.2020.126519

[31] D. Kubacka, M. Weiser, E. Spiecker, Early stages of high-temperature oxidation of Ni- and Co-base model superalloys: A comparative study using rapid thermal annealing and advanced electron microscopy, Corros. Sci. 191 (2021) 109744. https://doi.org/10.1016/j.corsci.2021.109744

[32] P. Pandey, S. Mukhopadhyay, C. Srivastava, S.K. Makineni, K. Chattopadhyay, Development of new γ′-strengthened Co-based superalloys with low mass density, high solvus temperature and high temperature strength, Mater. Sci. Eng. A 790 (2020) 139578. https://doi.org/10.1016/j.msea.2020.139578

[33] J.P. Moffat, N.G. Jones, P. Jackson, H.J. Stone, On the mechanism of carbide network oxidation and subsequent passivation interference in a high carbon-containing cobalt-based superalloy, J. Alloys Compd. 936 (2023) 168251. https://doi.org/10.1016/j.jallcom.2022.168251

[34] H.J. Im, W.S. Choi, K. Ryou, J.B. Seol, T.H. Kang, W. Ko, P. Choi, Enhanced microstructural stability of γ/γ′-strengthened Co-Ti-Mo-based alloys through Al additions, Acta Mater. 214 (2021) 117011. https://doi.org/10.1016/j.actamat.2021.117011

[35] C. Chu, C. Li, Y. Guan, Y. Liu, Microstructure-dependent coarsening behavior of γ′ precipitates in CoNi-based superalloys, Int. Commun. Heat Mass Transf. 140 (2022) 107396. https://doi.org/10.1016/j.intermet.2021.107396

[36] J. Chen, M. Guo, M. Yang, H. Su, L. Liu, J. Zhang, Phase-field simulation of γ′ coarsening behavior in cobalt-based superalloy, Comput. Mater. Sci. 191 (2021) 110358. https://doi.org/10.1016/j.commatsci.2021.110358

[37] W. Wu, U.R. Kattner, C.E. Campbell, J.E. Guyer, P.W. Voorhees, J.A. Warren, O.G. Heinonen, Co-Based superalloy morphology evolution: A phase field study based on experimental thermodynamic and kinetic data, Acta Mater. 233 (2022) 117978. https://doi.org/10.1016/j.actamat.2022.117978

[38] Z. Liang, M. Göken, U. Lorenz, S. Neumeier, M. Oehring, F. Pyczak, A. Stark, L. Wang, Influence of small amounts of Si and Cr on the high temperature oxidation behavior of novel cobalt base superalloys, Corros. Sci. 184 (2021) 109388. https://doi.org/10.1016/j.corsci.2021.109388

[39] Q. Gao, X. Zhang, Q. Ma, H. Zhu, H. Zhang, L. Sun, H. Li, Accelerating design of novel Cobalt-based superalloys based on first-principles calculations, J. Alloys Compd. 927 (2022) 167012. https://doi.org/10.1016/j.jallcom.2022.167012

[40] G. Cacciamani, G. Roncallo, Y. Wang, E. Vacchieri, A. Costa, Thermodynamic modelling of a six component (C-Co-Cr-Ni-Ta-W) system for the simulation of Cobalt based alloys, J. Alloys Compd. 730 (2018) 291-310. https://doi.org/10.1016/j.jallcom.2017.09.327

[41] Z. Yao, L. Bao, M. Yang, Y. Chen, M. He, J. Yi, X. Yang, T. Yang, Y. Zhao, C. Wang, Z. Zhong, S. Wang, X. Liu, Thermally stable strong <101> texture in additively manufactured cobalt-based superalloys, Scr. Mater. 242 (2024) 115942. https://doi.org/10.1016/j.scriptamat.2023.115942

Superalloys: Fundamentals and Applications
Materials Research Foundations 178 (2025) 81-94

Materials Research Forum LLC
https://doi.org/10.21741/9781644903698-5

Chapter 5

Iridium Base Refractory Superalloys

Ayanna Chanda[a], Leena Vij, Arti Jain[a*]

Department of Chemistry, Daulat Ram College, University of Delhi, Delhi-110007, India

Abstract

Iridium, a frequently overlooked metal, has emerged as a transformative force grasping onto the reins of high-temperature refractory superalloys. This chapter delves into the history, properties, manufacturing processes, and applications of iridium-based superalloys. Iridium's exceptional physical properties and corrosion resistance, coupled with its microstructure, has carved an exceptional place within the sphere of heat-resistant materials. Conventional techniques such as melting, purification and welding, alongside novel ones like powder metallurgy and deformation processing, have contributed to the manufacture of alloys with diverse applications. These range from aerospace to energy-related technologies, such as gas turbine engines, spark plugs, electrocatalysts, to emerging biomedical applications such as coronary stents and cancer treatment. Despite challenges such as brittleness, high cost and limited availability, the locked potential of this metal as an alloy is immense. Addressing these challenges and exploring newer combinations with other metals will further enhance the prospects. The future of Iridium based technology are endless, with the potential to be a driving force to drive economic growth and advance renewable energy solutions.

Keywords

High-Temperature, Microstructure, Corrosion Resistance, Aerospace, Melting

Contents

1.   Introduction

With the new advancements in technology, spurred by the advent of industrial revolution and beyond, superalloys have transitioned from a niche concept to the at-present cornerstone of modern engineering. These exceptional metallic products refer to a group of alloys with the remarkable high-temperature capability, while retaining their mechanical characteristics even under prolonged exposure to extreme temperatures. Driven primarily by the thermal requirements of jet engine turbines, which operate at temperatures close to their melting point, superalloys were initially developed. Fuelled by the combination of reliability and cost effectiveness, they have widespread adoption in diverse industrial sectors. This includes applications in petroleum refineries, chemical plants, military systems and other commercial fields. Furthermore, they provide considerable resistance to both mechanical degradation and environmental corrosion. Initially, they were categorised into three types.[1]

- Iron-based
- Nickel-based
- Cobalt-based

Recent decades have witnessed a concerted effort within the scientific community to surpass the limitation of the conventional superalloys. This endeavour focussed on improving efficiency and reducing emissions.[2] Refractory metals have come up as a sustainable innovation to this problem. A novel class of alloys based on the platinum group metals called the refractory superalloys were developed. These have a high melting temperatures, optimising their potential as structural materials. They address a key limitation inherent in the metallic heat resistant alloys with a BCC lattice, characterized by their high tendency to brittle fracture.[3]

Iridium with an exceptionally high melting point, stands out due its remarkable set of properties. It boasts the highest room temperature elastic modulus (570 GPa), a metric quantifying a substance's resistance to elastic deformation under stress[4]. Hence, it proves to be very stable against corrosion and oxidation. It was majorly employed in the manufacture of crucibles which endure high temperature conditions, as catalysts and spark plugs.[5] But the same reasons which make it highly efficient to be applied as a superalloy , provides a downside in difficulties in fabrication.[6] Under high tensile tests Ir and it's alloys also exhibit a tendency to brittle fracture, a property which has been worked upon to improve in the past decades. Doping with Cerium and Thorium has proved to be an effective alternative for this.[5] Given the metal's multifaceted and optimum attributes, its potential seems limitless.

Superalloys: Fundamentals and Applications   Materials Research Forum LLC
Materials Research Foundations 178 (2025) 81-94   https://doi.org/10.21741/9781644903698-5

2. Properties

The premier discovery and thereafter christening of this wonder metal can be traced back to the 1800s, when one Smithson Tennant found an insoluble black residue, while working with native platinum. Almost decades after this, Iridium found its applications in the nibs of gold pens conferring to it appreciable hardness. Alloying of Iridium followed much later in the 20th century with combinations of Rhodium and Ruthenium, to name some.[7]

Iridium shows great promise in refractory superalloys due to its resistance to oxidative degradation and stability at elevated temperature. Among the transition metals, Iridium supersedes with its high melting point of 2443 °C.[8] It is well suited for extreme thermal and mechanical treatment unlike other platinum alloys. The first attempts at melting Iridium were made by John George Children by the means of a powerful galvanic battery.[7]

The metal has exceptional mechanical properties, notably as the second-densest metal, after osmium with only slight difference. Investigation on the creep resistance, a solid's capability to resist slow deformation on exposure to increased levels of stress, across various temperatures showed a steady creep rate. They also exhibited a very high stress rupture strength and outstanding ductile strength for sheet as well as wire samples, even at enhanced thermal conditions.[9] Despite the high costs and challenging fabrication process, iridium's exceptional mechanical strength renders it as a prime candidate for applications subjected to extreme conditions.

Iridium has been longstanding the most corrosion resistant metal, with its properties extending to aerobic, water, acidic degradation, comprising of even aqua regia. It may however be dissolved by using molten sodium and potassium cyanides.[10] It is found in the oxidation states between -3 to +6, while being abundantly found in +3 and +4 states. IrO_2 is the only well characterised oxide and iridates such as K2IrO3 and KIrO3 can be prepared at higher temperatures. All halogens interact with Ir in the +3 oxidation state, but tetra-, penta- and hexafluorides are known. Hexachloroiridic acid is used as the industrial precursor and for purification of Iridium compounds.[11] Vaska's complex with its central Ir atom finds applications in its role as a homogenous catalyst and as a model compound to study reactions.[12]

In a concerted effort to prevent the brittle fracture of Ir, while maximising its abilities, alloying has historically proved to be the best way around this issue.

3. Composition

3.1 Microstructures and Phase Diagram

By antiquity, the term microstructures have been used to define a set of characteristics of a particular element which may be identified from the arrangement of the composite atoms and/or molecules. But with the more recent terminologies, microstructure is more often indicative of the particular defects residing in a material. Taking an example of microstructure polycrystals, which are aggregates of numerous crystals, we can take into account the defects. There are fine scale defects like dislocations as well as surface coatings and cracks which are classified under large scale defects.[13] Microstructures are a major influencing factor for the various properties conferred to the superalloys. They typically exhibit a dual phase, comprising a matrix phase and a strengthening phase. While the Iridium-based matrix phase contributes the majority of properties,

Superalloys: Fundamentals and Applications   Materials Research Forum LLC
Materials Research Foundations 178 (2025) 81-94   https://doi.org/10.21741/9781644903698-5

the strengthening phase, as the name suggests, plays a crucial role in enhancing the high temperature strength and creep resistance.[14,15]

As reported by Yamabe-Mitarai, Y. et al., Ir-based alloys with f.c.c. and $L1_2$ dual-phases pose similarities to Ni-based superalloys, leading to their naming as refractory superalloys. The morphology of these alloys were dependent on the lattice misfit between the two phases, which were directly influenced by the element that Iridium was alloyed with. It was established in this study that Ir-Nb alloys formed cuboidal precipitates with a 0.4% misfit. In contrast, Ir-Zr alloys exhibited plate-like morphology with a 2% misfit.[5]

Phase diagrams are essential tools for comprehending the thermodynamic behaviour of alloys, especially in the context of superalloys.These graphical representations depict the equilibrium phases present under varying sets of temperature, pressure and compostition. Understanding of phase diagrams is crucial for predicting phase transitions, such as melting point and boiling points. Two other vital properties obtained through phase diagrams are eutectic point and solubility limits. Eutectic point denotes the lowest temperature at which a mixture of two substances solidifies. Solubility limit refers to the maximum amount one element can be dissolved into another at a particular temperature.[16] This is especially utilised for optimizing alloy properties, valuable in industries for the purpose of metal casting and other manufacturing processes, ensuring quality assurance and precise control.

Figure 1 presents the experimentally determined phase diagram of the Ir-Nb-Ni-Al system, which corresponds to the microstructure described earlier. It reveals a three phase region comprising of f.c.c., $L1_2$-Ir_3Nb, and $L1_2$-Ni_3Al phases.[5]

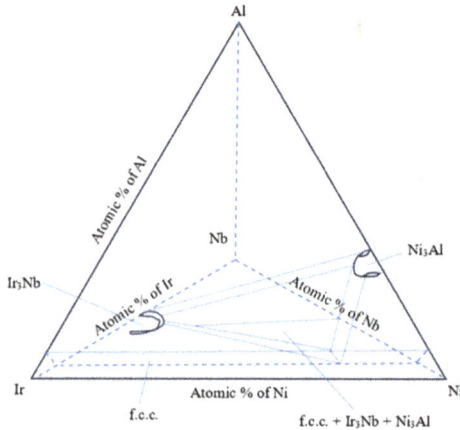

Fig. 1 Experimentally determined ternary diagram for Ir-Nb-Ni at 1300 deg Celsius. F.c.c. and L1 two-phase region is expanded upon addition of Ni. The two lines represent isothermal two phase region between Ir and Nb.

Superalloys: Fundamentals and Applications   Materials Research Forum LLC
Materials Research Foundations 178 (2025) 81-94   https://doi.org/10.21741/9781644903698-5

3.2   Role of Alloying Elements

Alloying elements primarily create lattice distortions, impeding dislocation, which are responsible for the high solid solution strengthening. Addition of Hafnium-Niobium solute mixture to Ir resulted in the anticipated formation of the fcc/L1$_2$ two phase structure. The microstructure contributed to significant improvements in compressive strengths and creep resistances.[17] Subsequent investigations revealed that precipitation strengthening of Ir with Aluminium and Tungsten further improves its mechanical properties at elevated temperatures. On addition of Tantalum alterations were found in the lattice and morphological parameters.[18]

DOP-26, an iridium alloy containing trace amounts of Thorium along with W and Al, has found widespread space power applications. The addition of Th enhances grain boundaries cohesion and forms Ir$_5$Th precipitates, leading to refined grain size. Metals like Ce, Y and Lu exhibit similar tendencies. Conversely, Si and P additions increase brittleness and resultingly, susceptibility to intergranular fracture.[19]

4.   Manufacture

Processing and manufacturing of Iridium alloys consists of multiple steps to obtain the final product.

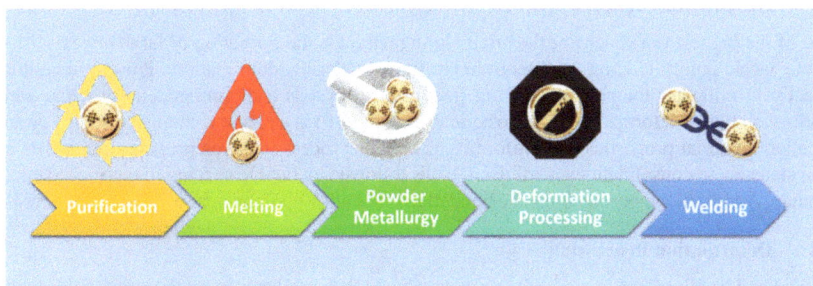

Fig. 2. Different processes involved in the manufacturing of Iridium alloys

4.1   Purification

Ir powder can only be obtained from secondary sources, such as by-products of Pt or Ni synthesis or scrap metal, which can contain significant impurities. These impurities have deleterious effects on the mechanical properties of Ir alloys. Therefore, purification is of utmost importance, ensuring the quality and performance of Ir-based materials.

Conventional purification methods involve using aqua regia to dissolve Ir oxides and precipitating it with ammonium chloride. While effective, this method can be costly and environmentally harmful. Solvent extraction method offers a more sustainable alternative. Electrodeposition or pyrometallurgical method are also viable options, which can remove both volatile and oxide-forming impurities as vapour or slag.

Superalloys: Fundamentals and Applications   Materials Research Forum LLC
Materials Research Foundations 178 (2025) 81-94   https://doi.org/10.21741/9781644903698-5

A recent and highly promising method has been found in the electron beam melting. This technique involves melting of compounds in vacuum to remove impurities. Studies have shown that this can achieve low impurity ratios and ensure ideal mixing and vaporisation.

By carefully selecting appropriate purification method, it is possible to obtain alloy grade Ir that is suitable for demanding applications from various industries.[6,20]

4.2   Melting

Electron beam melting is the most widely employed technique used to melt and alloy Ir with other metals. In addition, induction melting is also utilized, however the use of crucibles in this method might lead to excessive metal volatilisation. To mitigate this, alternative methods such as electron beam, plasma or arc melting are preferred.

Plasma melting, which operates under moderate pressure, is less effective than electron beam melting for Ir alloys. For smaller quantities of metals and alloys, button arc melting is commonly employed. At an industrial scale, the most prevalent method is Vacuum arc melting. This process employs inglots to control the porosity of the material, a critical aspect, as porosity is Dechallenging to manage post manufacturing. Ensuring minimal porosity during the melting and solidification process ensures high quality, defect-free alloys.[6]

4.3   Powder Metallurgy

One of the biggest drawbacks of the arc melting method is the formation of large crystals. These create weak points leading to intergranular fracture during deformation. Powder metallurgy effectively mitigates the risk by creating fine grained crystals and homogenous structure which works well under deformation. Hot Isostatic Pressing (HIP) is a recent technology, which ensures excellent material properties and uniformity.[21] The process involves pressing into billets and then sintering to consolidate the particles. This is followed by hot Deworking to produce the dense product. Additionally, porous Ir can be bonded to other alloy components to be used as filters.[22]

4.4   Deformation Processing

Iridium and its alloys can be difficult to process due to their tendency to crack during deformation. To prevent the same, processing must be performed at elevated temperatures as impurities may significantly influence the material's deformation behaviour. Annealing, which typically relieves internal stress, and alloying with specific elements can improve ductility and grain boundary properties of the microstructure. Initially, the preheat temperatures are high around 1500 K and hot extrusion and rolling may be used for working the alloys. The role of cold working and annealing typically is for refinement.

Manufacturing of sheets or plates is primarily performed using techniques like deep drawing, spinning, and pressing. Temperature control is of the essence to minimize cracking and ensure product quality. Hence it is established that the deformation behaviour of Ir is directly related to the microstructure of the samples.[6,23]

4.5   Welding

Welding is an essential process in the manufacturing of Ir-based components, such as of spark plug electrodes, nuclear fuel containers and crucibles. Arc welding, electron beam welding and laser welding are commonly used methods for joining Ir alloys.

Superalloys: Fundamentals and Applications   Materials Research Forum LLC
Materials Research Foundations 178 (2025) 81-94   https://doi.org/10.21741/9781644903698-5

Weld quality is influenced by various parameters, including current, arc length, speed, gas composition. This minimizes defects and ensures optimal performance. Techniques like arc deflection and oscillation are employed to refine the microstructure of the weld, improving its mechanical properties. Post manufacturing testing, such as dye penetrant and ultrasonic inspection evaluates the quality, by identifying defects and ensuring integrity.

Plasma arc welding is another method for making Ir alloys. It involves heating the workpiece using a constricted arc between the work piece and a non-consumable electrode.[6,24]

By effectively combining the various techniques and optimizing processes, it is possible to produce high-quality iridium-based alloys that meet the demanding requirements of the various applications.

5. Applications

Owing to its physical and chemical properties, Iridium has historically found multiple advantages, especially based on its strength under extreme conditions. Ir crucibles, one of the earliest applications, enabled the synthesis of oxide single crystals such as gadolinium gallium garnet and yttrium aluminium garnet.[25]

Ir with its high catalytic activities has also been utilised to for various binary and ternary alloy anodes, to perform oxidation reactions for the corrosive synthesis of methanol, ammonia and acetic acid.[26]

In response to the growing concerns about global warming and environmental pollution, there has been an increasing demand for vehicles with improved fuel efficiency and reduced emissions. To fulfil this, high performance spark plugs are needed. Iridium has become a novel alternative to the platinum ones which wear out easily. The iridium spark plugs achieve smaller discharge parts and extended service life, offering durability as well as environmental benefits. But Iridium by itself under high-speed driving leads to oxidation volatility and abnormal wear. To inhibit this, Rhodium is added which shows better resistance to wear at high temperatures.[27]

This metal along with its alloys has especially widely been used in the aerospace industry. The conventional usage of Ni-based heat-resistant alloys are soon becoming redundant owing to the fundamental limitation of its operating temperatures. With the constant demand in the aerospace industries for newer and more powerful engines, the usage of Nickel with it melting temperature only up to a mere 1450 °C, seems insignificant. There is a dire need for materials surpassing the current capabilities, a solution for which has come in the form of Ir superalloys. It successfully manages to fulfil the requirements of gas turbine engines with an operativity at temperatures as high as 1500°C. Furthermore, for the manufacture of space aircrafts, the fuel cladding of the thermoelectric temperature is made up of an iridium-based alloy. It's also resistant to the hypertoxic plutonium oxide which is filled inside.[28]

Radioisotope thermoelectric generators (RTGs), powered by plutonium dioxide, are essential for spacecraft propulsion. This heat source is encapsulated using Ir alloys, which provide a durable and efficient means for high temperature welding. The alloy DOP 26, due to the presence of Th, addresses the inherent ductility limitations of Ir, making it an ideal candidate for demanding outer planetary missions.[29] Radiation-cooled rockets, which are specifically designed to operate in high temperature conditions without conventional cooling methods, such as liquid propellants, traditionally utilise a niobium alloy. But the temperature limitations of niobium necessitate the usage of supplementary cooling methods, leading to significant energy and resource inefficiencies. A higher performance and cleaner alternative are found in the Ir- coated Rhenium system, which completely reduces or even eliminates the usage of fuel film cooling. The lifetimes of the combination of these two metals are prolonged due to the slow boundary diffusion rates of Re into Ir.[30]

Iridium-based electrocatalysts have emerged as a promising avenue for sustainable energy applications due to their unique redox properties. These catalysts demonstrate potential in various electrochemical reactions.

- Water splitting: Producing hydrogen and oxygen from water using renewable electricity.

- Fuel oxidation: Converting fuels like methanol and formic acid into electricity.

- Carbon dioxide reduction: Transforming carbon dioxide into valuable products such as hydrocarbons.

Heteroatom doping or alloying with transition metals has been shown to effectively modulate the electronic structure of Ir catalysts. For instance, the IrW alloy catalyst has been found to enhance proton adsorption energy and increase affinity for OH- under acidic conditions. This not only improves catalytic performance but also offers a strategy to reduce the reliance on and cost of iridium.

The potential of Ir-based catalysts to significantly impact industries such as energy, chemicals, gas capture, metallurgy, materials, and electronics manufacturing makes them a worthwhile area of exploration.[31]

Table 1 Novel applications of Iridium-based alloys and their advantages

Alloy	Application	Advantages	Ref
Pt-Ir	Picosecond laser micromachining of Pt-Ir Alloys for coronary stent applications	Limited material distortion which results in more precise and intricate dross-free cutting	[32]
Ir-Cu	Enhanced electrocatalysts for improved oxygen evolution and reduction reactions	Cost efficient and Effective	[33]
Ir based Heusler alloys (Ir_2CrSi, Ir_2CrGe, IrRhCrSi, and IrRhCrGe)	Data storage and spin electronic devices like magnetoresistive random access memory (MRAM), giant magnetoresistance (GMR), magnetic sensors.	Half metallicity and magnetism with exchange splitting	[34]
Ir added to Au-Ti binary alloy	Base for development of dental alloy for inlays, conventional crowns and bridges	Ir provided grain refining effects to the alloy	[35]

Re-Ir	Monolayer or multilayer coating for providing anti-sticking properties to glass during molding process	Provides interface resistance between the multiple layers	[36]
Re-Ir-Nb	Used as the release film for molding die to produce camera lenses	Provides high heat resistance, chemical stability, hardness and processability	[37]
Pt-Ir NP coatings	Reduce impedance in 3D neural electrodes	Improves efficiency of neural stimulation showing promising treatments like deep brain stimulation	[38]
Au-Ir	Excellent catalytic activity for glucose oxidation, making it ideal for glucose sensors and fuel cells	Synergistic effect provided by gold and iridium which enhances their properties	[39]
Ir-Y	Acid water hydrolysis to produce hydrogen from water Integrated with renewable energy sources would lead to sustainable hydrogen production	Shows higher catalytic activity and stability compared to other alloys and reduces overpotential required for oxygen evolution	[40]
Pd-Ir	Used in hydrogen storage alloys to modify the anode material	Enhances the kinetics of hydrogen sorption process in Ni-MH batteries	[41]
Ru-Ir	Oxidation of methane to methanol	Broadens the applicability of metal electrophilicity for C-H activation prediction	[42]
Ir-Ru	Modification of ballast water to transport aquatic creatures	Chlorination and disinfection of the water	[43]

6. Challenges

Table 2 Comparative Analysis of Melting Points and Densities of Refractory Metals

Metal	Melting Point [K]	Density [g/cm^3]
Rh	2237	12.4
Hf	2506	13.3
Ru	2607	12.1
Ir	**2719**	**22.42**
Nb	2750	8.57
Mo	2896	10.2
Ta	3290	16.4
Os	3306	22.57
Re	3459	20.8
W	3695	19.3

One of the defining characteristics of refractory metals is their melting temperature (T_m) which must be above 2200 °C.[44] By this definition, strictly, the elements stated in Table 2 are classified as refractory metals.[45] This goes to show that even though Ir has a very high melting point, metals like Tungsten, Tantalum, and Rhenium surpass it. This makes them better suited for ultra-high-temperature applications. Alloying with such metals would considerably make the material for efficient.

Iridium is one of the densest metals which while making it advantageous for aerospace or precision instrument components, leads to limitations in weight-sensitive applications.[46]

One of the most prominent disadvantages of Ir is its brittleness and susceptibility to intergranular fracture. This necessitates the usage of metals which are comparatively more flexible. Iridium is also very hard and often times becomes very difficult to roll or draw.

Iridium is one of the rarest metals found in the earth's crust, which depends on a variety of factors which influence its market dynamics. Nickel and platinum mines, predominantly found in South Africa and Russia, are the primary sources. Disputes in these regions due to geopolitical instability, labour disputes, or disrupted trade agreements, impact iridium production and supply. The continuous usage to meet the high demands to fulfil upcoming applications exacerbates the depletion of the present reserves. These interconnected factors disrupt the global market of the metal, limiting export and driving up prices. Careful assessment of these changes are important to understand the future trajectory of Ir's value.[47]

Fig 3 represents a graph exhibiting the increasing demand of Iridium through the years. After taking a slight hit during the pandemic years, the global market is once again on a steady rise.[48]

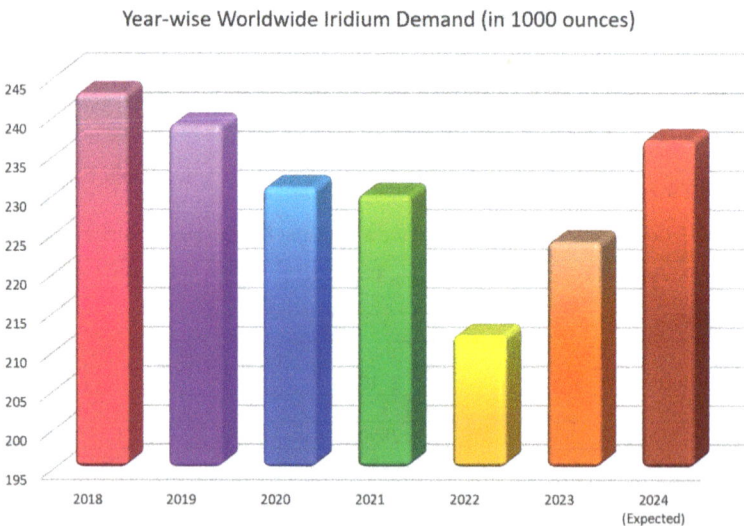

Fig. 3 Worldwide Iridium Demand per 1000 ounces from the years 2018-24

Superalloys: Fundamentals and Applications
Materials Research Foundations 178 (2025) 81-94

Materials Research Forum LLC
https://doi.org/10.21741/9781644903698-5

At present, there is not much existing data explaining the toxicity mechanisms of Ir exposure to the human body. An in-vivo study done conducted on the oral Ir exposure in rats caused nephrological and immunological imbalances.[49]

7. Summary

Iridium based superalloys are a class of high-temperature materials which have spurred upon a new age of alloys replacing the conventional Ni, Fe or Co-based alloys. The history of Iridium can be traced back to its discovery in the 1800s after which its exceptionally high melting point, density, creep resistance was explored. The mechanical properties were a direct reflection of its composition, namely microstructure and the representing phase diagrams, with f.c.c. and L1$_2$ dual-phases. Different alloying elements such as Al, Hf, Nb, Th, W overcome the brittleness, while strengthening and refining the grain size. These superalloys are manufactured through complex processes involving purification, melting, powder metallurgy, welding, and deformation processing. These techniques ensure the production of high-grade alloys that meet the demading requirements of the growing applications. These wide range of applications include spark plugs, gas turbine engines, biomedical devices and therapeutics, radioisotope thermoelectric generators, and electrochemical catalysts for hydrogen production using renewable sources of energy. However, this is encountered by various limitations, such as alternative metals with higher temperature endurance, weight-sensitive constraints, higher prices and lesser availability. Iridium's global market trends are influenced majorly by geopolitical instability, regulatory demands, limited supply and increased demand. Further development and investments for refractory superalloys, focussing on Iridium, while understanding its possible human toxicity, paves the path for multitude of innovations.

References

[1] H.A. Kishawy, A. Hosseini, Superalloys BT- machining difficult-to-cut materials: basic principles and challenges, in: H.A. Kishawy, A. Hosseini (Eds.), Springer International Publishing, Cham, 2019: pp. 97–137. https://doi.org/10.1007/978-3-319-95966-5_4

[2] D.B. Miracle, M.H. Tsai, O.N. Senkov, V. Soni, R. Banerjee, Refractory high entropy superalloys (RSAs), Scr. Mater. 187 (2020) 445–452. https://doi.org/https://doi.org/10.1016/j.scriptamat.2020.06.048

[3] I. Razumovskii, B. Bokstein, M. Razumovsky, Approaches to the development of advanced alloys based on refractory metals, Encyclopedia 3 (2023) 311–326. https://doi.org/10.3390/encyclopedia3010019

[4] S. Allard, Metals: Thermal and mechanical data, Elsevier, 2013.

[5] Y. Yamabe-Mitarai, Y.F. Gu, H. Harada, Two-phase iridium-based refractory superalloys, Platin. Met. Rev. 46 (2002) 74–81.

[6] E.K. Ohriner, Processing of iridium and iridium alloys, Platin. Met. Rev. 52 (2008) 186–197. https://doi.org/10.1595/147106708X333827

[7] L.B. Hunt, A history of iridium, Platin. Met. Rev. 31 (1987) 32–41.

[8] G. Xu, X. Chong, Y. Zhou, Y. Wei, C. Hu, A. Zhang, R. Zhou, J. Feng, Effects of the alloying element on the stacking fault energies of dilute Ir-based superalloys: A comprehensive first-principles study, J. Mater. Res. 35 (2020) 2718–2725.

Materials Research Forum LLC
https://doi.org/10.21741/9781644903698-5

[9] J. Merker, B. Fischer, D.F. Lupton, J. Witte, Investigations on structure and high temperature properties of iridium, in: Mater. Sci. Forum, Trans Tech Publ, 2007: pp. 2216–2221.

[10]   J. Emsley, Nature's building blocks: an AZ guide to the elements, Oxford University Press, USA, 2011.

[11]   K.P. Kühn, I.F. Chaberny, K. Massholder, M. Stickler, V.W. Benz, H.G. Sonntag, L. Erdinger, Disinfection of surfaces by photocatalytic oxidation with titanium dioxide and UVA light, Chemosphere 53 (2003) 71–77. https://doi.org/10.1016/S0045-6535(03)00362-X

[12]   Iridium, (n.d.). https://www.scientificlib.com/en/Chemistry/Elements/Iridium.html (accessed July 26, 2024).

[13]   K.J. Kurzydlowski, B. Ralph, The quantitative description of the microstructure of materials, CRC press, 1995.

[14]   G. Liu, L. Kong, S. Ruan, S. Birosca, Microstructure and phases structure in nickel-based superalloy IN713C after solidification, Mater. Charact. 182 (2021) 111566. https://doi.org/https://doi.org/10.1016/j.matchar.2021.111566

[15]   B. Gwalani, D. Choudhuri, K. Liu, J.T. Lloyd, R.S. Mishra, R. Banerjee, Interplay between single phase solid solution strengthening and multi-phase strengthening in the same high entropy alloy, Mater. Sci. Eng. A 771 (2020) 138620. https://doi.org/https://doi.org/10.1016/j.msea.2019.138620

[16]   S.J. Barnes, M.L. Fiorentini, Iridium, ruthenium and rhodium in komatiites: Evidence for iridium alloy saturation, Chem. Geol. 257 (2008) 44–58. https://doi.org/10.1016/J.CHEMGEO.2008.08.015

[17]   J.B. Sha, Y. Yamabe-Mitarai, Saturated solid-solution hardening behavior of Ir–Hf–Nb refractory superalloys for ultra-high temperature applications, Scr. Mater. 54 (2006) 115–119. https://doi.org/https://doi.org/10.1016/j.scriptamat.2005.08.038

[18]   X. Fang, R. Hu, J. Yang, Y. Liu, M. Wen, Microstructure evolution and mechanical properties of novel γ/γ′ two-phase strengthened Ir-based superalloys, Metals (Basel). 9 (2019) 1171.

[19]   Y. Liu, C.T. Liu, L. Heatherly, E.P. George, Effects of alloying elements on dendritic segregation in iridium alloys, J. Alloys Compd. 459 (2008) 130–134. https://doi.org/10.1016/J.JALLCOM.2007.04.252

[20]   E.K. Ohriner, Purification of iridium by electron beam melting, J. Alloys Compd. 461 (2008) 633–640. https://doi.org/10.1016/J.JALLCOM.2007.07.067

[21]   Z. Mao, W. Xian, G. Qin-Qin, H. Chang-Yi, Z. Jun, C. Hong-Zhong, Z. Gui-Xue, W. Xing-Qiang, Z. Xu-Xiang, W. Yan, Optimization of process parameters for hot isostatic pressing of Ir–Rh alloys based on first-principles calculations, Mater. Res. Express 9 (2022) 116513. https://doi.org/10.1088/2053-1591/AC9887

[22]   C. Xiang, H. Liu, Y. Huang, H. Tang, Preparation and microstructure of Ir-Zr-W alloys by powder metallurgy route, Rare Met. Mater. Eng. 38 (2009) 1132–1135. https://doi.org/10.1016/S1875-5372(10)60042-2

[23]   P. Panfilov, A. Yermakov, O. V. Antonova, V.P. Pilyugin, Plastic deformation of polycrystalline iridium at room temperature, Platin. Met. Rev. 53 (2009) 138–146. https://doi.org/10.1595/147106709X463318

[24]   Z. Amirsardari, Y. Vahidshad, M.A. Amirifar, A.H. Khodabakhshi, A. Dourani, N. Ghadiri Massoom, Method for preventing the contamination of iridium nanocatalyst during plasma arc welding procedure, J. Fail. Anal. Prev. 21 (2021) 1289–1296. https://doi.org/10.1007/S11668-021-01165-Z/FIGURES/9

[25]   J.R. Handley, Increasing applications for iridium, Platin. Met. Rev. 30 (1986) 12–13. https://doi.org/10.1595/003214086X3011213

[26]   W.P. Wu, Z.F. Chen, Iridium coating: processes, properties and application, Part I: Processes for protection in high-temperature environments against oxidation and corrosion, Johnson Matthey Technol. Rev. 61 (2017) 16–28. https://doi.org/10.1595/205651317X693606/CITE/REFWORKS

[27]   H. Osamura, N. Abe, Development of New Iridium Alloy for Spark Plug Electrodes, SAE Tech. Pap. (1999). https://doi.org/10.4271/1999-01-0796

[28]   J.R. Yang, X. Fang, Y. Liu, Z.-T. Gao, M. Wen, R. Hu, Microstructure evolution and mechanical properties of a novel γ' phase-strengthened Ir-W-Al-Th superalloy, Rare Met. 40 (2021) 3588–3597. https://doi.org/10.1007/s12598-020-01682-0

[29]   S.A. David, R.G. Miller, Z. Feng, Welding of unique and advanced alloys for space and high-temperature applications: welding and weldability of iridium and platinum alloys†, Sci. Technol. Weld. Join. 22 (2017) 244–256. https://doi.org/10.1080/13621718.2016.1222255

[30]   B.D. Reed, J.A. Biaglow, S.J. Schneider, Iridium-coated rhenium radiation-cooled rockets, (1997).

[31]   B. Huang, Y. Zhao, Iridium-based electrocatalysts toward sustainable energy conversion, EcoMat 4 (2022) e12176. https://doi.org/10.1002/EOM2.12176

[32]   N. Muhammad, D. Whitehead, A. Boor, W. Oppenlander, Z. Liu, L. Li, Picosecond laser micromachining of nitinol and platinum–iridium alloy for coronary stent applications, Springer 106 (2012) 607–617. https://doi.org/10.1007/s00339-011-6609-4

[33]   F. Wang, K. Kusada, D. Wu, T. Yamamoto, T. Toriyama, S. Matsumura, Y. Nanba, M. Koyama, H. Kitagawa, Solid-solution alloy nanoparticles of the immiscible iridium–copper system with a wide composition range for enhanced electrocatalytic applications, Angew. Chemie Int. Ed. 57 (2018) 4505–4509. https://doi.org/10.1002/ANIE.201800650

[34]   R. Paudel, S. KC, S. Adhikari, J.C. Zhu, S. Ahmad, G. Chandra Kaphle, D. Paudyal, Electronic and magnetic properties of iridium-based novel Heusler alloys, J. Magn. Magn. Mater. 555 (2022) 169405. https://doi.org/10.1016/J.JMMM.2022.169405

[35]   J. Fischer, Effect of small additions of Ir on properties of a binary Au–Ti alloy, Dent. Mater. 18 (2002) 331–335. https://doi.org/10.1016/S0109-5641(01)00058-6

[36]   X.Y. Zhu, J.J. Wei, L.X. Chen, J.L. Liu, L.F. Hei, C.M. Li, Y. Zhang, Anti-sticking Re-Ir coating for glass molding process, Thin Solid Films 584 (2015) 305–309. https://doi.org/10.1016/J.TSF.2015.01.002

[37]   Fukushima, Amorphous alloy, molding die, and method for producing optical element, (2013).

[38]   V. Ramesh, J. Johny, J. Jakobi, R. Stuckert, C. Rehbock, S. Barcikowski, Platinum-iridium alloy nanoparticle coatings produced by electrophoretic deposition reduce impedance

Materials Research Forum LLC
https://doi.org/10.21741/9781644903698-5

in 3D neural electrodes, ChemPhysChem 25 (2024) e202300623.
https://doi.org/10.1002/CPHC.202300623

[39]   A. Yu, S. Moon, T. Kwon, Y. Bin Cho, M.H. Kim, C. Lee, Y. Lee, Au-Ir alloy nanofibers
synthesized from Au-Ir/IrO2 composites via thermal hydrogen treatment: Application for
glucose oxidation, sensors and actuators. B Chem. 310 (2020) 127822.
https://doi.org/10.1016/J.SNB.2020.127822

[40]   X. Xiong, J. Tang, Y. Ji, W. Xue, H. Wang, C. Liu, H. Zeng, Y. Dai, H.J. Peng, T. Zheng,
C. Xia, X. Liu, Q. Jiang, High-efficiency iridium-yttrium alloy catalyst for acidic water
electrolysis, Adv. Energy Mater. 14 (2024) 2304479.
https://doi.org/10.1002/AENM.202304479

[41]   K. Hubkowska, M. Pająk, A. Czerwiński, The effect of the iridium alloying and hydrogen
sorption on the physicochemical and electrochemical properties of palladium, Mater. 2023,
Vol. 16, Page 4556 16 (2023) 4556. https://doi.org/10.3390/MA16134556

[42]   L. Yang, J. Huang, S. Dai, X. Tang, Z. Hu, M. Li, H. Zeng, R. Luque, P. Duan, Z. Rui,
Uniphase ruthenium–iridium alloy-based electronic regulation for electronic structure–
function study in methane oxidation to methanol, J. Mater. Chem. A 8 (2020) 24024–24030.
https://doi.org/10.1039/D0TA08350J

[43]   Y. Zhang, J. Guo, J. Ding, C. Li, Q. Zhao, A flow-through ti4o7 membrane electrode for
ballast water treatment: performance, mechanism, and comparison with ru-ir electrode,
(2024). https://doi.org/10.2139/SSRN.4776968

[44]   D.B. Miracle, M.H. Tsai, O.N. Senkov, V. Soni, R. Banerjee, Refractory high entropy
superalloys (RSAs), Scr. Mater. 187 (2020) 445–452.
https://doi.org/10.1016/J.SCRIPTAMAT.2020.06.048

[45]   National Center for Biotechnology Information, Melting point in the periodic table of
elements. https://pubchem.ncbi.nlm.nih.gov/ptable/melting-point/2024 (accessed September
16, 2024).

[46]   National Center for Biotechnology Information, Density in the periodic table of elements.
https://pubchem.ncbi.nlm.nih.gov/periodic-table/density/2024 (accessed September 16, 2024).

[47]   Iridium Prices– Historical Graph [Realtime Updates].
https://procurementtactics.com/iridium-prices/ (accessed September 17, 2024).

[48]   Iridium demand worldwide 2024 | Statista, (n.d.).
https://www.statista.com/statistics/585840/demand-for-iridium-worldwide/2024 (accessed
September 17, 2024).

[49]   I. Iavicoli, V. Leso, Iridium, Handb. Toxicol. Met. Fifth Ed. 2 (2022) 369–390.
https://doi.org/10.1016/B978-0-12-822946-0.00015-5

Superalloys: Fundamentals and Applications
Materials Research Foundations 178 (2025) 95-108

Materials Research Forum LLC
https://doi.org/10.21741/9781644903698-6

Chapter 6

Properties and Corrosion Behaviours of Different Types of Superalloys

Anirudh Pratap Singh Raman[1], Pallavi Jain[1*], Sunil Kumar Yadav[1], Prashant Singh[2]

[1]Department of Chemistry, Faculty of Engineering and Technology, SRM Institute of Science and Technology, Delhi-NCR Campus, Modinagar, India

[2]Department of Chemistry, Atma Ram Sanathan College, University of Delhi, Delhi India

* palli24@gmail.com, pallavij@srmist.edu.in

Abstract

This chapter delves into the properties and corrosion behaviours of different types of superalloys, specifically nickel-based, cobalt-based, and iron-based superalloys. These materials are crucial in power generation, high-temperature, aerospace, and automotive industries because of their exceptional mechanical properties and oxidation resistance. The chapter discusses the primary corrosion mechanisms affecting these superalloys, including general corrosion, crevice corrosion, intergranular corrosion, stress corrosion cracking (SCC), high-temperature oxidation, and hot corrosion. It also explores how factors such as alloy composition, microstructure, and environmental conditions influence corrosion resistance. Recent advancements in alloy development, including the incorporation of new elements and innovative processing techniques, are highlighted. The chapter concludes by identifying ongoing challenges and future research directions aimed at enhancing the corrosion resistance of superalloys, thereby broadening their applications and improving their performance in demanding environments.

Keywords

Nickel-Based, Cobalt-Based and Iron-Based Superalloys, Crevice Corrosion, Intergranular Corrosion, Stress Corrosion Cracking, Superalloys

Contents

1.   Introduction

Superalloys are high-performance materials renowned for their exceptional mechanical strength, resistance to thermal creep deformation, and stability at high temperatures. These characteristics make them indispensable in demanding applications such as aerospace, power generation, and chemical processing. However, despite their superior properties, superalloys are not immune to corrosion-a phenomenon that can significantly compromise their performance and longevity. Corrosion involves the slow deterioration of substances, typically metals, as a result of chemical interactions with their surroundings. For superalloys, which operate in extreme conditions, understanding and mitigating corrosion is crucial to ensuring their reliability and efficiency. This chapter delves into the various aspects of corrosion in superalloys, exploring the mechanisms, types, and factors that influence this detrimental process. High-temperature oxidation and abrasion are primary factors contributing to equipment failures in the hot sections of gas turbines, boilers, metallurgical furnaces, and petrochemical plants. Although superalloys have been engineered for high-temperature applications, they frequently encounter difficulties in meeting requirements for both elevated strength and resilience against erosion and corrosion [1-4].

Energy-generating systems frequently use low-quality fuel oils and fossil fuels containing complex mixes of molten sodium sulphate (Na_2SO_4) and vanadium pentoxide (V_2O_5) [5]. Turbine engines, which operate in extremely hot and severe conditions, are susceptible to substantial degradation of materials during operations. In today's gas turbine applications, a range of Ni, Co, and Fe based superalloys are used to survive these conditions. Alloy corrosion resistance is strongly dependent on the presence of elements play critical roles in the formation of protecting passivating oxide layers. However, the harshness of high-temperature molten salt environments presents a greater problem than aquatic solutions. In these conditions, the oxide covering may become non-protective, allowing the underlying alloy susceptible to corrosion. Corrosion of this kind often attacks the alloy's least noble metal component[6]. Corrosion processes have been developed as a consequence of researchers' efforts to comprehend the selective loss of elemental components from alloys under these circumstances [7]. As a result, the study of corrosion-related substances or oxides scale becomes vital in understanding and managing corrosion in these difficult environments.

In today's gas turbine applications, a variety of Ni, Co, and Fe based superalloys are employed to withstand these conditions. The corrosion resistance of alloys relies heavily on the presence of elements such as Al, Cr and Si, which play crucial roles in forming protective passivating oxide layers [8]. However, the harshness of high-temperature molten salt environments poses a greater challenge compared to aqueous solutions. In these environments, the oxide coating can become non-protective, leaving the underlying alloy vulnerable to corrosion. This corrosion typically targets the least noble metal constituent within the alloy [9]. Researchers have attempted to comprehend the selective removal of elemental constituents from alloys under such conditions, which has led to the formulation of corrosion mechanisms. As a result, study of products caused by corrosion or oxide levels becomes vital in understanding and managing rust in these tough environments [10,11].

In this chapter, we will look at the fundamental concepts of corrosion, including the electrochemical reactions that cause it and the environmental conditions that exacerbate it. Following that, we'll look at the many types of corrosion that affect superalloys, including oxidation, sulfidation, and hot corrosion. Each category provides unique issues that necessitate customised prevention and mitigation techniques. The impact of alloy composition, microstructure, and surface treatments on corrosion resistance will be addressed in detail. Advances in material science have led to the development of superalloys with higher corrosion resistance.

2.   Types of Superalloys

Superalloys are chosen for components subjected to extreme stress at temperatures greater than 540°C because of their higher strength, creep, retention, and resistance to oxidation at such high temperatures [12-15]. Superalloys can be classified into three types which are explained below.

2.1   Nickel-based Superalloys

The microstructure includes γ' phase, organized in the form of L12 structure, which is the source of high-temperature strength. Within the continuous fcc γ phase matrix, which produces narrow channels between the γ' cuboids, this γ' phase appears as cuboids with coherent interfaces. The dislocation movement is mostly restricted to the γ channels because to the strong stiffness and limited dislocation tolerance imparted by the L12 structure. The enhancement of creep resistance and high-temperature strength in these alloys was further bolstered by the restriction of oxidation. Under extreme thermal-mechanical conditions, three significant microstructural changes occurred: the development of dislocation networks, the coarsening of γ' phase cuboids, and the precipitation of topologically close-packed (TCP) phases, as illustrated in Figure 1 [16-18].

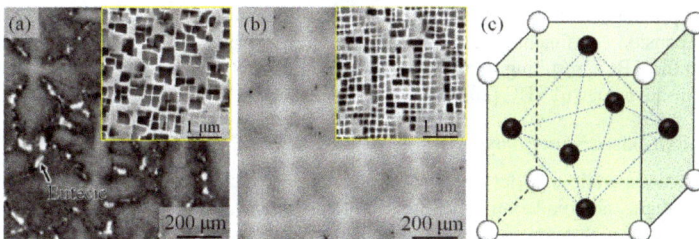

Figure 1 *Microstructures of Ni-based single crystalline superalloys [18]. (a) Microstructure of single crystalline DD22 superalloy in as-cast state; (b) Microstructure of the DD22 superalloy*

after γ' phase formation heat treatment; (c) Schematic of the A_3B type L12 structure of the γ' phase [11].

Since its inception in the 1940s as a nickel-based superalloy, the development of superalloys has been ongoing. This advancement has undergone four phases of microstructural change throughout its history. At first, superalloys were produced in the same way as other metals were: in wrought form. Problems with high production costs, poor success rates, and significant material waste were faced by these wrought superalloys, especially those used for turbine blades. During the second phase, which started in the 1950s, cast superalloys were introduced. These alloys offered many advantages, including faster production, cheaper costs, reduced material waste, and improved resistance to creep and high temperatures[14,19] (**Figure 2**).

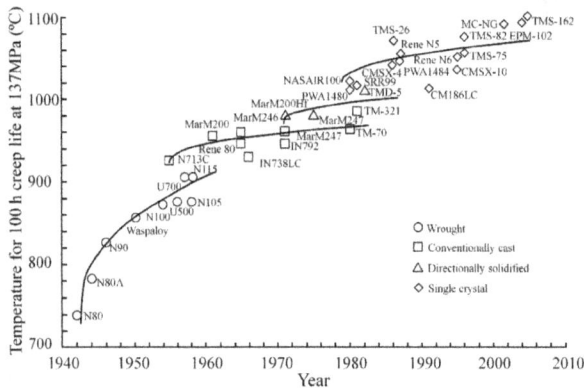

Figure 2 *Development histories of high-temperature properties of superalloys [11]*

Ni-based superalloys are often resistant to oxidation. However, in service, they rely on protective thermal barrier coatings (TBCs) as their principal means of preventing oxidation at high temperatures. TBCs typically have two layers: an exterior ceramic layer that protects the alloy from high temperatures, and an interior metal-based layer that produces an Al_2O_3 layer. These coatings are deposited on superalloys using either electron-beam physical vapour deposition or plasma spraying techniques. Because of their ceramic nature, TBC have drastically different thermal expansion coefficients than Ni-based superalloys. This disparity causes thermal stresses to occur in the TBCs, resulting in, splitting, and eventual rupture from the metal surface. This is the major disadvantage of TBC [20,21].

2.2 Cobalt-based Superalloys

Cobalt-based superalloys offer numerous advantages, including superb durability against corrosion, fatigue, and oxidation under elevated temperatures. These properties make them ideal for manufacturing high-temperature components like turbine guiding vanes in aeronautical engines and nozzles for engines powered by diesel. Traditionally, these alloys contain of a γ matrix and carbides, an important role in hardening. Nonetheless, it is important to note that mechanically, cobalt-based superalloys typically exhibit inferior performance at high temperatures compared to their nickel-based counterparts. Using directional solidification (DS) in an axial magnetic field, Zhang et al. designed Co-based superalloys composed of composite materials with properties of Co-8.8Al-9.8W-2Ta. Researchers looked at the relationship between magnetic field intensity and PDAS, or main dendrite arm spacing. Results showed that at various drawing speeds, increasing the magnetic field intensity initially increased and then lowered the PDAS of the samples [22] (**Figure 3**).

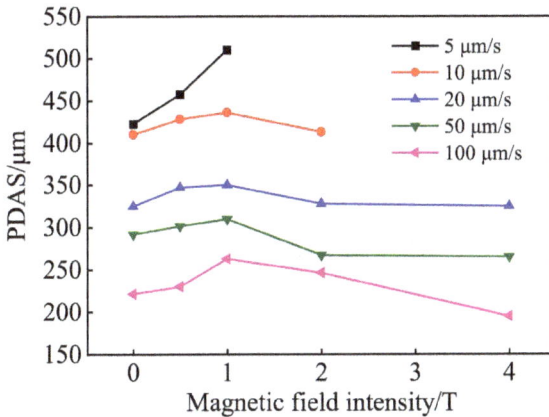

Figure 3 *The variation of PDAS with magnetic field intensity under different drawing speeds* *[22]*

Co-based superalloys are extensively utilized in corrosive environments and high-temperature because of their dynamic properties. In the aerospace industry, they are critical for manufacturing turbine guide vanes and combustor liners in aero-engines, where they endure extreme thermal and mechanical stress. These alloys are also pivotal in power generation, particularly in the hot sections of industrial gas turbines and steam turbines, where their ability to maintain strength and resist corrosion at elevated temperatures is essential [24–26]. The automotive industry benefits from Co-based superalloys in turbocharger components and diesel engine nozzles, which require high wear and thermal fatigue resistance. Additionally, their biocompatibility and durability make them suitable for medical implants and surgical instruments. In the oil and gas sector, these superalloys are used in downhole tools, valves, and pumps, where they withstand harsh operating conditions. The ongoing advancements in alloy compositions and coating technologies continue to enhance the performance and expand the applications of cobalt-based superalloys [27–29].

2.3 Iron-based Superalloys

It is widely acknowledged that nearly all moving parts are susceptible to wear, necessitating improvements in their wear resistance to prolong their lifespan. Considerable research and development efforts have focused on investigating a wide array of ferrous and non-ferrous alloys for this objective. Co-Ni based alloys are classified among the non-ferrous varieties within the spectrum of hard facing alloys accessible. These alloys are renowned for their exceptional durability against corrosion and abrasion, particularly in applications involving elevated

Superalloys: Fundamentals and Applications   Materials Research Forum LLC
Materials Research Foundations 178 (2025) 95-108   https://doi.org/10.21741/9781644903698-6

temperatures [30]. Fe-based alloys are valued for their affordability and versatility across various engineering applications [31]. Studies have shown that hard facing electrodes made of iron with varying amounts of carbon and Cr are very effective because to their great hardness and wear resistance [32]. Factors such as weldability, durability, wear mode, and cost-effectiveness play crucial roles in selecting the appropriate hardfacing alloy, as emphasized by Chandel [33]. Additionally, electrode selection is affected by the substrate material's and the deposited alloy's metallurgical compatibility. This is especially true in cost-conscious and application-friendly sectors, such as sugar manufacturing. Compared to more specialized materials, iron-based alloys are more popular due to the wide variety of alloy types they provide. In the minerals processing and associated sectors, eutectic alloys based on Fe-Cr-C compositions, which are white and contain high chromium, are used extensively for wear protection. Castings and hardfacings are two functions performed by these white irons [34,35]. They are usually made with a microstructure that is eutectic or slightly hypoeutectic, with primary austenite as its defining feature, for use in casting applications. Previous studies Researchers have stressed that achieving excellent wear resistance necessitates a significant proportion of M7C3 carbides within the austenitic microstructure. This particular structure is found in compositions ranging from subeutectic to eutectic and supereutectic, typically with a Cr/C ratio ranging from 5 to 8 [36,37]. Gahr [38] noted that while higher hardness levels generally enhance wear resistance, the morphological characteristics-including shape, size, and distribution of phases-significantly influence structural behavior, thereby affecting both wear and corrosion resistance. As environmental concerns continue to rise, there is an increasing demand for environmentally friendly hardfacing techniques. A more consistent and refined microstructure was shown to be an outcome of adding La_2O_3 to Fe-Cr alloys, according to research by Zhao et al. [39] The alloy's enhanced microstructure greatly boosted its strength, hardness, and resistance to wear. It is becoming more important to do research on coatings and treatments that lessen the hardfacing process's impact on the environment in terms of emissions, waste, and energy consumption, as well as the usage of hazardous compounds. Kim et al. states that rare earth oxides improve pitting corrosion resistance in super duplex stainless steel by decreasing the regions at interfaces that are more likely to start pitting corrosion [40].

3. Types of Corrosion

3.1 Oxidation

When a superalloy gets subjected to reactive materials in its surroundings, oxidation processes can occur, either improving or degrading its properties. A popular protection approach is to selectively oxidise one of the alloy's constituents to form a protective barrier. The oxide barrier is a slow-growing, passivating layer that adheres to the underlying alloy and is typically 1-20 μm thick. If the growth rate is too sluggish, a barrier of protection may not form constantly, whereas faster-growing scales are more likely to spallation. The most efficient protecting oxides are usually chromia (Cr_2O_3) or alumina (Al_2O_3). Chromia only provides resistance up to 1000-1100°C due to the generation of volatile CrO_3 at higher temperatures, but alumina is protective up to 1300-1400°C. As a result, superalloys designed for high-temperature applications develop alumina scales. The oxidation process are thermodynamically regulated by the system's Gibbs free energy (G), with the most stable state having the lowest G. The oxide scale is made up of the stable oxides that contribute the most to G reduction. Oxides that require greater oxygen partial pressures (pO) are found at the oxide-gas contact, whereas others can form beneath the surface. When a superalloy undergoes oxidation, the process can be controlled by diffusion, meaning the oxide layer forms continuously and adheres tightly to the metal surface. After an initial thin layer forms, further oxidation slows down because reactants need to diffuse through the oxide layer for the reaction to proceed. This selective oxidation is crucial for long-term protection of the alloy by passivating its surface. It relies on alloying elements diffusing rapidly to the interface between the alloy and oxide scale to create a continuous protective layer. Insufficient elemental concentration or slow diffusion can lead to internal oxidation, which can degrade the alloy's fatigue properties. The surface finish of the alloy also affects diffusion rates. Machining introduces subsurface defects that can accelerate external scale growth, influencing how oxidation progresses. Therefore, studies on oxidation need to use samples with carefully controlled surface finishes. Initially, during the oxide scale's early growth stage, all near-surface alloying elements form oxides. As elements crucial for the formation of fast-developing oxides diminish in the subsurface area, they are replaced by more sluggish and thermodynamically stable oxides. These new oxides gradually form a seamless and protective layer. During the phase of stable oxidation, selective oxidation takes precedence. Other oxide growth is inhibited by low oxygen partial pressure at the alloy-scale interface and by limited diffusion within the scale, which slows down oxide growth beyond the protective layer. During thermal cycling, stresses from differing thermal expansion coefficients between the scale and alloy can cause oxide scale spallation. Regrowth of the oxide layer may be challenging as the subsurface alloy region becomes depleted of oxidizing elements [41,42].

3.2 Corrosion in Molten State

Superalloys can undergo corrosion, particularly under extreme conditions like elevated temperatures in molten salts. Meier categorized hot corrosion into two types: Type I (high temperature) and Type II (low temperature). Type I corrosion occurs when the deposited salt mixture becomes liquid upon initial deposition due to high temperatures, while Type II corrosion starts with the salt mixture in a solid phase, transitioning to a liquid phase upon exposure to elevated temperatures [43]. Moreover, during high-temperature corrosion, superalloys go through a two-step deterioration process: a beginning phase, and then a later, more strenuous phase of dissemination [44]. The rate of degradation of superalloys at the initial stage is comparable to that which would occur in the absence of deposits, as stated by Pettit and Meier. Nevertheless, during the propagation stage, deposits greatly diminish the protective characteristics of oxide scales, resulting in substantially increased rates of corrosion. Corrosion and electrochemical reactions are both accelerated when these deposits change into conductive liquids at high enough temperatures.The corrosion mechanisms observed in aqueous solutions also apply to molten electrolytes. Therefore, molten salts may also undergo corrosion processes such pitting, galvanic corrosion, intergranular corrosion, and uniform surface corrosion, all of which are common in watery settings [45].

3.2.1 Molten Nitrate salt

Molten nitrate salts are widely used as thermal energy storage and heat transfer fluids in concentrated solar power (CSP) facilities and other industrial applications. Molten nitrate salts have low viscosity, allowing for easy pumping and circulation in thermal systems. Their low vapor pressure at high temperatures reduces the risk of evaporation, ensuring long-term safe and efficient operation. Additionally, these salts demonstrate excellent chemical stability and are compatible with various materials used in industrial equipment, including stainless steel and other corrosion-

Materials Research Forum LLC
https://doi.org/10.21741/9781644903698-6

resistant alloys. This compatibility helps lower maintenance costs and prolongs equipment lifespan. Furthermore, molten nitrate salts are relatively inexpensive and widely available, making them attractive for large-scale industrial use[46,47]. McConohy and Kruizenga investigated the changes in the behaviour of Ni-based HA230 and In625 alloys in a mixture percentage of 60:40 of $NaNO_3$ and KNO_3 at 600°C and 680°C for up to 4000 hrs. The result analysis showed that the corrosion rate is minimal for both the alloys at 600 °C but there is a considerable increase at 680°C. The findings also revealed that measured thermophysical properties have no major impact on the functioning of solar power facilities.In a salt combination comprising 40% KNO_3 and 60% $NaNO_3$ at 390°C, Fernandez compared the hot corrosion of 304 SS, T22 steel, and an alumina-forming austenitic (AFA) stainless steel to that of other steels. They used electrochemical impedance spectroscopy (EIS) to assess corrosion behaviour. Furthermore, they compared the chloride, sulphate, and nitrite concentrations of the salt before and after the experiments; they discovered little changes from the initial concentrations, which were insignificant in relation to the corrosion of AFA OC4 [48]. Cheng et al. evaluated the effects of chromium incorporation on the high-temperature corrosion-resistant properties of Cr-Mo steels in molten $LiNO_3$-$NaNO_3$-KNO_3 at 550°C for three different time spans. The results showed that the amount of weight loss of the steels reduced with increasing chromium content, which corresponded to much lower corrosion rates at higher chromium concentrations [49].

3.2.2 Molten Fluoride Salt

These are highly appreciated as heat transfer media due to their outstanding features like low viscosity, high thermal conductivity, high heat capacity, low vapour pressure. However, in many molten salt environments, the passive coating on alloys becomes unstable. Once these films evaporate, dissolution selectively attacks the least noble alloying elements [50]. Three main corrosion processes have been postulated for materials immersed in static molten fluoride salts: intrinsic, galvanic, and impurity-induced. Scientists have looked at adding trace quantities of certain metals to molten salts as one method to increase their corrosion resistance. By way of illustration, Li et al. studied the effects of adding yttrium to a superalloy in molten FLiNaK for 620 hours at 850°C. A Y-rich layer was formed, which stopped the dissolution of chromium and iron, and the correct quantity of yttrium was found to increase corrosion resistance. Negative effects were seen, nevertheless, with high yttrium levels [51]. Likewise, Cheng et al. studied the corrosion of SS316L and Hastelloy-N when zirconium was added to static molten FLiNaK at 850°C for 1000 hours. The inclusion of zirconium considerably improved both metals' corrosion resistance by producing a Zr-Ni intermetallic compound layer that prevented outward dissolution [52]. It is crucial to note that corrosion processes in molten salts can be complicated and vary greatly, typically due to a lack of comprehensive parametric experiments that span a wide range of temperatures and contamination concentrations.

3.2.3 Molten Chloride salt

Molten chloride salts, consisting primarily of metal chlorides like NaCl, $MgCl_2$ and magnesium chloride ($MgCl_2$), are gaining significant attention for their high-temperature applications in industries involving superalloys. These salts exhibit remarkable thermal stability, high conductivity, and low viscosity, making them ideal for use as heat transfer fluids in advanced energy systems such as CSP plants [53]. In oxygen-deprived environments, Hastelloy alloys exhibited notably reduced corrosion rates when exposed to molten chloride salts, even at temperatures as high as 800°C, validating their suitability for ternary molten chloride salt applications. A specific ternary eutectic blend of molten chloride salts consisting of 50.6% LiCl, 22.5% KCl, and 26.9% CsCl has been suggested for use in liquid salt cooling operations [54]. Hofmeister et al. investigated the corrosion resilience of alloys using a low-carbon stainless steel, a Ti-stabilized high-carbon stainless steel, and a Ni-based superalloy (CMSX-4) [55]. In the realm of electrochemistry, molten chloride salts serve as effective electrolytes for electrorefining and electroplating processes, which are crucial for producing and maintaining the integrity of superalloys used in aerospace, power generation, and other high-performance applications. However, the high corrosiveness of chlorides at elevated temperatures poses a substantial challenge, necessitating the development of superalloys with enhanced corrosion resistance or the application of protective coatings [56]. Research efforts are increasingly focused on understanding the thermophysical properties and chemical interactions of molten chlorides to improve their compatibility with structural materials. Innovations in material science, such as the formulation of new superalloys and the exploration of advanced coatings, are essential to mitigate corrosion issues and extend the operational lifespan of components exposed to these harsh environments.

4. Recent Advances and Future Directions

Recent advances in the field of corrosion in superalloys have focused on enhancing their resistance to degradation in aggressive environments. Researchers are exploring advanced alloy compositions, such as nickel-based and cobalt-based superalloys, to improve their corrosion resistance while maintaining mechanical integrity at elevated temperatures. Innovations in surface treatments, including the development of novel coatings and protective layers, aim to mitigate corrosion effects in aerospace, power generation, and industrial applications. Furthermore, there is a growing interest in understanding the fundamental mechanisms of corrosion in superalloys through advanced characterization techniques like electron microscopy and spectroscopy. These methods provide insights into the interaction between alloy microstructures and corrosive agents, enabling targeted improvements in material design. Future directions in the field involve integrating computational modeling and simulation to predict corrosion behavior under diverse operational conditions accurately. This approach will facilitate the development of tailored corrosion-resistant superalloys with enhanced performance and durability, addressing the evolving demands of high-stress, high-temperature environments in various industrial sectors

Conclusion

In conclusion, the study of corrosion in superalloys reveals a complex interplay of material composition, environmental factors, and advanced engineering solutions. Superalloys, notably nickel-based and cobalt-based variants, stand out for their robust mechanical properties and resistance to high-temperature degradation, making them indispensable in critical industries such as aerospace, power generation, and automotive sectors. Recent advancements in alloy design and surface treatments have significantly enhanced their corrosion resistance, paving the way for safer and more efficient operational practices. Looking forward, continued research efforts will focus on overcoming remaining challenges, such as improving resistance to aggressive corrosion environments and extending the lifespan of protective coatings. Integrating cutting-edge analytical techniques and computational modelling will be pivotal in predicting and mitigating corrosion effects under diverse operational conditions. By addressing these challenges, the future holds promising opportunities to expand the application scope of superalloys, foster innovation in material science, and meet the stringent demands of next-generation industrial technologies. In essence, the journey towards corrosion-resistant superalloys underscores a commitment to advancing materials engineering for sustainable and high-performance solutions in the face of evolving global challenges.

References

[1]Q. Tan, K. Liu, J. Li, S. Geng, L. Sun, V. Skuratov, A review on cracking mechanism and suppression strategy of nickel-based superalloys during laser cladding, J. Alloys Compd. 1001 (2024) 175164. /https://doi.org/10.1016/j.jallcom.2024.175164

[2]A. Tiwari, D.K. Singh, S. Mishra, A review on minimum quantity lubrication in machining of different alloys and superalloys using nanofluids, J. Braz. Soc. Mech. Sci. Eng. 46 (2024) 112. https://doi.org/10.1007/s40430-024-04676-6

[3]M. Kopec, Recent advances in the deposition of aluminide coatings on nickel-based superalloys: A synthetic review (2019–2023), Coatings. 14 (2024). https://doi.org/10.3390/coatings14050630

[4]G. Prashar, H. Vasudev, A. Mehta, 12 - Application of bimodal stuructured thermal sprayed coatings in power plant components, in: R.K. Gupta, A. Motallebzadeh, S. Kakooei, T.A. Nguyen, A. Behra, Advanced Ceramic Materials, Elsevier (2024) pp. 273–287. https://doi.org/10.1016/B978-0-323-99620-4.00012-9

[5]J. Gonzalez-Rodriguez, S. Haro, A. Martinez-Villafane, V. Salinas-Bravo, J. Porcayo-Calderon, Corrosion performance of heat resistant alloys in Na2SO4–V2O5 molten salts, Mater. Sci. Eng. A. 435–436 (2006) 258–265. https://doi.org/10.1016/j.msea.2006.06.138

[6]I. Gurrappa, Y. Injeti, A. Gogia, The behaviour of superalloys in marine gas turbine engine conditions, JSEMAT. 01 (2011). https://doi.org/10.4236/jsemat.2011.13022

[7]Z. Wang, Y. Yan, Y. Wu, Y. Zhang, X. Zhao, Y. Su, L. Qiao, Recent research progress on the passivation and selective oxidation for the 3d-transition-metal and refractory multi-principal element alloys, npj Mater. Degrad. 7 (2023) 86. https://doi.org/10.1038/s41529-023-00410-0

[8]S. Gu, H. Gao, H. Pei, C. Zhang, Z. Wen, Z. Li, Z. Yue, Degradation of microstructural and mechanical properties with serviced turbine blades, Mater. Charact. 182 (2021) 111582. https://doi.org/10.1016/j.matchar.2021.111582

[9]J.Y. Hu, S.S. Zhang, E. Chen, W.G. Li, A review on corrosion detection and protection of existing reinforced concrete (RC) structures, Constr. Build. Mater. 325 (2022). https://doi.org/10.1016/j.conbuildmat.2022.126718

[10]   S. Kamal, R. Jayaganthan, S. Prakash, High temperature cyclic oxidation and hot corrosion behaviours of superalloys at 900°C, Bull. Mater. Sci. 33 (2010) 299–306.

[11]   H. Long, S. Mao, Y. Liu, Z. Zhang, X. Han, Microstructural and compositional design of Ni-based single crystalline superalloys: A review, J. Alloys Compd. 743 (2018) 203–220. https://doi.org/10.1016/j.jallcom.2018.01.224

Materials Research Forum LLC
https://doi.org/10.21741/9781644903698-6

[12]   H. Fecht, D. Furrer, Processing of nickel-base superalloys for turbine engine disc applications, Adv. Eng. Mater. 2 (2000) 777–787. https://doi.org/10.1002/1527-2648(200012)

[13]   P. Auburtin, T. Wang, S.L. Cockcroft, A. Mitchell, Freckle formation and freckle criterion in superalloy castings, Metall. Mater. Trans. B. 31 (2000) 801–811. https://doi.org/10.1007/s11663-000-0117-9

[14]   J.S. Benjamin, Dispersion strengthened superalloys by mechanical alloying, Metall. Trans. 1 (1970) 2943–2951. https://doi.org/10.1007/BF03037835

[15]   S.J. Rettig, J. Trotter, The crystal and molecular structure of B,B-bis-(p-flurophenyl)boroxazolidine, (p-FC$_6$H$_4$)$_2$BO(CH$_2$)$_2$NH$_2$, Acta. Cryst. B30 (1974) 2139-2145.

[16]   E. V Kozlov, E.L. Nikonenko, N.A. Koneva, Energy of planar defects in the Ni3Al phase: Theory and experiment, Bull. Russ. Acad. Sci.: Phys. 71 (2007) 198–202. https://doi.org/10.3103/S1062873807020128

[17]   S. Xiang, S. Mao, H. Wei, Y. Liu, J. Zhang, Z. Shen, H. Long, H. Zhang, X. Wang, Z. Zhang, X. Han, Selective evolution of secondary γ′ precipitation in a Ni-based single crystal superalloy both in the γ matrix and at the dislocation nodes, Acta. Mater. 116 (2016) 343–353. https://doi.org/10.1016/j.actamat.2016.06.055

[18]   Y. Luo, Y. Zhao, S. Yang, J. Zhang, D. Tang, Effects of Ru on microstructure and stress rupture property of Ni-based single crystal superalloy DD22, Hangkong Cailiao Xuebao/JAM. 36 (2016) 132–140. https://doi.org/10.11868/j.issn.1005-5053.2016.3.014

[19]   A. Devaux, E. Georges, P. Héritier, Development of new C&W superalloys for high temperature disk applications, Adv. Mater. Res. 278 (2011) 405–410. https://doi.org/10.4028/www.scientific.net/AMR.278.405

[20]   R.J. Bennett, R. Krakow, A.S. Eggeman, C.N. Jones, H. Murakami, C.M.F. Rae, On the oxidation behavior of titanium within coated nickel-based superalloys, Acta. Mater. 92 (2015) 278–289. https://doi.org/10.1016/j.actamat.2015.03.052

[21]   M. Bensch, J. Preußner, R. Hüttner, G. Obigodi, S. Virtanen, J. Gabel, U. Glatzel, Modelling and analysis of the oxidation influence on creep behaviour of thin-walled structures of the single-crystal nickel-base superalloy René N5 at 980 °C, Acta. Mater. 58 (2010) 1607–1617. https://doi.org/10.1016/j.actamat.2009.11.004

[22]   Z. Zhang, G. Tang, X. Peng, F. Sun, S. Liu, Y. Xi, A review of the microstructure and properties of superalloys regulated by magnetic field, J. Mater. Res. Technol. 30 (2024) 9285-9317. https://doi.org/10.1016/j.jmrt.2024.05.189

[23]   J. Yu, Y. Hou, C. Zhang, Z. Yang, J. Wang, Z. Ren, Effect of high magnetic field on the microstructure in directionally solidified Co-Al-W lloy, Jinshu Xuebao/Acta. Metall. Sinica. 53 (2017) 1620–1626. https://doi.org/10.11900/0412.1961.2017.00165

[24]   B. Gao, L. Wang, Y. Liu, X. Song, S.Y. Yang, A. Chiba, High temperature oxidation behaviour of γ′-strengthened Co-based superalloys with different Ni addition, Corros. Sci. 157 (2019) 109–115. https://doi.org/10.1016/j.corsci.2019.05.036

[25]   S. Yu, C. Chen, S. Xu, T. Hu, S. Shuai, J. Wang, J. Zhao, Z. Ren, Effect of static magnetic field on the molten pool dynamics during laser powder bed fusion of Inconel 718

superalloy, Int. J. Therm. Sci. 197 (2024) 108851.
https://doi.org/10.1016/j.ijthermalsci.2023.108851

[26]   J. Sato, T. Omori, K. Oikawa, I. Ohnuma, R. Kainuma, K. Ishida, Cobalt-base high-temperature alloys, Science. 312 (2006) 90–91. https://doi.org/10.1126/science.1121738

[27]   Z. Zhang, H. Huang, Z. Zhang, Y. Wang, B. Zhu, A review of the microstructure and properties of superalloys regulated by magnetic field, J. Mater. Res. Tech. 30 (2024) 9285–9317. https://doi.org/10.1016/j.jmrt.2024.05.189

[28]   R. Sivakumar, B.L. Mordike, High temperature coatings for gas turbine blades: A review, Surf. Coat. Technol. 37 (1989) 139-160. https://doi.org/10.1016/0257-8972(89)90099-6

[29]   N. Lape, C. Dudfield, A. Orsillo, Environmentally friendly power sources for aerospace applications, 181 (2007) 353–362. https://doi.org/10.1016/j.jpowsour.2007.11.045

[30]   O.O. Zollinger, J.E. Beckham, C. Monroe, What to know before selecting hardfacing electrodes, Weld. J. 77 (1998) 39–43. https://www.scopus.com/inward/record.uri?eid=2-s2.0-0032002762&partnerID=40&md5=1f2d639372f8234ef85ea5dcc6560139

[31]   H. Berns, A. Fischer, Microstructure of Fe-Cr-C hardfacing alloys with additions of Nb, Ti and, B, Metallography. 20 (1987) 401–429. https://doi.org/https://doi.org/10.1016/0026-0800(87)90017-6

[32]   A.V. Byeli, V.A. Kukareko, A.A. Kolesnikova, S.K. Shykh, Structure-based selection of surface engineering parameters to improve wear resistance of heterogeneous nickel- and iron-based alloys, Wear. 255 (2003) 527–534. https://doi.org/10.1016/S0043-1648(03)00170-4

[33]   R.S. Chandel, Hardfacing consumables and their characteristics for mining and mineral processing industry, Indian. Weld. J. 34 (2001) 26–34.
https://www.informaticsjournals.com/index.php/IWJ/article/view/39327

[34]   A. Neville, F. Reza, Erosion-corrosion of cast white irons for application in the oilsands industry, NACE - Int. Corros. Conf. Ser. (2007) 76781–767812.
https://www.scopus.com/inward/record.uri?eid=2-s2.0-69249120458&partnerID=40&md5=229fe2ec7e7c2d66561251fbcfc1eba7

[35]   J. Liu, L. Wang, J. Huang, PTA clad (Cr, Fe)7C3/γ-Fe in situ ceramal composite coating, Int. J. Miner. Metall. Mater. 13 (2006) 538–541. https://doi.org/10.1016/S1005-8850(06)60109-6

[36]   J. Singh, S.S. Chatha, B.S. Sidhu, Abrasive wear characteristics and microstructure of Fe-based overlaid ploughshares in different field conditions, Soil and Tillage Research. 205 (2021) 104771. https://doi.org/10.1016/j.still.2020.104771

[37]   E. Zumelzu, I. Goyos, C. Cabezas, O. Opitz, A. Parada, Wear and corrosion behaviour of high-chromium (14–30% Cr) cast iron alloys, J. Mater. Process. Technol. 128 (2002) 250–255. https://doi.org/10.1016/S0924-0136(02)00458-2

[38]   K.H.Z. Gahr, Microstructure and Wear of Materials, (1987) 1-560.

[39]   Y. Zhao, W. Meng, P. Wang, C. Du, X. Wang, Effect of lanthanum oxide addition on microstructure and wear performance of iron-chromium alloy manufactured by laser direct deposition additive manufacturing, Materials. 15 (2022). https://doi.org/10.3390/ma15093234

[40]   S.-T. Kim, S.-H. Jeon, I.-S. Lee, Y.-S. Park, Effects of rare earth metals addition on the resistance to pitting corrosion of super duplex stainless steel – Part 1, Corros. Sci. 52 (2010) 1897–1904. https://doi.org/10.1016/j.corsci.2010.02.043

[41]   M. Stratmann, The atmospheric corrosion of iron — A discussion of the physico-chemical fundamentals of this omnipresent corrosion process: Invited review, Ber. Bunsenges. Phys. Chem. 94 (1990) 626–639. https://doi.org/10.1002/bbpc.19900940603

[42]   Y. Ma, H. Hu, D. Northwood, X. Nie, Optimization of the electrolytic plasma oxidation processes for corrosion protection of magnesium alloy AM50 using the Taguchi method, J. Mater. Process. Technol. 182 (2007) 58–64. https://doi.org/10.1016/j.jmatprotec.2006.07.007

[43]   G.H. Meier, A review of advances in high-temperature corrosion, Mater. Sci. Eng. A. 120–121 (1989) 1–11. https://doi.org/10.1016/0921-5093(89)90712-0

[44]   F.S. Pettit, Oxidation and hot corrosion of superalloys, JOM (1984), 651-687. https://www.tms.org/superalloys/10.7449/1984

[45]   N.S. Patel, V. Pavlík, M. Boča, N.S. Patel, V. Pavl, High-temperature corrosion behavior of superalloys in molten salts: A review, Crit. Rev. Solid State. 42 (2016) 83–97. https://doi.org/10.1080/10408436.2016.1243090

[46]   B. Dudda, D. Shin, Effect of nanoparticle dispersion on specific heat capacity of binary nitrate salt eutectic for concentrated solar power applications, Int. J. Therm. Sci. 69 (2013) 37–42. https://doi.org/10.1016/j.ijthermalsci.2013.02.003

[47]   H.L. Zhang, J. Baeyens, J. Degrève, G. Cacères, Concentrated solar power plants: Review and design methodology, Renew. Sustain. Energy. Rev. 22 (2013) 466–481. https://doi.org/10.1016/j.rser.2013.01.032

[48]   G. McConohy, A. Kruizenga, Molten nitrate salts at 600 and 680°C: Thermophysical property changes and corrosion of high-temperature nickel alloys, Sol. Energy. 103 (2014) 242–252. https://doi.org/10.1016/j.solener.2014.01.028

[49]   W.-J. Cheng, D.-J. Chen, C.-J. Wang, High-temperature corrosion of Cr–Mo steel in molten $LiNO_3$–$NaNO_3$–KNO_3 eutectic salt for thermal energy storage, Sol. Energy Mater. Sol. Cells. 132 (2015) 563–569. https://doi.org/10.1016/j.solmat.2014.10.007

[50]   Y.L. Wang, Q. Wang, H.J. Liu, C.L. Zeng, The effect of the microstructure on the corrosion behavior of N5 superalloy in a molten $(Li\{,\}Na\{,\}K)F$ eutectic salt, RSC Adv. 5 (2015) 32755–32760. https://doi.org/10.1039/C5RA04755B

[51]   X. Li, S. He, X. Zhou, P. Huai, Z. Li, A. Li, X. Yu, High-temperature corrosion behavior of Ni–16Mo–7Cr–4Fe superalloy containing yttrium in molten LiF–NaF–KF salt, J. Nucl. Mater. 464 (2015) 342–345. https://doi.org/10.1016/j.jnucmat.2015.05.007

[52]   W.J. Cheng, R. Sellers, M. Anderson, K. Sridharan, C.-J. Wang, T. Allen, Zirconium effect on the corrosion behavior of 316L stainless steel alloy and hastelloy-N superalloy in molten fluoride salt, Nucl. Technol. 183 (2013) 248–259. https://doi.org/10.13182/NT12-125

[53]   M. Liu, M. Belusko, N.H.. Tay, F. Bruno, Impact of the heat transfer fluid in a flat plate phase change thermal storage unit for concentrated solar tower plants, Sol. Energy. 101 (2014) 220–231. https://doi.org/10.1016/j.solener.2013.12.030

[54]   K. Vignarooban, P. Pugazhendhi, C. Tucker, D. Gervasio, A.M. Kannan, Corrosion resistance of Hastelloys in molten metal-chloride heat-transfer fluids for concentrating solar power applications, Sol. Energy. 103 (2014) 62–69. https://doi.org/10.1016/j.solener.2014.02.002

[55]   M. Hofmeister, L. Klein, H. Miran, R. Rettig, S. Virtanen, R.F. Singer, Corrosion behaviour of stainless steels and a single crystal superalloy in a ternary LiCl–KCl–CsCl molten salt, Corros. Sci. 90 (2015) 46–53. https://doi.org/10.1016/j.corsci.2014.09.009

[56]   A. Abramov, V. Karpov, A. Zhilyakov, S. Belikov, V. Volkovich, I. Polovov, O. Rebrin, Corrosion resistance of nickel-based alloys in salt and metal melts containing REE, *AIP Conf. Proc.* 1886 (2017) 020029. https://doi.org/10.1063/1.5002926

Superalloys: Fundamentals and Applications   Materials Research Forum LLC
Materials Research Foundations 178 (2025)109-130   https://doi.org/10.21741/9781644903698-7

Chapter 7

Comprehensive Insights on Corrosion in Superalloys: Testing, Protection Methods, Influencing Factors, and Mechanisms

Aalia Asghar[1], Nadia Akram[1]*, Khalid Mahmood Zia[1], Zumaira Siddique[1], Atta-ul-Haq[1]

[1]Department of Chemistry, Government College University Faisalabad, Faisalabad-38000, Pakistan

nadiaakram@gcuf.edu.pk

Abstract

Super alloys can be defined as high-performance materials developed for maintaining great mechanical properties and resisting corrosive attack at considerable temperatures, making their application in aerospace, power generation and chemistry industries. Based on the chemical analysis, corrosion characteristic and protective measures of the super alloy are also investigated in this review. There are three subgroups of super alloys known namely as nickel group, cobalt group and iron group with the nickel based super alloys taking the largest percentage due to high heat resisting ability and oxidation resisting ability. The originalities of super alloys arise from their composition which combines chromium, aluminum and titanium, as well as refractory metals like tungsten. These elements assist in the formation of Steady State precipitates like the γ'-precipitates Ni_3 (Al, Ti) which are useful in strengthening at high working temperatures as well as resisting creep age. There are different types of corrosion, which affect super alloys operating at high temperatures and stressing conditions such as high temperature oxidation, hot corrosion, hydrogen induced embrittlement and stress corrosion cracking. Kinetics and mechanism of corrosion depend very much on the presence of protective oxide layers such as Cr_2O_3 and Al_2O_3. To overcome the problem of corrosion several techniques of surface treatment and coating are used. These, anodizing and plating, thermal barrier coatings (TBCs) and environmental barrier Coatings (EBCs) are some of the most developed future trends in super alloy are directed toward enhancing the staking of alloy constituents, innovative manufacturing processes such as direct metal laser sintering and advanced coating systems. Current investigation is continued to overcome problems in environmental issues, economic factors, and technological issues of super alloys in corrosion protection.

Keywords

Metals, Super Alloys, Corrosion, Factors Affecting Super Alloys, Corrosion Mechanism

Contents

1. Introduction

Super alloys or high-performing alloys are specialized materials that are used in harsh environments or at high temperatures. They mostly consist of high temperatures, great mechanical loads, and severe corrosive atmospheres characteristic of aerospace, power plant, and chemical production applications. The term super alloy was first used in 1940s referring materials that showed enhanced performance even at high temperature, especially in Jet engines [1]. Superalloys are generally classified into three main categories based on their primary alloying element: There are three primary classes of solid oxide fuel cells, namely nickel-based, cobalt-based, and iron-based. Out of all the super alloys, Nickel based super alloys, namely Inconel and hastily dominate the market given their high input tolerance to heat and resistances to oxidation.

Super alloys based on cobalt, such as Haynes 188, are highly valuable for their hot corrosion and strength characteristics at high temperatures. Superalloys are mostly based on iron and are actually modified stainless steels and are used where cost is a paramount factor [2]. An important feature of superalloys lies in their ability to maintain strength properties in stage II oxidation conditions. This is mainly done by precipitation of a secondary phase, for instance gamma prime (γ') in Ni based super alloy which restrict motion of dislocation and thus increases resistance to creep [3]. Besides, that the contents of refractory ingredients such as tungsten, molybdenum and rhenium improves high-temperature property by stabilizing the structure of the steel. Another important feature is their ability of resistance to oxidation and corrosion. Superalloys commonly contain high percentages of chromium, aluminum and might also contain silicon, which upon exposure to high temperatures develop passive oxide layers on the surface. These oxide layers form a layer that shields the other layers from degrading any further[4]. Also, superalloys have good phase stability, which is the constancy of its structure and characteristics when used for a long period at high temperatures. This stability is desirable for components like turbine blades in jet engines where a materials failure would be disastrous [2]. Superalloys play a critical role on high temperatures service applications and their significance cannot be overemphasized. They are now used extensively in several employment areas because of their mechanical strength, resistance to oxidation, and phase stability. Some of the most common and well-known uses of superalloys are in aerospace industries such as in manufacturing jet engine parts. The actual working temperatures of jet engines range above 1200 degrees centigrade and the greater degree of temperature means a more efficient jet engine. Superalloys are used in fabricating turbine blades, discs, and other parts for such engine because of their high temperature strength and oxidation resistance. For instance, Nickel based super aluminum such as Inconel 718 and Rene 41 are well known for the blades used in turbines due to their ability to withstand creep and thermal stability [3]. In power generation, superalloys apply in the manufacture of gas turbines for electricity generation. These turbines can work at a higher temperature than the normal temperature due to the application of high temp resistant super alloys, and thus the thermal efficiency and emission are known to be higher. According to the findings of a study conducted by the Electric Power Research Institute, the addition of new super alloys to the GTs can help increase its efficiency by 3 percent or increase fuel conservation, besides helping reduce greenhouse emissions [5].Super alloys are also used in chemical processing industry where equipment is placed in corrosive environment at high temperatures. A number of characteristics make it suitable for such an environment, for instance it has good pitting, stress corrosion cracking and oxidation resistance [6].

Thirdly, super alloys have significant roles to play as nuclear reactors advance and new designs are considered. Such reactors function at far higher temperatures than typical reactors, and so the material must cope with long-term elemental and thermal degradation and sustained radiation. For example, superalloys including Alloy 617 and Hastelloy N are in consideration for reactor internals, and heat exchangers as well [7]. The unique characteristics of superalloys render them inevitable wherever high temperature use is involved in different sectors of the economy. Their propensity to retain their mechanical strength, anti-oxidation and anti-corrosion features and guarantee phase stability at high subjection make them reliable and efficient for components used in aerospace, power generators, chemical processing and nuclear reactor.

2. Types of Superalloys

Nickel-Based Superalloys

This is because nickel-based superalloys are mainly an alloy of nickel in which nickel accounts for more than 50 percent of the whole composition. Some of the well-known alloying elements include chromium, cobalt, aluminum, titanium, and molybdenum, which can be added to improve certain properties. Chromium enhances oxidation resistance by developing oxides layer while Aluminum and Titanium are involved in making the gamma prime phase Ni_3 (Al, Ti) crucial for precipitation hardening of the alloy. γ Phase: The primary matrix, a face-centered cubic (FCC) structure.

γ' Phase: Precipitates of Ni_3(Al, Ti), which are coherent with the γ matrix and provide substantial resistance to dislocation movement.

Refractory elements such as tungsten and rhenium addition, assist to pin the microstructure and enhance the mechanical characteristics at high temperature. These include but are not limited to Nickel-based superalloys which find wide application in aerospace field, mainly in the production of turbine blades, discs, and other parts of jet engines. They can be used at temperatures over 10000c and ensure that the materials used have high mechanical strength and resistance to oxidation and corrosion. For instance, Inconel 718, a nickel-based superalloy, has gained immense popularity due to its high creep resistance and fatigue strength; thus used in the manufacture of the components in gas turbines [1].

Nickel based superalloys in the power generation sector are used in gas turbines for the production of electricity. They also have better tolerance to high temperatures that increase the thermal efficiency of turbines and result in lower consumption of fuel and emission of gases. The Electric Power Research Institute (EPRI) estimates that newly developed nickel based superalloys could improve efficiency by 3% potentially resulting in significant cost and environmental gains [3].

Cobalt-Based Superalloys

Cobalt based superalloys mainly contain cobalt, a substantial portion of chromium, tungsten, nickel and molybdenum. Chromium offers oxidation protection while tungsten and molybdenum build up high temperature strength. The microstructure of cobalt-based superalloys includes cobalt FCC matrix and carbide phases like M23 C6 and M6C which offer high strength to these alloys by preventing the movement of dislocation [2]. FCC Matrix: The primary cobalt-rich face-centered cubic structure.

An important characteristic of cobalt based superalloys is their capability of retaining their mechanical properties at high service temperatures. They have carbides and solid solution strengtheners like tungsten and molybdenum which give high thermal fatigue and hot corrosion resistance making these alloys ideal for high operating temperature [1]. Cobalt based superalloys are most commonly used in the gas turbines especially in the hot gas path of the turbines where temperatures are very high.

Materials like Haynes 188 and MAR-M-509 are widely applied for the manufacturing of turbine vanes and combustion chambers because of the high hot-corrosion resistance and higher strength at elevated temperatures [6]. Cobalt-based superalloys find application in the production of orthopedic implants and prosthetics in the medical field because they possess desirable biocompatibility and high mechanical strength. For instance, Co-Cr-Mo alloy used in hip and knee replacements offers durability and corrosion resistance in the long run [7].

Iron-Based Superalloys

Iron based super alloys are based on stainless steels to which have been added substantial amounts of nickel and chromium for high temperature performance. Additional elements such as molybdenum, vanadium and niobium may also be included to enhance the strength or corrosion resistance of these alloys. Iron-based superalloys have unique microstructures developed basically from austenitic (FCC) and ferritin (BCC) phases together with carbide and intermetallic particles for precipitation strengthening [2]. Austenitic (FCC) Phase: The first phase responsible for the strength and flexibility. Ferritin (BCC) Phase: It offers heat and load bearing properties when the working temperature is at its peak. Carbides: Such as M23C6 and M6C, which can improve high-temperature strength through getting precipitation strengthening. Intermetallic Compounds: It enhance the total properties of the alloy and its stability.

Materials Research Forum LLC
https://doi.org/10.21741/9781644903698-7

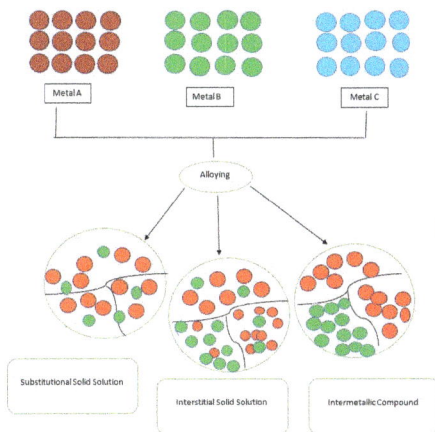

Fig 1: Intermetallic compounds

Iron-based super alloy containing Chromium interacts with O_2 in service and forms a stable oxide scale on the outside surface, therefore increasing the sample's resistance to oxidation and corrosion. Some of the added elements include nickel that increases the toughness and ductility of the material while others such as niobium and vanadium help in the formation of stable carbides that increase the high-temperature strength [3]. Applications of these iron based superalloys include situations where cost is very important the operating temperatures are comparatively lower than forged nickel and cobalt based ones. These alloys are especially applied in industrial gas turbines, steam turbines, as well as heat exchangers. For instance, alloy 800H is a Fe-based superalloy which is used in steam generator tubing in nuclear power plants because of its ability to resist oxidation and creep strength at high temperatures [6]. The chemical processing industry also requires iron-based superalloys like the Alloy 20 for use in conditions where they come into contact with sulfuric acid since the material is quite resistant to corrosion. These alloys offer targeted solutions for applications that need moderate high temperatures and good environmental corrosion resistance [2].

Titanium-Based Superalloys

Titanium based super alloys are primarily constituted of a standard constituent of titanium, aluminum, vanadium and molybdenum bearing extremely small proportions of other additives to increase their fortitude and corrosion resistance. Titanium based super alloy contains desired properties of strength, ductility and toughness by pin pointing the dominating alpha phase (hexagonal closed packed) and beta phase (body center closed packed) structures and by stabilizing it with alloying additions [8]. Alpha Phase (HCP): This also shows a good ability to resist creep as well as having good creep's stability at these high temperatures. Beta Phase (BCC): Strengthens the ductile and tough characteristics of the material of the alloy.

Intermetallic Compounds: Similar to Ti$_3$Al that improves high-temperature strength and stability of material. Aluminum and vanadium act as strong solid solution strengtheners and also as stabilizers of the stable phases that improve high-temperature characteristics. The presence of combined molybdenum and other substances even enhances the features of these alloys in titanium based superalloys are widely employed in many aerospace applications and more specifically in airframe and engine parts where the strength to weight ratio is very important. For instance, Ti-6Al-4V is employed in the production of turbine blades, compressor disks, and structural parts of airplanes because of the material's highly desirable mechanical characteristics and corrosion tolerance.

The Aerospace Industries Association confirmed that penetration of titanium based superalloys for construction of aircrafts has risen significantly on account of their low densities and high strength which help in enhancing the mileage and overall efficiency [8]. Titanium-based superalloys find application in the medical sector in orthopedic implants and dental prostheses since they are bio-compatible and highly resistant to corrosion. Some of the titanium alloys that are used include Ti-6Al-4V and Ti 6-7AlNb that can enable the hip and knee replacements last longer without negative reactions to the body [9].

3.   Factors Affecting Corrosion

Impact of High-Temperature Oxidation on Superalloys

Oxidation at high temperatures is one of the major contributors to the degradation of the superalloy performance and their life. Superalloys are vulnerable to oxidations at high temperatures and therefore forms oxide layers on them upon exposure to the surrounding environment. Altogether, this oxidation process can be either constructive or destructive depending on the type of oxide layer developed.

Protective Oxide Layers: In many nickel based superalloys, the development of an adherent, continuous oxide layer typically including chromium oxide (Cr$_2$O$_3$) or aluminum oxide (Al$_2$O$_3$) can cover the base metal complement and hinder the future oxidation and corrosion. These oxides having diffusion power and they also work as protector by not allowing the oxygen and other injurious material to enter into the formation of an alloy [4]. The degree of these protective layers determines the reliability in operating jet engine and gas turbines where components are subjected to temperatures above 3000 F.

Detrimental Oxidation: Oxidation leads to the formation of non-s protective layers among them being nickel oxide NiO that tend to spall off under thermal cycling resulting to fresh alloy surface being exposed to oxidation. It is possible for this spalling to be a progressive process, causing a continual loss of material and damage to the mechanical properties of the material. For instance, in service conditions that subject the component to thermal cycling, stress variation in the oxide layers causes them to spall off from the surface and increases the oxidation rate and shortens the life of the component [10].

Effects of Sulfidation and Chlorination

Super alloys are usually used in environments having high levels of sulfur and chlorine mainly in processes such as chemical processing and electric power generation. One has seen that Sulfidation and Chlorination are the common types of corrosion which impact heavily on the corrosion of superalloys.

Sulfidation: Sulfidation is the process by which superalloys dissolve sulfur-containing species, producing surface precipitates such as nickel sulfide, NiS, or cobalt sulfide. These sulfides are usually less stable than oxides and hence cause faster material degradation. Sulfur compounds present in fuels and industrial gases are capable of stimulating sulfidation particularly at high temperatures. For instance when gas turbines were run on fuels containing sulfur, sulfidation causes formation of non-protective sulfide scales which in turn accelerates corrosion and decrease the life.

Chlorination: Chlorination entails action between superalloys and chlorine-containing molecules resulting into complimentary formation of chlorides which are comparatively volatile and less protective in comparison to oxides. Superalloys compromise in marine surroundings or any production facilities where chlorine is used and this causes Chlorination hence, pitting and localized corrosion [11]. For example, when used in high-temperature industrial applications where chlorine-containing gases are generated, nickel-based superalloys alloy Ni.Cl with chlorine to form $NiCl_2$ that has tendency to volatilize and erode rapidly [6].

Role of Environmental Contaminants

Different types of salts, acids, and various industrial pollutants are responsible for escalating the overall corrosion rate of superalloys and in conditions that involve high temperature.

Salts: Sodium and potassium seem to have a role with the heavy hot corrosion form of the salt deposit crystal. NaCl in the marine environment can lead to the formation of molten salts at high temperatures and can flux protective oxide layers to cause very fast oxidation and sulfidation in the material. This kind of hot corrosion is experienced in marine gas turbines and industrial boilers [12].

Acids: For instance, chemical processing industries contain environments with low pH that could result in localized corrosion and pitting in superalloys. Typical industrial acids are sulfuric acid, H_2SO_4, and hydrochloric acid, HCl, which dissolve the oxide layers and in turn cause further higher material loss. Superalloys functioning in such conditions must exhibit higher tolerance to the corrosive effect that the acidic environment has on it [13].

Industrial Pollutants: The environmental contaminants that are likely to affect the formation of superalloys are sulfur dioxide (SO_2) and nitrogen oxides (NOx), which can cause corrosion. Such pollutants can create acidic deposits when mixed with moisture; an action that produces sulfuric and nitric acids that may corrode super alloy parts [14].

Influence of Mechanical Stresses and Strain

These mechanical stresses increase the load or strain and influence the corrosion processes and in turn leads to the occurrences of SCC and Fatigue corrosion interactions.

Stress Corrosion Cracking (SCC) Stress is one of the most important causes that can result in SCC in all super alloys and the tensile stress in severe mediums is much more critical. Vibrations other mechanical loads combined with environment that promotes corrosion, increase the rate at which cracks are initiated and grow leading to structural collapse. For example when exposed to tensile stresses and highly corrosive environment as in the case of gas turbine blades, SCC developed and significantly reduced the component's lifespan [15].

Fatigue-Corrosion Interactions: super alloys undergo cyclic loading in corrosive environments, they are manufactured to degenerate faster in fatigue failure. A corrosive agent can start the crack at the stress concentration and cyclic stresses can then cause expansion of the crack resulting to early failure. This interaction is of most concern in aerospace and power generation sector which comprises components that are susceptible to mechanical loads besides being exposed to corrosive[16].

Thermal Stresses: Corrosion behavior of superalloy is also affected by thermal stresses due to temperature gradients that occur along with the temperature field. Fluctuations in temperature pose a significant problem to materials as it triggers the phenomenon of thermal expansion and contraction thereby causing the creation of micro cracks on the surface of protective oxide layers. These micro cracks increase the vulnerability of the underneath material to oxidation and corrosion and hence shorten its lifecycle [17].

Table 1: Factors affecting corrosion in superalloys

Factor	Description	Impact on Superalloys
High-Temperature Oxidation	Oxidation occurs when superalloys react with oxygen at high temperatures, forming oxide layers on the surface.	Protective oxides (e.g. Cr_2O_3, Al_2O_3) prevent further oxidation and corrosion. Non-protective oxides (e.g. NiO) can spall off, leading to continuous material loss and degradation.
Sulfidation	Sulfidation involves the reaction of super alloys with sulfur-containing compounds, forming sulfides.	Formation of non-protective sulfides (e.g. NiS, CoS) accelerates material degradation. Common in environments with sulfur-containing fuels and gases.
Chlorination	Chlorination involves the reaction with chlorine-containing compounds, forming volatile chlorides.	Formation of volatile chlorides (e.g. $NiCl_2$) leads to significant material loss. - Common in marine environments and chemical processing plants.
Environmental Contaminants	Includes salts, acids, and industrial pollutants that interact with superalloys.	Salts (e.g. NaCl) can cause hot corrosion by fluxing protective oxide layers. Acids (e.g., H_2SO_4, HCl) cause localized corrosion and pitting. Industrial pollutants (e.g. SO_2, NOx) form corrosive acids.
Mechanical Stresses and Strain	Mechanical stresses and strain can exacerbate corrosion processes through stress corrosion cracking and fatigue-corrosion interactions.	SCC: Tensile stress in corrosive environments causes crack initiation and propagation. Fatigue-corrosion: Cyclic loading in corrosive environments accelerates fatigue failure.

4. Mechanisms of Corrosion in Superalloys

Mechanisms of Oxidation in Superalloys

Metals usually accelerate and corrode at high temperatures the high temperature oxidation is one of the major challenge areas in super alloy as are High temperature application. This implies that when the superalloys are exposed to oxidizing environments then oxide layers are formed on its surface. These oxide layers can further be characterized as being either beneficial or harmful, depending on their formation and growth mechanisms. Passively oxides, including chromium oxide (Cr_2O_3) and aluminum oxide (Al_2O_3), form clear, tenacious layers on the surface of the alloying elements that protect them from further chemical transformation by keeping out oxygen and other aspect of clad [10]. But nonproductive oxides such as nickel oxide (NiO), can spall off under thermal cycling conditions to constantly expose fresh alloy surface leading to continuous materials loss and reduction in mechanical properties [17].

It has also determined that individual alloying elements have a pivotal role to play in improving the oxidation characteristics of superalloys. Out of all the transition metals, both chromium and aluminum are significant as they produce Cr_2O_3 and Al_2O_3respectively that are very passivating. More recently, reactive elements like tungsten, molybdenum, and rhenium follow boosting high-temperature mechanical properties and stabilizing the microstructure to increase the stability of protecting oxide layers [3].

Understanding Hot Corrosion

Hot corrosion is a kind of high temperature corrosion in which the corrosion takes place in the contact area with molten salt like sodium sulfate (Na_2SO_4).

Type one hot corrosion which takes place at temperature between 900°C and 1100°C concerns the formation of molten salts which flux protective oxides and bring fast oxidation and sulfidation [12]. Type II hot corrosion takes place between 700 °C and 900 °C and involves the formation of porous and non-adherent oxide scales, causing localized penetration and pitting.

Hot corrosion is the conversion of the alloy with molten salts leading to the formation of oxide and non-protective scales. These parameters include salt distribution, temperature, and composition of the alloy, among others, affect the extent of hot corrosion. Sodium sulfate and vanadium pentoxide in the environment can raise the corrosion rate due to their action, which include the depression of the melting point of protective oxides as well as dissolution of the oxides [17].

Hydrogen Embrittlement

Hydrogen embrittlement can be expressed as a condition in which parts of a material become brittle and fail suddenly due to hydrogen absorption. Hydrogen atoms penetrate into the alloy metal and get trapped along with grain boundaries and other micro defects, which destroy the metallic bond and make the material more brittle and prone to crack under load [18].

This is because hydrogen atoms are small and can readily penetrate the metal lattice. Afterwards, they can move to places with increased stress or defects and initiate decohesion and cracking. The normal adsorption sites for hydrogen are grain boundaries and other areas of low metallic bonding which in turn increases the stress required to fracture the material. This leads to intergranular or tran's granular fracture depending on the texture of the alloy and the stress conditions prevailing at that time [19]. It is also possible to employ several strategies in order to minimize the level of hydrogen embrittlement of these particular super alloys. In a bid to develop a HE resistant super alloy, certain aspects can be taken into consideration that includes the kind of elements featured in the alloy and in some proportion in the micro structure [20]. It has also been discovered that the application of coatings and plating at the surface of the material can equally work as a barrier against hydrogen. Measures such as cathodic protection and application of hydrogen-entry resistant coatings help in minimizing hydrogen uptake [21]. Further, proper control of the

Hydrogen content in the process environment or utilizing hydrogen absorption users can go a long way to reduce chances of embrittlement [18].

Stress Corrosion Cracking (SCC)

Stress corrosion cracking (SCC) is one of the corrosion types that develop from tensile stress as well as the corrosive conditions. SCC usually occurs in tensile stresses which are caused by the mechanical load or th ermal expansion and the influences of specific corrosive agents, including chloride, sulfate, and caustic solutions [15].

Tensile stress is in a position to pull the crack forward and to cause crack extension and crack initiation. Stresses can be induced mechanically, due to thermal expansion or as a result of manufacture. The corrosiveness promotes the generation of localized corrosion sites over which cracks tend to develop. The susceptibility of superalloys to SCC is influenced by the microstructure and alloy content of the subject material of all the microstructures that play a role in the determination of the cracking of the alloy, the microstructural factors that have been considered include grain size, distribution of precipitates, and the kinds of alloying elements [2].

SCC can result in component failure and a dramatic reduction in the remaining service life of structures, as well as greater operational expenses due to inspection and maintenance. The performance of superalloy components in corrosive environments can therefore only be guaranteed through careful consideration of SCC mechanisms and practical techniques such as material selection, stress reduction measures and environment control measures [22].

It is therefore important to understand the modes of corrosion in superalloys: oxidation, hot corrosion, hydrogen embrittlement, and SCC with a view of improving their performance in high temperature environments. It is important to understand these corrosion mechanisms and how to avoid them through proper alloy design, surface engineering, and environmental control to help extend the lifespan of superalloy components in hostile services.

5.   Corrosion Testing in Superalloys

Superalloys require corrosion testing so that their performance and durability in high temperature and using conditions can be effectively evaluated. In this type of testing, lab and practical methods are used to expose the superalloys to situations that are like real life and study their corrosion behavior.

5.1   Laboratory Methods for Testing Corrosion Resistance

Static and Dynamic Testing Methods

In static testing methods superalloy samples are subjected to corrosive environment for a given time without subjecting to any mechanical force. These tests make it easier to determine the performance of the material concerning uniform corrosion and pitting in controlled environments. Some of the most frequent static tests are immersion tests where samples are exposed to corrosive media and salt spray tests that involve exposure of samples to chloride containing atmosphere to sample corrosion resistance [23]The other category of accelerated testing methods involve the use of mechanical load while the material is undergoing environmental exposure. These tests mimic practical working environments where the components are loaded mechanically while being exposed to corrosive agents. Dynamic tests include the rotating beam test and the cyclic bending test which are used to measure the rate of stress corrosion cracking (SCC) and fatigue-corrosion interactions [24].

High-Temperature Exposure Tests

Thermal cycling tests prove to be very useful in determining the capability of the super alloys to resist high temperature oxidation and hot corrosion tests. Such tests include hot chamber tests in which specimens are exposed to thermal environments and conditions similar to actual use in industrial gas turbines as well as other jet engines. The samples are normally exposed to thermal cycling up to some specific cycles to create conditions similar to practical applications. To determine the material oxidation properties and thickness and quality of the protective oxide layers, weight gain measurements, metallography, and oxide scale characterization may be made where necessary [25].

Electrochemical Testing Techniques

Cathodic and anodic methods of electrochemical tests help in studying the processes and rates of corrosion in superalloys. These methods include applying an electrical current across the material to be tested and measuring the resulting current to make conclusions regarding the electrochemical behavior of the material. Some of dynamic electrochemical techniques that maybe employed include potentiodynamic polarization, electrochemical impedance spectroscopy and cyclic voltammogram. Since potentiodynamic polarization tests give information about the tendency of the material to corrode, corrosion potential and the corrosion current density allow determination of the corrosion potential. EIS is employed to determine the ability of the material to prevent charge transfer as well as the stability of the protective oxide layers. Cyclic voltammetry is applicable for investigations of the process of formation and dissolution of passive films on the surface of super alloys [26].

5.2   Techniques for Evaluating Corrosion in Practical Applications

Field Exposure Tests

One of the most common corrosion tests is field exposure tests where actual superalloy samples are exposed to actual working conditions to determine their corrosion characteristics IMA. These tests give important information about the material's performance in practical conditions of use and the consequences arising from changes in temperature, humidity and contaminants. Field exposure tests are normally carried out in an industrial setting such as in factories and power plants, and in marine conditions to check on the ability of the material in terms of resistance to oxidation or hot corrosion and other forms of degradation [23].

Nondestructive Evaluation (NDE) methods

Superalloy components are sensitive materials that may not withstand invasive techniques of evaluating corrosion damage; therefore, it is crucial to use nondestructive evaluation methods. These techniques are practiced for in-service inspection and maintenance of structures to guarantee the reliability of key components.

Some of the more widely used NDE techniques are ultrasonic testing, radiographic inspection, and current testing. Ultrasonic testing employs sound waves of high-frequency to observe internal defects as well as evaluate the thickness of oxide coatings. Radiography makes use of X-rays or gamma rays to depict internal flaws and corrosion extents. Eddy current testing relies on electromagnetic induction to identify crack or flaws on the surface as well as on a shallow penetration [27].

In-Situ Monitoring Techniques

Morphological analysis techniques give real-time information on the corrosion processes of superalloys while in operation. These techniques include mounting sensors and data acquisition equipment, which tracks the state of the material and signs of corrosion.

Some of the typical methods mounted for in-situ corrosion monitoring include use of corrosion sensors that can detect physical characteristics such as electrochemical potential, temperature and humidity. Fiber optic sensors and acoustic emission sensors are also applied for sensing the changes in the material properties that occur due to corrosion. Onsite monitoring plays a crucial role of alerting on the need for maintenance and avoiding major system breakdowns because it updates on the condition of the material [28].

There are some laboratory techniques and practical assessment methods used in corrosion testing of superalloys with the aim of determining the material's ability to resist various forms of degradation. The static and dynamic testing methods, the tests at high-temperature exposure, and the electrochemical techniques are beneficial in understanding the corrosion and corrosion rates. Field exposure tests, NDE methods, and in-situ monitoring techniques helps to prove the durability and durability of the superalloy parts in applications.

6. Protective Measures for Superalloys

Coatings for superalloys are of great importance to increase their durability and performance in conditions that are characterized by corrosion, oxidation, or any other detrimental effects. Several forms of surface treatments and coatings are used to prolong the lifespan of the superalloys as well as to optimize their performance in application areas including aerospace, power generation, and chemical processing.

Surface Treatments to Prevent Corrosion

Anodizing and Plating: There are two techniques of surface treatment that can be used to enhance the corrosion behavior of superalloys and these include anodizing as well as plating. Anodizing is an electrochemical technique where the surface of the metal is converted to oxides to shield the metal from corrosion. This process is particularly appropriate for aluminum that contains superalloys since anodic oxide can improve oxidation and wear resistance [20].

Plating is the process of building up a layer of a particular metal on the surface of the superalloy. Some of the most frequently used plating metals are nickel, chromium, and gold to give various degrees of protection. Nickel plating for example has a good resistance to corrosion and wear, useful in parts that are likely to operate in corrosive conditions [20].

Shot Peening and Laser Shock Peening: Laser shock peening is one of the mechanical surface treatment processes that has been increasingly applied to enhance fatigue and corrosion properties of the super alloys. Shot peening is the process of firing spherical shots onto the surface of the alloy so as to build up layers of residual compressive stresses thus enhancing the ability of the component to resist formation of new cracks and growth of existing ones [29].

Superalloys: Fundamentals and Applications   Materials Research Forum LLC
Materials Research Foundations 178 (2025)109-130   https://doi.org/10.21741/9781644903698-7

Laser shock peening, conversely, employs an intense laser pulse to create pressure waves that cause the formation of compressive residual stress in the material. This method gives higher depth of compressive stress than conventional shot peening and is useful for enhancing the high cycle fatigue strength of the super alloys [29].

Thermal Barrier Coatings (TBCs): Thermal barrier coatings are the latest coating systems that are applied to super alloys with the goal of protecting them from high temperatures, and thermal degradation. The principles constituents of TBCs are the ceramic layer such as YSZ and the metallic layer such as MCrAlY (in which M is nickel, cobalt or iron) [30].

Diffusion Coatings: Super alloys are coated with one or more elements and then alloyed into the base metal to create protective layers commonly known as diffusion coatings. One of the most widely used diffusion coating processes is aluminizing which involves the deposition of a layer of aluminum into the surface. This layer subsequently oxidizes to form a protective and dense alumina (Al_2O_3) scale that offer excellent oxidation durability [31].

Other examples are chromizing and siliconizing, where chromium or silicon diffuses into the surface to form a chromium or silicon oxide layer. These coatings are more effective for enhancing high temperature corrosion characteristics of super alloys in high corrosive environments [31].

Overlay Coatings: Overlays are typically thick layers formed on better-navigated superalloys with the intention of guarding them against the perils of oxidation, corrosion, and thermal attack. The most of these coatings are derived from high-temperature alloys such as Ni, Co. or Fe based MCrAlY or platinum aluminide alloy that proves both high temperature stability and resistance to corrosion [32].

Thermal spraying, electron beam physical vapor deposition for overlay coating application and plasma spraying. Such coatings are of immense importance to increase the service life of critical components of gas turbines, jet engines and applications involving high temperature [32].

Environmental Barrier Coatings (EBCs): They are called environmental barrier coatings (EBCs) and are aimed at protecting the superalloys from the aggressive environment like high humidity and corrosive gases for example. They are quite useful in the defense of silicon carbide ceramics and the CMC's used in gas turbines and other elevated temperature [33].

EBCs often involve layers of silicon carbide (SiC) or other oxidation-resistant ceramics since these offer protection against the ingress of moisture and corrosive species. These coatings are essential to deliver desirable attributes of superalloys and CMCs and keep them performing under harsh conditions [33].

Thus, corrosion or oxidation protection of the super alloys in the elevated temperature condition is very significant and is provided by either surface treatments or different coating technologies. Surface treatments such as anodizing, plating, shot peening and laser shock peening improves the surface features of superaltars whereas, modern coating systems include TBCs diffusion coating, overlay coating and EBCs providing improved thermal and environmental protection. These protective measures are all essential to prolong the service life and service efficiency of super alloy parts in aerospace, electric power industry and high temperature services.

7.   Examples of Corrosion Failures in Superalloys

Corrosion failures in superalloys exhibit potential consequences in different critical applications related to numerous industries such as industrial gas turbines, jet engines, and chemical processing plants. This forms the basis of studying these failures through case histories and metallurgical investigations, in order to draw conclusions that would help enhance the performance of the material and reduce or eliminate the prospects of such failure occurrences in the future.

Industrial Gas Turbines: Components of industrial gas turbines are designed to withstand high temperature, mechanical loads and corrosive atmosphere. A classic example is the proctographic analysis of a turbine blade made of a nickel-based super alloy that failed in a gas turbine for power application. A high degree of oxidation and hot corrosion on the blade surface were observed, which led to the formation of non-adherent oxide scales and sulfide deposits [34]. Metallographic examination showed that the passive oxide film had shed off exposing the metal to further corrosion. It was revealed that the fuel was containing sulfur which served as the cause of the hot corrosion.

Engines Jet: The components of the jet engines are exposed to even harsher conditions compared to those for industrial gas turbines, with very high and fluctuating thermal stresses. A good example of corrosion failure is the case where the turbine blades of a jet engine that has been manufactured by cobalt-based superalloy corroded. The root cause of failure was identified as hot corrosion of type I where sodium sulfate creates a molten phase at the blade surface [35]. These deposits stirred up the protective oxide layer thus resulting to fast oxidation and formation of non-sacrificial sulfides. Based on the failure analysis it was found that the engine was working in an environment full of sulfur and sodium, which speeds up the corrosion process.

Chemical Processing Plants: Superalloys are in chemical processing plants where they work under conditions that involve chemical attack as well as high temperature. Chloride induced corrosion was also underlining in a case study which dealt with the failure of a heat exchanger tube made from an iron based superalloy. Tube in a plant processing hydrochloric acid shows the highest level of pitting and intergranular corrosion [36]. Chloride ions were observed to have attacked the inner surface of the protective oxide layer thus causing localized corrosion near the grain boundaries. This they said was due to wrong material choice or lack of it and the presence of residual stresses after production.

Metallurgical Analysis: Possibly the most important factor for comprehending the failure modes of super alloys is metallurgical evaluation. This process involves investigating the morphological, elemental and mechanical properties of the crippled part for the purpose of establishing the root cause of corrosion. SEM, element mapping by EDS and XRD are some of the major analytical techniques used in the study [37]. For instance, in the case of the gas turbine blade failure, the SEM analysis carried out on the samples highlighted the fact that there were chromium-depleted zones below the oxide layer. Concentration of sulfur and sodium were identified by EDS and the failure was attributed to the hot corrosion.

Environmental and Operational Factors: Corrosion response of superalloys also depends on environment and operational conditions. In the case of jet engines, the metal components corrosive failure was attributed to high operating environment sulfur and sodium content. These elements produced bitter and reactive molten salts that dissolved the barrier oxide layers thus leading to fast oxidation and sulfidation of the material [35].

Superalloys: Fundamentals and Applications   Materials Research Forum LLC
Materials Research Foundations 178 (2025)109-130   https://doi.org/10.21741/9781644903698-7

Other factors that could affect the operational performance of the point of care testing equipment include temperature variations and mechanical stresses applied to the instrument. In the chemical processing plant case, secondary stress arising from manufacturing process left residual stresses on the components that led to initiation and growth of intergranular corrosion. The increases in both the aggressive chloride concentrations and residual stresses produced a suitable environment for localized corrosion and probable failure [36].

Measures like enhancing the chemical composition of the superalloy, enhancing the protective film or coating, and eliminating factors that may lead to corrosion are key in avoiding corrosive failures in the superalloy. Besides, proper maintenance, inspection, and Monitoring are also important to know the early stage of corrosion and avoiding disasters.

Thus, identification of corrosion failure in superalloys through case studies and metallurgical analysis is crucial to enhance material reliability and eliminate such occurrences. There are severe conditions in industrial gas turbines, jet engines, and chemical processing plants where superalloys are used.

8.   Successful Approaches to Mitigate Corrosion

Improved Alloy Design and Composition

Role of Alloying Elements: It is usually observed that superalloys are characterized by improved corrosion properties due to alloying additions. This is because the common alloying additives include Chromium (Cr), Aluminum (Al), and Silicon (Si) among others since they form oxide layers on the surface. It oxidizes to a stable layer of Cr_2O_3 that endows chromium with high resistance to oxidation and corrosion. Aluminum is reacts with oxygen in a process that forms Al_2O_3 which is a good passivating oxide that provides high temperature stability. Although it is used in smaller capacity, silicon reacts to produce silicon oxide (SiO_2) layers that are beneficial in oxidizing atmosphere [4].

The incorporation of refractory elements such as tungsten, molybdenum and rhenium also improves high temperature strength and resistance of the superalloys. These contribute to reinforcement and enhancement of the microstructure in as much as the general performance of the material under conditions of stress [3].

Development of New Super alloy Grades. Another constant topic is the production of new generations of super alloy grades to improve the corrosion and mechanical characteristics. For instance, earlier nickel base superlatives with increased percentage of chromium and aluminum increase their tolerance for oxidation hence improves their service period. Inconel 718 and Rene 41 are far more complicated super alleys which are specifically used for working in high temperature circumstances such as in gas turban and jet engine [1].

Current research direction involves the creation of superalloys with nanostructures since they exhibit desirable mechanical properties together with strength against environmental effects. These new grades are expected to be capable of operating at higher temperatures and more severe corrosive conditions as a development over the original material to increase the life of critical components [2].

Advanced Coating Technologies

The next generation of coatings is in the process of being created to combat high temperature oxidation and corrosion. These coatings are designed to form stable, adherent oxide layers that protect the underlying superalloy from environmental degradation. One example is the development of thermal barrier coatings (TBCs) with improved thermal insulation properties and enhanced adhesion to the substrate [30].

High-surface-area to volume ratio is a characteristic of nanostructured coatings because of their specific microstructures providing improved protective characteristics. These coatings offer enhanced anti-oxidation and hot corrosion protection that makes them perfect for use in gas turbines and other high temperature applications [38].

The application techniques of protective coatings play an important role in the success of protective coatings. Thermal spraying can therefore be defined as the use of molten or partially sintered materials to deposit a layer on the substrate material. EB-PVD is a vacuum deposition process that allows deposition of uniform, dense and high purity coatings with a good spall strength. Plasma spraying is a method where the coating material is atomized by a plasma jet and then projected at high speed to form a layer of protection on the substrate surface [39].

Enhanced Surface Treatments

Some of the advanced surface engineering techniques being considered today aims at enhancing the corrosion and mechanical characteristics of the superalloys. Methods like laser surface alloying, ion implantation, and surface Nano crystallization are seen to improve the properties of the superalloys in hostile environments.

Laser surface alloying is a process of treatment of a surface layer by means of a high-energy laser beam to melt the surface layer and mix it with alloying elements. It forms a new surface layer with enhanced Vickers hardness, wear resistance, and corrosion resistance. Ion implantation is defined as the action of introducing alloying elements into the surface layer using high energy ions and enhancing the surface features. Nano crystallization or, in other words surface mechanical attrition treatment (SMAT), strengthens the grain size of the surface layers and increases its physical mechanical properties and resistance to corrosion and various environmental impacts [40].

All are good and as well have their demerits. Anodizing and plating are very effective in this context of the protection against corrosion, but has limitations with high heat treatment. Both shot peening and laser shock peening enhance the fatigue limit but neither imparts a surface layer which can prevent oxidation of the surface. TBCs are known to provide excellent thermal insulation but have poor resistance to thermal cycling and are prone to spallation. Diffusion coatings are good for oxidation but may not be easy to apply uniformly onto the surface. Overlay coatings and environmental barrier coatings are strong coatings that can provide protection but may be expensive and must be applied carefully.

It is important to appreciate these benefits and limitations as they help identify the right protective measure for certain applications. Surface treatments and coatings are usually applied in conjunction in order to get the best performance and life from superalloy parts [2].

Therefore, effective ways of combating the problem of corrosion in superalloys involve the development of new chemical compositions and new and improved coatings and surface treatment.

Superalloys: Fundamentals and Applications   Materials Research Forum LLC
Materials Research Foundations 178 (2025)109-130   https://doi.org/10.21741/9781644903698-7

This is by refining the chemical makeup of the alloy and the creation of the enhanced types of superalloy that make the valuable material more intrinsically resistant to corrosion. Other types of coatings and novel methods provide additional protection and retain the performance of each superalloy part in severe environments.

9.   Future Trends and Challenges

Anticipated Advances in Superalloy Development

The prospects of continued improvement in the performance of superalloys are therefore grounded on the continued research and innovation in the chemistry of the alloys and in the manufacturing technology. Materials scientist are already on the task of developing new alloys to improve the high temperature, high corrosion and mechanical characteristics of the material. High entropy alloys HEAs consisting of five or more elements with nearly equiatomic concentrations of each metallic element is one of the areas of interest. HEAs also have other characteristics found in other classes of material such as high strength, good oxidation and high thermal stability which could make them appropriate as potential next generation or next generation advanced super alloys [41]. Another important line of research is the improvement of the composition of the alloys to provide the best conditions for the formation of oxide films. For example, by raising the aluminum and chromium content in nickel based superalloy, the formation of Al_2O_3 and Cr_2O_3 scales can be enhanced thus improving the oxidation performance of the alloy. Furthermore, the incorporation of additional elements like Yttrium and Cerium will help in increasing the bond strength of these oxide layers [2]. Alloys in aerospace applications, it is therefore imperative to develop new manufacturing processes for super alloys. Some of the greatest changes being observed in the make of super alloy components are through the use of Additive Manufacturing also known as 3D Printing. AM provides a greater amount of control over microstructure and content, which can facilitate the creation of parts with superior Mechanical property and corrosion resistance [42].

Other developments are the powder metallurgical processes for manufacturing super alloy products with refined, homogeneous microstructures. Such techniques as hot isostatic pressing (HIP) and spark plasma sintering (SPS) that comes under powder metallurgy reprocesses the super alloys to enhance the density and mechanical characteristics. These techniques also allow for the additions of the nanoscale precipitates responsible for the high temperature strength and oxidation resistance [43].

Emerging Coating Technologies for Enhanced Corrosion Resistance

Nanocomposite coatings are attracting the interest of many researchers due to their improvement in corrosion protection and mechanical features. These coatings are made up of Nano particles that form a compact layer to prevent corrosive species from attacking the surface. The enhanced physical turnover area means that a large surface area is in contact with the substrate, which improves adhesion and reduces the probability of developing defects that provide pathways for corrosion [38].

Nanostructured TBCs fabricated from YSZ exhibit better thermal insulating property and resistance to oxidation than the conventional TBCs. These coatings are specifically useful when applied to the super alloys utilized in the gas turbines and the jet engines since they can enhance component operational life and efficiency [38]

Self-healing smart coatings go a long way to enhance corrosion protection in various ways. These coatings are capable of self-healing effects resulting from mechanical abrasion or any other hostile environment and retain their protective attributes in the process. Usually self-healing methods include smart healing agents, in form of microcapsules or substances dispersed in the coating matrix [44].

An example of some emerging repair strategies is the creation of polymer-based coatings containing microencapsulated healing agents. When the coating is damaged, the capsules break and the healing agent is released which then polymerizes to seal the cracks. These coatings have potential for the enhancement of the service life of superalloy component in severe conditions [44].

Challenges in Corrosion Management

The environmental regulations describe below are likely to create opportunities as well as threats in the production of corrosion-resistant superalloys. The rules and regulations for cutting down emissions and impacts on the environment has put pressure to use materials that can handle the tough and diverse conditions effectively. For instance, outright regulations that demand low sulfur and nitrogen oxide emissions require superalloys that are not easily corroded by these elements [45].

There is also a need to design the friendly environment coatings and surface treatment to enhance conformation to environmental laws. This also includes removal of carcinogenic substances such as hexavalent chromium/ widely used in corrosion prevention, but with adverse effects to the environment and human health. It remains a major challenge for researchers to find new coatings that will offer similar performance to the existing coatings that are currently being regulated but which will meet the set requirements [45].

Other concerns that are majorly perceived as challenges include economic and technological barriers in developing and incorporating new generation superalloys and protective coatings. This means that the cost of the raw materials, the cost forces sing and other quality controls can be very high, and thus, the cost of producing the superalloy components can be prohibitive[46].

Technological factors depend on the current technologies that the company uses and the available technologies to produce products and offer services. Apparently, additive manufacturing is still hindered by some issues concerning reproducibility and stabilization of the fabricated parts. One technological challenge that the advanced coatings must also overcome is the ability to deposit them with good uniformity and good adhesion to structures with complicated shapes [42].

To overcome these limitations, current and future research is aimed at designing conducive manufacturing processes and economical techniques used in the coating process. Due to such challenges, superalloy development cooperation between various industries, academia, and governing bodies should be enhanced.

As for the prospects of further advancements in superalloys one should talk about the new generations of alloy compositions, new manufacturing methods, and new coatings. Advanced coatings such as nanostructured and smart coatings are directional techniques to achieve improved corrosion protection and increased service life of superalloy parts. However, some problems regarding environmental standards, cost effectiveness, and the existing technology remain critical inquiries that require research and development solutions.

Superalloys: Fundamentals and Applications   Materials Research Forum LLC
Materials Research Foundations 178 (2025)109-130   https://doi.org/10.21741/9781644903698-7

Conclusion

Superalloys, usually referred to as materials or components that possess elevated-temperature strength, oxidation, and corrosion resistance, are widely used in several applications that include aerospace, power generation, and chemical processing industries. This chapter was aimed of providing an understanding of all the comprehensive facets of superalloy and their performance, with regards to their material composition, corrosion behavior, testing procedures and protection, and lastly the advancement and future prediction. It was centered on the influence of alloying additions including chromium, aluminum and refractory on corrosion stability and mechanical processes. Lastly, general information on four types of corrosion like high temperature oxidation, hot corrosion, hydrogen embrittlement and stress corrosion cracking was reviewed, concentrating on how these four corrosion types influence the working life and stability of super alloy parts. Both laboratory and field corrosion techniques contribute to the determination of the overall susceptibility of superalloys to environmental corrosion. Superalloys are protected from corrosion by using various surface treatment and coating techniques which comprise of anodizing, plating, shot peening, laser shock peening, TBCs, diffusion, overlay, and EBC. Future trends in superalloy are included on themes such as improved alloy composition, manufacturing technology including additive manufacturing technology and advanced coatings including nanostructured and self-healing coatings. This gives corrosion management a vital significance towards the durability, dependability and functionality of the superalloy parts. These must be to increase the component life, maintain its functionality, meet the environmental standards and affordability. Corrosions fail inhibitions lead to low frequent maintenance and replacement hence improving the operational costs and safety. Moreover, utilization of a material and adequate preservation of the mechanical properties of the superalloys contribute to the effectiveness and durability of crucial systems. Continuous research and development work are imperative to overcome limitations and exploit most of the superalloys sufficiently in demanding applications. The need to develop a holistic strategy for combating corrosion of superalloys include; recognizing the modes through which corrosion takes place, conducting intrusive tests, and applying the best protective measures as useful process in countering corrosion.

References

[1] C.T. Sims, N.S. Stoloff, W.C. Hagel, Superalloys II, Wiley New York 1987.

[2] R.C. Reed, The superalloys: fundamentals and applications, in: Cambridge university press 2008.

[3] T.M. Pollock, S. Tin, Nickel-based superalloys for advanced turbine engines: chemistry, microstructure and properties, J. Propuls. Power. 22 (2) (2006). p. 361-374. https://doi.org/10.2514/1.18239

[4] B. Pint, Experimental observations in support of the dynamic-segregation theory to explain the reactive-element effect, Oxid. Met. 45 (1996) p. 1-37. https://doi.org/10.1007/BF01046818

[5] J. Smith, et al., Thermal barrier coating validation testing for industrial gas turbine combustion hardware, J. Eng. Gas Turbines Power. 138 (2016) p. 031508. https://doi.org/10.1115/1.4031448

Superalloys: Fundamentals and Applications   Materials Research Forum LLC
Materials Research Foundations 178 (2025)109-130   https://doi.org/10.21741/9781644903698-7

[6] C. Giggins, F. Pettit, Corrosion of metals and alloys in mixed gas environments at elevated temperatures, Oxid. Met. 14 (1980) pp. 363-413. https://doi.org/10.1007/BF00603609

[7] T. Gibbons, I.G. Wright, A review of Materials for gas turbines firing syngas fuels, 2009. https://doi.org/10.2172/970884

[8] F. Froes, Titanium: physical metallurgy, processing, and applications, ASM international, 2015. https://doi.org/10.31399/asm.tb.tpmpa.9781627083188

[9] M.H. Sullivan, On the development of thermally grown oxide in high water vapor environments for thermal barrier coating systems, University of California, Irvine 2013.

[10] D.J. Young, High temperature oxidation and corrosion of metals, Second ed., Vol. 1. Elsevier 2008. https://doi.org/10.1016/S1875-9491(08)00001-X

[11] A. Zahs, M. Spiegel, H. Grabke, The influence of alloying elements on the chlorine-induced high temperature corrosion of Fe-Cr alloys in oxidizing atmospheres, Mater. Corros. 50 (1999) p. 561-578. https://doi.org/10.1002/(SICI)1521-4176(199910)50:10<561::AID-MACO561>3.0.CO;2-L

[12] R.A. Rapp, Hot corrosion of materials: a fluxing mechanism?, Corros. Sci. 44 (2002) p. 209-221. https://doi.org/10.1016/S0010-938X(01)00057-9

[13] N. Shi, et al., Biomass-inspired semiconductor photocatalysts for solar degradation of organics, Curr. Org. Chem. 19 (2015) p. 521-539. https://doi.org/10.2174/1385272819666150115000005

[14] K.J. Tan, T.A. Hatton, Electrochemically mediated sustainable separations in water, in: G. Szekely, D. Zhao, (Eds.), Sustainable Separation Engineering: Materials, Techniques and Process Development, 2022 p. 1-62. https://doi.org/10.1002/9781119740117.ch1

[15] R.E. Ricker, et al., Chloride ion activity and susceptibility of Al alloys 7075-T6 and 5083-H131 to stress corrosion cracking, Metall. Mater. Trans. A, 44 (2013) pp. 1353-1364. https://doi.org/10.1007/s11661-012-1500-2

[16] X.S. Zhang, Kinetics of O2 reduction on oxide-covered Ni-Cr-Mo alloys, The University of Western Ontario., Canada, 2012.

[17] N. Birks, G.H. Meier, F.S. Pettit, Introduction to the high temperature oxidation of metals, Second ed., Cambridge university press., 2006. https://doi.org/10.1017/CBO9781139163903

[18] S. Lynch, Hydrogen embrittlement phenomena and mechanisms, Corros. Rev. 30 (2012) p. 105-123. https://doi.org/10.1515/corrrev-2012-0502

[19] R.P. Gangloff, Hydrogen-assisted cracking. Comprehensive structural integrity, 6 (2003) p. 31-101. https://doi.org/10.1016/B0-08-043749-4/06134-6

[20] S.D. Cramer, et al., ASM Handbook, ASM international Materials Park, Ohio. 13 (2003).

[21] W.F. Smith, J. Hashemi, Foundations of Materials Science and Engineering. Mcgraw-Hill Publishing 2006.

[22] R. Rebak, Stress corrosion cracking (SCC) of nickel-based alloys, in: Stress Corrosion Cracking, Elsevier, 2006 p. 273-306. https://doi.org/10.1533/9780857093769.3.273

[23] F. MG, A Critical Analysis of Pitting Corrosion, Corros. Eng. 8 (1959) p. 298-307. https://doi.org/10.3323/jcorr1954.8.7_298

[24] D. Buhrmaster, N. Wilson, Military Aircraft and Associated Equipment, in: Supplement to Corrosion Tests and Standards: Application and Interpretation, Second ed., 2022, pp. 393-397. https://doi.org/10.1520/MNL202NDSUP20190028

[25] P. Kofstad, High Temperature Corrosion, Elsevier Applied Science, London, 1988, pp. 382-385.

[26] F. Mansfeld, Electrochemical methods of corrosion testing, in: Corrosion: Fundamentals, Testing, and Protection, Association of corrosion engineers, 2003, pp. 446-462. https://doi.org/10.31399/asm.hb.v13a.a0003644

[27] J. Blitz, Electrical and Magnetic Methods of Non-Destructive Testing. Second ed., Springer Science & Business Media 1997 pp. XI, 216. https://doi.org/10.1007/978-94-011-5818-3

[28] G.S. Frankel, Electrochemical techniques in corrosion: Status, limitations, and needs. J. ASTM. Int. 5 (2009) 101241. https://doi.org/10.1520/JAI101241

[29] S.G. Irizalp, N. Saklakoglu, 1.14 laser peening of metallic materials, in: Comprehensive Materials Finishing, Elsevier Oxford., 2017, pp. 408-440. https://doi.org/10.1016/B978-0-12-803581-8.09160-8

[30] D. Clarke, C. Levi, Materials design for the next generation thermal barrier coatings. Ann. Rev. Mater. Res. 33 (2003) pp. 383-417. https://doi.org/10.1146/annurev.matsci.33.011403.113718

[31] R. Darolia, Thermal barrier coatings technology: critical review, progress update, remaining challenges and prospects, Int. Mater. Rev. 58 (2013) pp. 315-348. https://doi.org/10.1179/1743280413Y.0000000019

[32] Y. Liu, J. Shi, Y. Wang, Evolution, Control, and mitigation of residual stresses in additively manufactured metallic materials: A review, Adv. Eng. Mater. 25 2023. pp. 2300489. https://doi.org/10.1002/adem.202300489

[33] C.G. Levi, Emerging materials and processes for thermal barrier systems, Curr. Opin. Solid State Mater. Sci. 8(1) (2004) pp. 77-91. https://doi.org/10.1016/j.cossms.2004.03.009

[34] C. Zhou, Y. Song, Oxidation and hot corrosion of thermal barrier coatings (TBCs), in: H. Xu, H. Guo (Eds.), Thermal Barrier Coatings, Second ed., Elsevier 2011, pp. 193-214. https://doi.org/10.1533/9780857090829.3.193

[35] R. Viswanathan, An investigation of blade failures in combustion turbines, Eng. Fail. Anal. 8 (2001) 493-511. https://doi.org/10.1016/S1350-6307(00)00043-1

[36] P.A. Schweitzer, Corrosion of linings & coatings: cathodic and inhibitor protection and corrosion monitoring, First ed., CRC press, 2006, pp. 568. https://doi.org/10.1201/9780849382482

[37] J.R. Davis, Metals Handbook Desk Edition. Second ed., ASM International, 1998. https://doi.org/10.31399/asm.hb.mhde2.9781627081993

[38] H. F. Chen, et al., Recent progress in thermal/environmental barrier coatings and their corrosion resistance, Rare Metals, 39 (2020). pp. 498-512. https://doi.org/10.1007/s12598-019-01307-1

[39] L. Pawlowski, The Science and Engineering of Thermal Spray Coatings, First ed., John Wiley & Sons, 2008, pp. 597-626. https://doi.org/10.1002/9780470754085

[40] P. Dearnley, A review of metallic, ceramic and surface-treated metals used for bearing surfaces in human joint replacements, Proc. Inst. Mech. Eng. Part H: J. Med. Eng. 213 (1999) pp. 107-135. https://doi.org/10.1243/0954411991534843

[41] E.P. George, D. Raabe, R.O. Ritchie, High-entropy alloys, Nat. Rev. Mater. 4 (2019) pp. 515-534. https://doi.org/10.1038/s41578-019-0121-4

[42] W.E. Frazier, Metal additive manufacturing: A review, J. Mater. Eng. and performance, 23 (2014) pp. 1917-1928. https://doi.org/10.1007/s11665-014-0958-z

[43] I. Chang, Y. Zhao, Advances in Powder Metallurgy: Properties, Processing and Applications, First ed., Elsevier, 2013. https://doi.org/10.1533/9780857098900

[44] W.Y. Jeng, Characterization and Dielectric Properties of Pure and re-substituted bi4ti3o12 and bi5crti3o15 aurivillius ceramics, 2018.

[45] R.W. Revie, Corrosion and Corrosion Control: An Introduction to Corrosion Science and Engineering, John Wiley & Sons, 2008. https://doi.org/10.1002/9780470277270

[46] G.H. Koch, et al., Corrosion Cost and Preventive Strategies in the United States, United States, Federal Highway Administration, 2002.

Superalloys: Fundamentals and Applications
Materials Research Foundations 178 (2025)131-142

Materials Research Forum LLC
https://doi.org/10.21741/9781644903698-8

Chapter 8

Limitations of Superalloys and Future Research

Aalia Asghar[1], Nadia Akram[1]*, Khalid Mahmood Zia[1], Muhammad Saeed, Akbar Ali[1]

[1]Department of Chemistry, Government College University Faisalabad, Faisalabad-38000, Pakistan

* nadiaakram@gcuf.edu.pk

Abstract

Superalloys, majorly nickel based, are widely used in applications where materials are required to operate at high temperature. However, there are certain challenges that still stand and hamper the growth of superalloys in aerospace and power generation industries even though they hold the major market share for their application. This review sought to look at the current state, opportunity and challenges as well as the future outlook of superalloys from materials chemistry standpoint. Again, superalloys are characterized by a few disadvantages such as; high cost, need complex manufacturing processes and environmental unfriendly. The cost problem results from varied prices of uncomplicated components for instance nickel besides the embraced usage of rare costlier elements for instance rhenium. Superior properties of the γ/γ' microstructure of super alloys come at the cost of difficult manufacturing and recycling processes. There are few environmental considerations: high energy usage is used in production and there are problems with recyclability. Exploring these challenges is area of future studies and development of new strategies for carrying out the research. They include oxide dispersion strengthened (ODS) super alloys for fusion reactor using 3-D printing technology for material minimization and refractory high-entropy alloys (RHEAs) for hypersonic flight purposes. Measures that are in the works include creation of protective smart coatings against environmental erosion, coming up with rhenium free super alloys and enhancing efficiency in the recovery via electrochemistry. Recent publications have revealed improvements in super alloy design using machine learning thus expectant property predictions and beneficial composition modifications. Yet, the field is at the crossroads for which the priorities lie in the improvement of performance, at the cost of economic and environmental impacts. The prospects for superalloys are defined by the need to create parts from materials that are stronger, more resistant to temperature and are ecologically friendly to be used in the frames of the further anticipated advancement of manufacturing technologies and application.

Keywords

Super Alloy Market, Challenges, Cost and Complexity, Resource Availability, Applications of Super Alloys, Future Progress

Contents

1.   Limitations of Superalloys and Future Research Directions

1.1   Current Scenarios

The present situation of super alloys is understood by their vital utilization in various industries but the uses and demands of these super alloys are still under pressure. In the atmosphere sector, nickel based super alloys forms between a half of the total weight of advanced aircraft engines, and the market for these materials is expected to grow to $9.3 billion [3].

Nevertheless, this dominance comes with a lot of controversy. Although superalloys make it possible to work at higher temperatures and increase fuel efficiency [7]. Stresses that further improvements in performance are gradually becoming less significant, and that it takes more and more effort to achieve the same level of improvement with each 1°C increase in temperature. In the power generation industry, super alloys are used in gas turbine parts, where thermal efficiency is as high as 60% in combined cycle power plants [11]. Argue that using super alloys hampers the shift to renewable energy sources for the same reason that these materials are not fit for cyclic loading conditions that characterize wind and solar thermal power generation.

Super alloys are widely used in downhole tools and offshore platforms of the oil and gas industry because of their high resistance to corrosion. However, note that the industry's move towards more complex, or rather aggressive, wells is progressing faster than the advancement in super alloy technology, with existing materials unable to cope with hydrogen sulfide contents of more than 10% by volume. For turbocharger components in the automotive industry, the use of super alloys has increased by 15% per year in the last ten years [3]. But this trend is slowly changing as the automobile industry moves towards electric cars which do not require the high temperature application of super alloys to the same extent [11].

Biomedical applications offer a new market for super alloys especially in orthopedic applications because of their compatibility with human tissues and their mechanical strength. However, they also point out the disadvantage of the high elastic modulus of super alloys compared to the human bone Such multiple applications demonstrate the versatility of super alloys although the present and future trends indicate a fairly competitive market with super alloys losing their supremacy due to changing industrial requirements, environmental considerations and new materials [1].

Superalloys: Fundamentals and Applications
Materials Research Foundations 178 (2025)131-142

Materials Research Forum LLC
https://doi.org/10.21741/9781644903698-8

1.2   Limitation of Superalloys

Super alloys have been key in the development of HTPG applications especially in the aerospace and power generation industries. However, these materials have several drawbacks that limit their applicability and efficiency in high-temperature applications. For instance, although super alloys offer excellent strength and creep resistance at high temperatures, their maximum working temperatures have stagnated at 1100°C which is only 85% of their melting This thermal ceiling is mainly attributed to the fact that the γ' phase which is the strengthening phase in these nickel based super alloys is relatively unstable at ultra-high temperatures More [13].

It is crucial to mention that the density of nickel-based super alloys, which is in the range of 8.0-9.0 g/cm³, is a significant disadvantage for weight-critical applications. Thus, the corrosion properties of super alloys, though higher than those of conventional alloys, remain insufficient for highly sulfide ring and chlorination conditions as evidenced by accelerated hot corrosion tests where material loss rates grew by up to 40 % compared to standard oxidation conditions. From an economic point of view, the authors note that the high and fluctuating prices of critical alloying elements, including rhenium and ruthenium, which have increased by more than 300% in the last ten years, pose a significant risk to the development of advanced super alloys [14].

These limitations point to the need for new ideas in the design and processing of super alloys. Furthermore, there are those who believe that a gradual optimization of the material composition is possible and sufficient while others propose that revolutionary shifts in the material systems, e.g. refractory high entropy alloys, are required to bypass the fundamental constraints of super alloys. These challenges must be tackled in future studies by paying attention not only to the increased temperature potential but also to the creation of more sustainable and affordable materials that can satisfy the requirements of modern advanced engineering systems [18].

The disadvantages of super alloys, especially the cost and the difficulty in manufacturing, are major barriers to the further application and improvement of super alloys. These factors have become more important than ever before as the need for better materials in high stress conditions has risen [7].

According to the literature, one of the key challenges facing the super alloy industry is the cost, including the fact that the price of nickel, a key component in super alloys, has been known to fluctuate by as much as 500% in five years, affecting manufacturing costs and market conditions. Furthermore, the use of some special and critical materials like rhenium which can be up to 6 wt. % in single crystal super alloys has caused a 60% rise in the raw material costs over the past two decades [15]. This cost is further amplified by the high energy consumption needed to produce super alloys through the manufacturing processes.

Vacuum induction melting and subsequent remitting that are important to provide the required chemical homogeneity can make up to 40% of the total costs. Super alloy makers' proponents have defended these costs arguing that super alloys offer superior performance [17] yet others note that the incremental benefits no longer justify the skyrocketing costs especially when compared to new super materials such as high entropy alloys [22].

This has remained a big problem especially given that the superalloy industry is faced with very many cost related issues that puts a jeopardy the sustainability as well as the competitiveness of the sector. The key constituent components of this issue involve highly fluctuating metallic pricing, especially in terms of nickel that has been known to fluctuate by up to 500 percent within a time period of just five years. This large fluctuation has not only implications on the cost modeling of manufacturing, but also adds a level of variability to the marketplace that resonates in the aerospace and energy industries that are primary uses of superalloys [26].

The demand for unassociated raw materials including rhenium metals increases more complexities to the economic challenges facing manufacturers in production of these modern superalloys. In the advanced single crystal superalloys, the rhenium content has gone up to 6 wt. % and the raw material cost rose a staggering 60% over two decades. This trend provokes a number of profound questions about the economic feasibility of steady consummation of such in supplied resources, particularly referring to such geopolitical circumstances, which can compromise supply chains [16].In this case, several of these manufacturing processes prove adequate for the high costs of superalloy. Post forming remelting procedures which are important in attaining the desired purity and uniformity of the microstructure can represent as much as 40% of the total manufacturing costs. Although these processes are crucial to achieve the high quality level required by the stringent applications, they are expensive barriers that hinder the wider use of superalloys in applications with lower performance requirements.

As argued for superalloys, despite their expensive costs in their production due to their excellent mechanical qualities and heat resistance, the rhodium coated superalloys are more appropriate for use. However, this perspective has recently been criticized by researchers such who argue that while there have been constant enhancements in the properties that surround organizations, the costs have also been escalated to an exponential level. This critical viewpoint becomes pertinent especially to new projects as more apparently superior material such as high entropy alloys seem to provide similar performance at probably cheaper costs [27]. Super alloy industry stands at a crossroad now, where attaining higher and higher level of performance is opposed to the economic consideration. This conventional wisdom of accepting escalating cost for marginal improvement in properties does not seem sustainable in the long run. Therefore, there is a need for fresh thinking and novel solutions toward designing and fabricating these alloys with significantly lower production costs while maintaining performance.

Thus, in conclusion, we have to state that although the theme of superalloys is still urgently relevant for many key applications, the industry needs to face the operational and, especially, economic issues to continue its development. The way through this is probably going to entail a materials approach, process approach and a change in how we use performance materials based on the economic downturn. Unless the superalloy industry adopts such a rescue mechanism, the industry will gradually disappear from the competitive market [12].

Another major constraint is the characteristic complexity of the super alloys. The fine structure that is responsible for the high performance of super alloys like the γ/γ' phase is sensitive to composition and heat treatment conditions. This is to prove that small differences of 0.1wt% in some alloying elements may cause the creep life to reduce by 20%, which This complexity carries through to the manufacturing process such as the directional solidification process for single crystal super alloys, which has a typical yield of only 70 percent compared to conventional casting techniques [25].

Additionally, the problem of recycling super alloys is another issue since it is impossible to recycle super alloy scrap into primary aerospace applications due to the presence of multiple tightly controlled alloying elements, which leads to ecological concerns in the context of the growing

Superalloys: Fundamentals and Applications   Materials Research Forum LLC
Materials Research Foundations 178 (2025)131-142   https://doi.org/10.21741/9781644903698-8

focus on the environmental impact of aerospace production. Complexity is evident in repair and maintenance operations where techniques such as diffusion brazing are applied, setting higher lifecycle costs and poor field reparability [21].

Super alloy supporters counter that these restrictions are simply the price that has to be paid for superior performance, but skeptics such as point out that the business has gone as far as it can go. They claim that future developments in high temperature materials may necessitate a shift from the existing super alloy strategy to compositionally and structurally less complex systems capable of delivering similar performance at lower costs [23].

Superalloys exhibit many challenging characteristics not only in their formulation and manufacturing but also disposal cycle also offers a major challenge. On a micro scale, the γ/γ' two phase system which is the basis of superalloys' superior characteristics profound strict requirements on the alloying composition and the technological treatments. For example [6] showed how even minor changes of 0.1 wt. % in specific alloying elements can be enough to cause a creep life decrease of 20 % in superalloys, which illustrates how narrowly balanced superalloy compositions are. This sensitivity is further magnified during manufacturing especially in single crystal superalloy manufacturing. In the same regards, pointed that the directional solidification process which is essential in the formation of the required microstructure endures only 70% success ratio which is considerably lower than that of the standard casting process and greatly contributes towards the cost of superalloy manufacture [3].

The complexity cascade does not stop here; the organization must also factor sustainability and a lifecycle for a product. [15] further pointed that recycling of superalloys is a reality with a daunting process simply because of the complexity of composition of the purchased scrap with some components containing very strictly controlled alloying materials making the recycling of the superalloy scrap with merits for the first stage aerospace applications economically unviable. This limitation is especially daunting given the increasing trend towards circular economy in the aerospace industry. [8] the authors calculated the recycling efficiency of nickel-based superalloys to be 30 percent with 70 percent of the material down cycled through less demanding products indicating loss of critical materials.

Stakeholders in the development of superalloys may counter these factors as acceptable trade-offs for achieving superior performance; however, recent studies indicate that the superalloy field may be reaching a significant milestone. postulated that each new generation of superalloys is being developed at an exponentially higher cost for each successive percent increase in high temperature service life. Similar to this, observed the growth in the rate of temperature capability in superalloys level out to about 4°C per year in the last twenty years than the 15°C per year seen in 1970s and 1980s.

Some people including say that the future belongs to compositionally and microstructurally lightweight materials, namely, refractory high entropy alloys. This they pointed out that some of the RHEAs can have as much specific strength as nickel based superalloys up to a temperature of 1600 C while at the same time may have ease of processing. However, in their recent study, pointed out that despite the promising RHEAs, they continue to pose numerous problems including their ability to resist oxidation and stability at high temperatures, out of some of the studied compositions, only 5% can be useful for high temperature structural applications [17].

This conflict between radicalization and traditionalization concerning the future of superalloys and their possible replacements is not unique to superalloys, but rather a general conflict in materials science between the paradigm of continued improvement of existing materials and the search for completely new materials. According the next generation materials for high temperature application may pose a major science and engineering challenge as the fundamental concepts of alloy design may need to shift from targeting high performance at high temperature alone, to combining performance with workability, cost, and environmental considerations and implications. This view is increasingly common, as evidenced by a Walston Survey in which 65 percent of aerospace material scientists argued that a major departure from conventional superalloys can be expected within the next twenty years [19].

The industry is facing pressure to increase performance, lower costs and reduce environmental impacts; all of which highlight the drawbacks of using superalloys. There remains considerable debate as to what the way ahead is, based more particularly between the evolutionist models for developing new superalloy systems and the revolutionist school of thinking on a completely new one [25].

Table 1: challenges faced by the superalloy industry

Challenge	Description	Impact	References
Cost	Nickel price fluctuations (up to 500% in 5 years). Use of special materials (e.g. Rhenium). High energy consumption in manufacturing.	Affects manufacturing costs and market conditions. 60% rise in raw material costs over two decades. Vacuum induction melting can make up to 40% of total costs.	[1], [2], [3]
Complexity	Sensitive γ/γ' phase structure. Complex manufacturing processes. Difficult recycling.	0.1wt% composition change can reduce creep life by 20%. Single crystal production yield only 70%. Recycling efficiency of only 30%.	[4], [5], [6]
Performance vs. Cost	Diminishing returns on performance improvements. Competition from new materials.	Incremental benefits may not justify rising costs. High entropy alloys offer similar performance at potentially lower costs.	[7], [8]
Environmental Concerns	Poor recyclability. High energy consumption in production.	Difficulty in achieving a circular economy. Increased focus on environmental impact in aerospace production.	[9], [10]
Supply Chain Issues	Dependence on rare materials. Geopolitical factors affecting supply	Vulnerability to supply disruptions. Potential for compromised supply chains.	[11], [12]
Future Sustainability	Debate between evolutions vs. revolution in alloy design. Need for new approaches.	Potential shift away from conventional super alloys within 20 years. Requirement for balancing performance, workability, cost, and environmental considerations.	[13], [14]

1.3   Environmental and Sustainability Issues

The environmental and sustainability problems of super alloys such as resource scarcity and recycling difficulties have emerged as critical concerns over the last few years, and they question the future of super alloys in a wide range of industries.

Super-alloys are known to have a high environmental cost; super alloy manufacturing can consume up to ten times more energy than the production of conventional steel, with one kilogram of nickel-based super alloy requiring over 300 kWh of energy.

It is an energetically demanding process that has a high impact on climate change, estimating that the super alloy industry produces about 0.5% of the world's CO_2 emissions. In addition, the extraction and refining of the constituent elements and especially nickel and cobalt have been found to be causing serious environmental pollution. According to Mud (2010) the amount of nickel mining alone emits more than 1 million tons of SO_2 every year and causes acid rain and other effects to the ecology within mining areas [6].

Resource depletion is another major factor that is likely to hinder the sustainability of superalloys. A recent study estimates that economically recoverable resources of some key elements, such as rhenium, that are used in some of the most advanced super alloys, may last for only 40 years at the current rates of consumption. This is especially problematic for materials such as hafnium which is noted to have no suitable alternatives in many super alloy applications, despite estimated reserves only being sufficient for another 70 years of mining at current rates. This impending shortage has given rise to geopolitical issues; it has been estimated that more than 80% of the supply of certain critical super alloy elements comes from three countries only, which is a cause for concern regarding supply chain risks and price fluctuations [22].
Recycling challenges

Thus, recycling issues only add to the sustainability concerns of super alloys. Although recycling of super alloys is theoretically possible, the Due to the many element and compound components of super alloys, the separation and purification process is challenging and requires a large amount of energy [9]. In addition, note that the use of tramp elements and impurities in recycled super alloys results in variations in the material properties, which may deter manufacturers from using high levels of recycled content in critical structures.

The current super alloy industry has been criticized for having a poor response to these questions. Pro advocates like and contend that these issues can be managed design by design, and incremental improvements in alloying and fabrication techniques can help offset several of these concerns while asserting for a leap in the choice of material systems with high entropy alloys or intermetallic that can provide similar or possibly better properties to that of existing alloys but with far improved environmental footprint and recyclability. The controversy points out the necessity of developing new ideas for designing super alloys with efficiency performance characteristics consistent with environmental and sustainability constraints [20].

2.   Future Research Directions

In the coming years, elaborate studies are cultivated to be conducted on super alloys not only to overcome the present shortcomings but to diversify its uses and come up with preventive measures reasonably enough for the encountered problems. This field is still growing, as those researchers continue to develop new and interesting ways to increase performance, sustainability, and decrease cost [18].

Much work has been done highlighting the potential of oxide dispersion strengthened (ODS) super alloys for fusion reactor applications because of their irradiation resistance. However, they note that a considerable amount of work should be devoted to the study of such long term irradiation effects and compatibility issues with the coolant [14].

3.   Application

In terms of applications, one of the most promising is the use of super alloys for additive manufacturing processes.CSS point out that AM technique might radically change the production of intricate components from super alloys, with the material waste rate as low as 10%. However, note that the AM involves extremely high rates of solidification that may cause no equilibrium states in the microstructure of the alloy in question. The future studies have to be more dedicated to the fine tuning of the AM parameters and novel alloys designed for these techniques [10].

Another future use is in hypersonic flights suggest that RHEAs, which is a subgroup of super alloys, may be useful in the temperatures over 2000 0 C and oxidation conditions met in hypersonic vehicles. But, the authors recognized that there are important issues regarding density reduction and oxidation resistance that need to be solved, meaning that new compositions and protection coatings have to be investigated [19].

Additive manufacturing of superalloys is considered as one of the most challenging and innovative areas of material, science, and engineering. [24] AM techniques could unlock conventional manufacturing of highly complex superalloy parts Dro using up to 90% less of the material than used by conventional manufacturing processes. This massive cut in wastage is not only enterprise related implications but also address some of the contemporary environmental issues affecting the manufacturing industries. However, problems are encountered when using AM for superalloys. [15] point out that high cooling rates typical for AM processes result in nonequilibrium microstructures that may negatively affect the properties of the alloy. This concern rises the issue of how to maintain the benefits that AM brings while at the same time protecting the unique properties of superalloys.

Another promising field is the usage of superalloys in hypersonic flight is another perspective direction. [22] Similarly consider RHEAs, a category of superalloys, as materials capable of withstanding the operating conditions above 2000°C and oxidative conditions present in hypersonic vehicles. This proposition is especially important considering the rise in demand for hypersonic technology in both commercial and defense industries. The authors also understand that further improvements have to be made, with density reduction and oxidation resistance being the most challenging areas. Such concerns point to the potential for additional investigation into new composites as well as for better encapsulation techniques that may help to close the existing gap between current technological developments and the needs of hypersonic flight.

Today, the use of superalloys is being discussed for future generation power in nuclear energy production. In a related study, discuss the opportunity of using oxide dispersion strengthened (ODS) superalloys in fusion reactor confines because of their high irradiation tolerance. Superalloys' application in nuclear could help create advance nuclear power plants, helping solve the issues with the present nuclear energy technology However, the authors state that much work

Superalloys: Fundamentals and Applications   Materials Research Forum LLC
Materials Research Foundations 178 (2025)131-142   https://doi.org/10.21741/9781644903698-8

should be done toward head irradiation effects and compatibility with the coolant systems. This notice is intended as a word to the wise as to the challenges of procuring and modifying material for nuclear use and the measures necessary to guarantee that it is safe and will perform to specifications [27].

However, it is imperative that these applications are considered with a realistic approach; superalloys have a bright future within various fields, but they have different strengths and weaknesses still need to prove themselves. The problem areas identified by point to the fact that much has to be achieved in the superalloys' modification and development to effectively support these future uses. However, more emphasis should be placed on assessing the sustainability of employing these superior performing materials in volume applications. Subsequent studies have to address not only issues related to the technical characteristics of superalloys but also potential ways to reduce the cost of their manufacturing to allow their usage in broader amounts of applications [26].

4. Prevention method

Centre-measures against identified problems are a major focus of future research activities. Among these goals, protection of components from environmental degradation is prominent with emphasis being made on combating hot corrosion and oxidation. Pettit and recommend the use of adaptive smart coatings in which lifetimes of components may be increased by up to 50% depending on fluctuating changes in their environment. But to their knowledge, the application of these coatings has been a major challenge because it is difficult to incorporate these with the presently used super alloy systems without loss of the mechanical properties of the material.

For resource depletion issues, researchers are looking at ways to remove or at least minimize key elements in the super alloys. This show great potentials in developing rhenium free super alloys with similar creep resistance to those containing rhenium. Yet, the authors note that it is currently impossible to apply the same obtained performance to all properties, which is why further studies of other strengthening mechanisms are needed [11].

It is also proposed that recycling challenges are also present and are being met through following novel methods there is an attempt for proposing a new electrochemical method to selectively recover valuable elements from super alloy waste in order to boost the recycling efficiency ranging up to 30%. Nonetheless, the authors also explain the process as still highly energy consuming and that it needs improvement for making the process industrially feasible [9].

machine learning and artificial intelligence become intriguing approaches in super alloy design based on computational methods such as Neural Networks are shown to be capable of predicting the super alloy properties to levels of precision similar to experimental methods and therefore show promising aspects in the development of the alloy. But as they note, these models are inevitably as good as the data they've been trained on, thus highlighting the importance of experimental confirmation [1].

However, some researcher's state that advancements in these directions are gradually slowing down and warn that further metabolism of the traditional super alloy concept can be considered only within creating fundamentally new alloys based on radically different principles, such as multiprincipal systems. They state that they provide improved compositional freedom and may

Superalloys: Fundamentals and Applications   Materials Research Forum LLC
Materials Research Foundations 178 (2025)131-142   https://doi.org/10.21741/9781644903698-8

possess better characteristics; however, it is agreed that much work is still required to develop them to a technology maturity level [5].

In conclusion, it can be summarized that super alloy research has huge opportunity in the future, but in the meanwhile also has great challenges. The field has to find that approach, where the focus on improvement of performance is not overshadowed by emergent environmental and economic issues. It is implicit that the next generation of super alloys will have to be stronger, more temperature resistant but also sustainable, while at the same time are capable to meet emerging manufacturing technologies. This will need a combined effort and out of the box thinking from interdisciplinary professionals in order to maintain relevance of super alloys in technology advancements.

References

[1] R. Darolia, Development of strong oxidation and corrosion resistant nickel-basee superalloys: critical review of challenges, progress and prospects, Int. Mater. Rev. 64 (2019) 355-380. https://doi.org/10.1080/09506608.2018.1516713

[2] K. Moeinfar, F. Khodabakhshi, S.F. Kashani-Bozorg, M. Mohammadi, A.P. Gerlich. A review on metallurgical aspects of laser additive manufacturing (LAM): Stainless stells, nickel superalloys, and titanium alloys, J. Mater. Res. Technol. 16 (2022) 1029-1068. https://doi.org/10.1016/j.jmrt.2021.12.039

[3] R.R. Srivastava, M.S. Kim, J.C. Lee, M.K. Jha, B.S. Kim, B. S. (2014). Resource recycling of superalloys and hydrometallurgical challenges, Journal of Materials Science, 49 (2014) 4671-4686. https://doi.org/10.1007/s10853-014-8219-y

[4] A.N. Jinoop, C.P. Paul, K.S. Bindra, Laser-assisted directed energy deposition of nickel super alloys: a review, Proc. Inst. Mech. Eng. Pt L: J. Mater. Des. Appl. 233 (2019), 2376-2400. https://doi.org/10.1177/1464420719852658

[5] A. Kracke, A. Allvac, (2010, October). Superalloys, the most successful alloy system of modern times-past, present and future, In Proceedings of the 7th International Symposium on Superalloy. 71 (2010) pp. 13-50 https://doi.org/10.7449/2010/Superalloys_2010_13_50

[6] W. Betteridge, S.W.K. Shaw, Development of superalloys, Mater. Sci. Technol. 3 (1987), 682-694. https://doi.org/10.1179/mst.1987.3.9.682

[7] A. Thakur, S. Gangopadhyay, (2016). State-of-the-art in surface integrity in machining of nickel-based super alloys, Int. J. Mach. Tools Manuf. 100 (2016) 25-54. https://doi.org/10.1016/j.ijmachtools.2015.10.001

[8] R. Dasgupta, A look into Cu-based shape memory alloys: Present scenario and future prospects, J. Mater. Res. 29 (2014), 1681-1698. https://doi.org/10.1557/jmr.2014.189

[9] W. Xia, X. Zhao, L. Yue, Z. Zhang, A review of composition evolution in Ni-based single crystal superalloys, J. Mater. Sci. Technol. 44 (2020) 76-95. https://doi.org/10.1016/j.jmst.2020.01.026

[10] H. Long, S. Mao, Y. Liu, Z. Zhang, X. Han, Microstructural and compositional design of Ni-based single crystalline superalloys-A review, J. Alloys Compds. 743 (2018) 203-22. https://doi.org/10.1016/j.jallcom.2018.01.224

Materials Research Forum LLC
https://doi.org/10.21741/9781644903698-8

[11] A. Bandyopadhyay, K.D. Traxel, M. Lang, M. Juhasz, N. Eliaz, S. Bose, Alloy design via additive manufacturing: Advantages, challenges, applications and perspectives, Mater. Today. 52 (2020) 207-224. https://doi.org/10.1016/j.mattod.2021.11.026

[12] M. Kuntoğlu, E. Salur, M.K. Gupta, S. Waqar, S., Szczotkarz, G. Vashishtha, G.M. Krolczyk, A review on microstructure, mechanical behavior and post processing of additively manufactured Ni-based superalloys, Rapid Prototyping Journal. 9 (2024) 1890-1910. https://doi.org/10.1108/RPJ-10-2023-0380

[13] Y.T. Tang, C. Panwisawas, J.N. Ghoussoub, Y. Gong, J.W. Clark, A.A. Németh, R.C. Reed, Alloys-by-design: Application to new superalloys for additive manufacturing, Acta Mater. 202 (2021) 417-436. https://doi.org/10.1016/j.actamat.2020.09.023

[14] H.A. Kishawy, A. Hosseini, H.A. Kishawy, A. Hosseini, Superalloys. Machining Difficult-to-cut Materials: Basic Principles and Challenges, Springer, 2019, 97-137. https://doi.org/10.1007/978-3-319-95966-5_4

[15] E. Yasa, O. Poyraz, Powder bed fusion additive manufacturing of Ni-based superalloys: applications, characteristics, and limitations, in: Additive Manufacturing Applications for Metals and Composites, IGI Global. 2020, pp. 249-270 https://doi.org/10.4018/978-1-7998-4054-1.ch013

[16] C.T. Sims, A contemporary view of nickel-base superalloys, JOM. 18 (1966) 1119-1130. https://doi.org/10.1007/BF03378505

[17] M. Perrut, P. Caron, M. Thomas, A. Couret, A. High temperature materials for aerospace applications: Ni-based superalloys and γ-TiAl alloys, C. R. Phys.19 (2018) 657-671. https://doi.org/10.1016/j.crhy.2018.10.002

[18] H. Daiy, Y. Najafi, Z.D. Ragheb, H.R. Abedi, A review study on thermal stability of psowder-based additively manufactured alloys, J. Alloys Compds. (2023) 171384. https://doi.org/10.1016/j.jallcom.2023.171384

[19] D. Li, P.K. Liaw, L. Xie, Y. Zhang, W. Wang, Advanced high-entropy alloys breaking the property limits of current materials, J. Mater. Sci. Technol. (2024) 186. https://doi.org/10.1016/j.jmst.2023.12.006

[20] S.U.N. Yuan, Q.I.N. Xindong, W.A.N.G. Shiyang, H.O.U. Xingyu, Z.H.A.N.G. Hongyu, X.I.E. Jun, Y.U. JinJiang, Research status and future perspectives on superalloy fusion welding, Chin. J. Eng. 46 (2024), 1065-1076.

[21] D. Li, P.K. Liaw, L. Xie, Y. Zhang, W. Wang, Advanced high-entropy alloys breaking the property limits of current materials, J. Mater. Sci. Technol. (2024) 186 https://doi.org/10.1016/j.jmst.2023.12.006

[22] T. Sonar, M. Ivanov, E. Trofimov, A. Tingaev, I. Suleymanova, An overview of microstructure, mechanical properties and processing of high entropy alloys and its future perspectives in aeroengine applications, Mater. Sci. Energy Technol. 7 (2024) 35-60. https://doi.org/10.1016/j.mset.2023.07.004

[23] H. Hamdi, H.R. Abedi, Thermal stability of Ni-based superalloys fabricated through additive manufacturing: A review, J. Mater. Res. Technolo. 30 (2024) 4424-4476. https://doi.org/10.1016/j.jmrt.2024.04.161

[24] M.P. AS, V.S. Kumar, A review: fabrication techniques of the hastelloy (super alloy) composites and it impacts on the properties, Mater. Today Proc. (2024)

[25] T.M. Pollock, et al. Additive manufacturing of nickel-based superalloys: challenges and opportunities, J. Mater. Sci. 51 (2016) 12-34.

[26] O.N. Senkov, et al. Refractoryhigh-entropy alloys for hypersonic applications, J. Mater. Sci. 53 (2018) 1445-1460.

[27] W. Xu, et al. Additive manufacturing of superalloys: Microstructure evolution and property enhancement, Mater. Sci. Eng. A. 721 (2018) 126-136.

[28] S.J. Zinkle, L.L. Snead, Designing radiation resistance in materials for fusion energy, Annu. Rev. Mater. Res. 44 (2014) 241-267. https://doi.org/10.1146/annurev-matsci-070813-113627

Superalloys: Fundamentals and Applications
Materials Research Foundations 178 (2025)143-199

Materials Research Forum LLC
https://doi.org/10.21741/9781644903698-9

Chapter 9

Superalloys for Nuclear Plants

Fadhli Muhammad[1], Dinni Nurhayani[1]*, Tria Laksana Achmad[1], Anisa Fitri[2],
Siti Khodijah Chaerun[1]*

[1]Department of Metallurgical Engineering, Faculty of Mining and Petroleum Engineering,
Institut Teknologi Bandung, Indonesia

[2]Department of Materials Engineering, Institut Teknologi Sumatera, Indonesia

*skchaerun@itb.ac.id; skchaerun@gmail.com; dinni.nurhayani@gmail.com

Abstract

This review examines the critical role of superalloys in nuclear power plants, focusing on their unique properties, manufacturing processes, and applications. Nickel-based, cobalt-based, and iron-based superalloys are discussed, highlighting their high-temperature strength, corrosion resistance, and radiation tolerance. The chapter explores the microstructure and strengthening mechanisms of these alloys, as well as their manufacturing techniques, including casting, powder metallurgy, and advanced surface treatments. Applications in reactor components, turbines, and heat exchangers are detailed, along with case studies demonstrating their performance. The review also addresses current challenges, technological advancements, and future prospects, including the potential of superalloys in next-generation nuclear reactors and emerging sustainable production methods.

Keywords

Superalloys, Nuclear Power Plants, Manufacturing Processes, High-Temperature Strength, Corrosion Resistance, Next-Generation Reactors

Contents

1. Introduction

The construction and operation of nuclear power plants must prioritize public safety and environmental protection. The role of materials science is critical in reactor design, manufacturing, and operation, as selecting materials that can withstand the demanding conditions within a reactor is essential to prevent degradation and ensure safe, efficient system performance. Nuclear reactors, particularly water-cooled ones, face unique challenges due to the presence of high-temperature water, mechanical stress, vibration, and intense neutron radiation. The materials used in these reactors must be resilient enough to endure such conditions without compromising performance or causing system failures. The next-generation Gen IV reactors introduce even greater material challenges due to their advanced designs and operational conditions.

The materials used in nuclear power plants are very diverse. Each component in this plant also works in different working environments and loadings. These conditions make the study of material degradation in nuclear power plants complex. There are more than 25 kinds of alloys, including low alloy steels (LAS), nickel superalloys, and stainless steel, which are used in the primary and secondary systems of pressurized water reactors (PWR) and require welding. Similar complexities arise in boiling water reactors (BWRs) and pressurized heavy water reactors (PHWRs). In addition to the reactor materials, a variety of other materials are employed in the construction of concrete containment vessels, instrumentation, cabling, buried piping, and other supporting systems. Forecasting the performance and longevity of materials used in reactors is challenging due to varied service conditions, where loads and material degradation over time show significant variability. The degradation of components, structures, and systems in reactors is complex and occurs through multiple pathways.

1.1 Overview of superalloys and their significance in nuclear applications

Superalloys are alloys designed to maintain their performance under various loads and stresses at extreme working temperature environments, such as in reactors. This makes superalloys an important material in the manufacture of nuclear reactor components. Choosing materials that are both durable and reliable is vital to ensuring the safe and efficient operation of nuclear power plants. Superalloys generally use nickel, cobalt, or iron as the base metal, and then add alloying elements to obtain the desired properties in the superalloy, including refractory elements such as molybdenum, tungsten, and chromium. One important strengthening mechanism at the microstructure level in superalloys is the presence of precipitates, which are dispersed in the matrix and increase the durability and strength of the superalloy so that the material can maintain its structural integrity in extreme service conditions. Superalloys are also usually alloyed with alloying elements that are able to produce a protective oxide layer, such as aluminum. These oxide layers will actively protect the materials from wear and oxidation and make the superalloys suitable for nuclear reactor working environments. Superalloys are utilized in numerous components of nuclear power plants. In gas-cooled reactors, they are employed in turbine blades and vanes that are subjected to elevated temperatures and stress during electricity production. In liquid metal-cooled reactors, superalloys are utilized in components that interact with the coolant to prevent corrosion. Additionally, they are utilized in nuclear fuel assemblies to guarantee the structural integrity and safety of the reactor core.

The deployment of nickel-based superalloys in sophisticated reactor designs is contingent upon their capacity to withstand high temperatures. Nevertheless, the utilization of nickel-based alloys in PWRs and BWRs has been constrained due to their relatively high neutron absorption cross-section, which diminishes power generation efficiency and renders components radioactive. Additionally, the cost of nickel as an alloying element represents a substantial financial burden. Various components in the manufacture of water reactors, such as attachment welds, bolts, and nozzles, use nickel-rich superalloys, like Alloy 600 and Alloy 690. Although these materials can work under extreme conditions, such as in nuclear reactors, material failures influenced by the environment can occur, such as fatigue, stress corrosion cracking, and fracture [1]. A detailed overview of these alloys, their chemical composition, and their applications in water reactors is provided in Table 1.1. Additional components manufactured from nickel-based alloys are listed in Table 1.2.

Superalloys: Fundamentals and Applications
Materials Research Foundations 178 (2025)143-199

Materials Research Forum LLC
https://doi.org/10.21741/9781644903698-9

Table 1.1 Several Ni-based superalloys utilized in nuclear reactor applications [1]

Alloy	Composition (%)							Usage
	Ni	Cr	Al	Fe	Ti	Nb	Mo	
Inconel 600	75	16	0.2	8	0.3	-	-	BWR, PWR, PHWR
Inconel 625	61	22	0.3	<5	0.3	3.5	9	BWR
Inconel 690	61	29	0.5	9	0.5	-	-	PWR, PHWR
Inconel 718	53	18	0.6	19	0.9	5	2.5	BWR, PWR, PHWR
Inconel 750	72	15.5	0.7	7	2.5	1	-	BWR, PWR, PHWR
Incoloy 800	33	21	0.4	>39.5	0.4	-	-	BWR, PWR, PHWR

Table 1.2 Some components fabricated from Ni-based superalloy [1]

Reactor Components	Nickel-based Superalloy Designation
Pressure vessel attachment pads	Alloy 182
Control rod penetrations	Alloy 600
Pressure vessel nozzles	Alloy 182 and 82
Safe ends	Alloy 600
Fuel rod spacers	Alloy X-750
Jet pump beams	Alloy X-750

The nuclear industry faces ongoing challenges with environmental degradation, particularly when materials succumb to stress-induced corrosive damage. To combat these issues, metallurgical engineers and scientific investigators have pioneered enhanced material compositions, focusing specifically on elevating the chromium concentrations in newer alloy formulations. A notable advancement in this field is exemplified by the widespread adoption of Alloy 690 in modern pressurized water reactor systems (PWRs), which has largely superseded its predecessor, Alloy 600, due to its superior resistance to environmental deterioration. Researchers are studying nickel-based superalloys for liquid-sodium-cooled fast-breeder reactors since nickel's capture cross-section for fast neutrons is low compared to other materials.

1.2   Brief history of superalloy development for nuclear plants

Gas turbine engines for jet propulsion during World War II influenced superalloy development. These alloys were initially intended to endure the jet engine temperatures but quickly adopted for various high-temperature applications, including nuclear power plants. With the emergence of nuclear power in the mid-20th century, the industry faced new material challenges. Reactor core temperatures significantly surpassed those in traditional power plants, leading to frequent component failures when using conventional materials. The introduction of nickel-based superalloys, such as Inconel and Hastelloy, marked a significant advancement. These alloys, known for their superior high-temperature strength, creep resistance, corrosion resistance, and resilience to radiation-induced damage, enabled the production of more durable and reliable components for nuclear reactors. Intensive research on irradiation effects during the 1960s and 1970s resulted in the development of alloys like Incoloy 800 and Inconel 625, which were optimized for nuclear environments. The discovery of precipitation-hardening mechanisms further improved performance, leading to the creation of superalloys like Nimonic and René 41.

As the nuclear industry progressed, the demand for more advanced superalloys grew. The introduction of Advanced Gas-Cooled Reactors (AGRs) and fast breeder reactors required materials capable of withstanding even harsher conditions. In response, alloys like Hastelloy N were developed to meet these demands. Concurrent advancements in alloy composition, manufacturing techniques, and microstructural control improved oxidation resistance, corrosion resistance, and overall damage tolerance. In recent years, the focus has shifted towards developing superalloys for next-generation nuclear reactors, such as molten salt and gas-cooled fast reactors, which require materials that can perform under higher temperatures and more corrosive conditions. Advanced manufacturing technologies, including additive manufacturing, are being explored to produce complex components with tailored properties. The progression from early jet engine alloys to today's advanced superalloys reflects continuous innovation in materials science. Sustaining nuclear power facility reliability and protection protocols hinges upon the persistent dedication to improving material resilience and performance characteristics.

2. Properties of superalloys

Superalloys are known for their exceptional strength, corrosion and oxidation resistance, and their ability to endure elevated temperatures. These characteristics are crucial for nuclear power plants, where components must perform reliably under extreme conditions. Nickel-based superalloys, in particular, are commonly utilized in nuclear reactors because of their excellent creep resistance and capability to operate at temperatures exceeding 1200°C.

2.1 Physical properties

The physical attributes of superalloys depend upon the composition of the alloy and the corresponding compounds generated from this combination. Nickel-based alloys are primarily strengthened through two mechanisms: solid solution strengthening and precipitation strengthening. Solid solution-strengthened alloys gain their strength from substitutional solid solution hardening, whereas precipitation-strengthened alloys are reinforced by the addition of elements like titanium and aluminum, which form small particles of Ni_3Ti and Ni_3Al. As an example, Table 1.3 provides an overview of the physical properties of selected intermetallic compounds present in different superalloys.

Table 1.3. Physical and mechanical properties of several intermetallic constituents of superalloy [2]

Compounds	Melting Point (°C)	Density (g/cm³)	Room Temperature Yield Strength (GPa)	Max. Operating Temperature (°C) Creep Resistance	Corrosion Resistance
Ni_3Al	1390	7.5	179	1000	1150
NiAl	1640	5.86	294	1200	1400
Fe_3Al	1540	6.72	141	600	1100
FeAl	1250	5.56	261	800	1200
Ti_3Al	1600	4.2	145	760	650
TiAl	1460	3.91	176	1000	900

Nickel-based superalloys exhibit melting points typically ranging from 1300 to 1450°C. Their densities vary from 8.0 to 9.0 g/cm³, offering favorable mass efficiency for rotating components. However, their thermal conductivity is relatively low (15-30 W/mK), while electrical conductivity is moderate, lower than pure nickel. The characteristics of nickel-based superalloys confer excellent thermal stress resistance and a relatively small coefficient of thermal expansion during operation.

Compared to cobalt-based and iron-based superalloys, nickel-based superalloys have advantages in physical and mechanical properties, especially for applications in extreme conditions, such as high temperatures in reactors. Cobalt-based superalloys boast even higher melting points, often exceeding 1450°C. Their densities are generally similar to nickel-based alloys, but both thermal and electrical conductivity are typically lower. Iron-based superalloys, on the other hand, have the lowest melting points among the three categories, typically below 1400°C. Their densities are also lower (7.5 to 8.0 g/cm³) compared to nickel- and cobalt-based alloys. Iron-based superalloys also have greater thermal diffusion capacity, electrical conductivity, and thermal expansion coefficients.

2.2 Mechanical properties: High-temperature strength and creep resistance

Scientists and engineers design superalloys as materials that can maintain high strength under elevated temperatures, which can be measured through various parameters such as yield strength, creep strength, rupture strength, or fatigue resistance. While superalloys are commonly employed in aerospace and chemical industries, the operational conditions between the nuclear and aerospace sectors differ significantly. The nuclear industry often utilizes coolants such as water, helium, molten salts, or liquid metals, whereas the aerospace industry functions in air or space environments. Regarding aerospace applications, elevated strength and resistance to creep deformation are critical for nickel-based alloys. In contrast, nuclear environments prioritize resistance to corrosion, stress corrosion cracking (SCC), and irradiation-assisted SCC (IASCC), given the aqueous or moist conditions [3].

Nickel-based superalloys are designed to have certain physical and mechanical properties that meet their service conditions, including in nuclear reactors, through alloying and processes to obtain the desired properties, including strength. There are three main strengthening mechanisms in nickel-based superalloys: (i) solid solution strengthening with the addition of alloying elements like Mo, Cr, Ti, Fe, Nb, and Al; (ii) coherent precipitation strengthening, where γ' and γ" phases are formed; and (iii) incoherent precipitation strengthening, which involves the formation of larger, non-integrated particles, such as carbide phases (typically chromium) [3–5].

Particles' size, quantity, and distribution of the precipitates in the microstructure impacted the characteristics of alloys that are strengthened by precipitation hardening mechanism. Alloys that are hardened through precipitation include X-750 and 718. Carbides frequently develop and accumulate along grain boundaries during high-temperature treatments [4,6]. Conversely, the formation of precipitate particles as gamma prime (γ') and gamma double prime (γ") occurs inside the grains and along the grain boundaries. Intragranular precipitates significantly contribute to the enhancement of the alloy's strength, and their size and density can be adjusted via aging treatments. Overaging, however, can degrade the ordered precipitate structure, leading to the formation of larger η- and δ-phases. The γ'-phase is especially critical for maintaining high yield and creep strength in nickel-based superalloys, even at elevated temperatures. In some cases, yield strength (YS) may even increase as the temperature rises [6,7].

The increase in YS at elevated temperatures in certain superalloys is attributed to dislocation interactions within the ordered γ'-phase. Dislocations can split γ' precipitates into pairs, creating antiphase boundaries (APBs). The presence of anti-phase boundaries (APBs) increases the critical resolved shear stress required for deformation. However, when a subsequent dislocation with the same Burgers vector and slip plane passes through the material, the APB is eliminated, lowering the energy needed for further plastic flow. The degree of coupling between the paired superdislocations - whether strong or weak - is dependent on the size and distribution of the γ' strengthening precipitates within the material microstructure. At high temperatures, the probability of dislocations undergoing cross-slip onto {010} planes increases, resulting in the development of Kear-Wilsdorf locks. This mechanism is contingent upon the presence of γ' precipitates, which are cut by dislocations. The optimal hardening occurs at a specific particle size where the coupling between the paired superdislocations is between weak and strong.

Other theoretical framework proposes the occurrence of single dislocations, where the tension magnitude in the dislocation line will drive the propensity to either cut through or bypass a precipitate. Smaller precipitates are more likely to be cut, while larger ones are typically bypassed through the Orowan looping mechanism. The shift from cutting to looping is influenced by the size of the precipitate and the energy required to penetrate larger ones. For a given volume fraction, larger precipitates with greater spacing tend to favor Orowan looping over cutting. When Orowan looping prevails, the degree of material strengthening is primarily controlled by the precipitate size and spacing. The Orowan mechanism plays a significant role in engineering alloys like Inconel X-750, as indicated by the presence of dislocation tangles near γ' precipitates. This is likely due to the alloy's low γ' volume fraction, which makes cutting less likely. Additionally, a lower γ' volume fraction prevents the occurrence of unusual temperature-dependent behaviors.

Solution strengthening, crystallite size reduction, and the controlled formation of dispersed particles are mechanisms that affect the mechanical performance of superalloys, especially the nickel-based superalloys. In order to obtain the desired properties and match the designation of the material, the material is generally subjected to thermomechanical treatment, which will modify the grain size, as well as the formation of γ and γ' phases in solution enhancement and precipitation hardening. Adjusting the variations and amounts of alloying elements like titanium, aluminum, niobium, and tantalum can significantly impact strength, hardness, creep resistance, and other mechanical and physical characteristics of the superalloy. Alloys rich in these elements tend to facilitate strengthening through the controlled formation of dispersed particles within the metal matrix. Conversely, alloys with lower concentrations of such elements rely solely on solution strengthening.

2.3   Chemical properties: corrosion resistance in nuclear environments

A critical component used in the cooling system of Generation IV nuclear power plants is designed to endure high temperatures. The cooling system in nuclear reactors may use either gas or molten salts as the cooling medium. Helium is the most commonly used gas due to its inert properties, while molten salts, typically a mixture of LiF, NaF, and KF (FLiNaK), are frequently

Superalloys: Fundamentals and Applications   Materials Research Forum LLC
Materials Research Foundations 178 (2025)143-199   https://doi.org/10.21741/9781644903698-9

employed as well. The interaction between the superalloy material and the environment under both cooling conditions has been extensively studied.

In a helium gas environment, various impurity gases, such as hydrogen (H_2), methane (CH_4), carbon monoxide (CO), carbon dioxide (CO_2), and water vapor (H_2O), can potentially react with the material. One potential oxidation reaction, involving CO_2 gas, is represented by equation 1 [8].

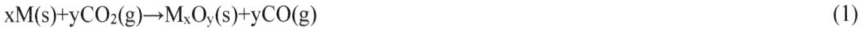

$$xM(s)+yCO_2(g)\rightarrow M_xO_y(s)+yCO(g) \tag{1}$$

In addition to oxidation reactions, decarburization reactions may also occur when carbon is present in the superalloy used, according to the reaction represented by equation 2 [8].

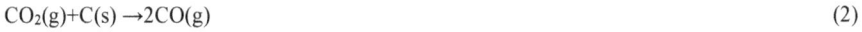

$$CO_2(g)+C(s)\rightarrow 2CO(g) \tag{2}$$

The occurrence of both reactions is contingent upon the partial pressure of CO_2 and CO gases. The greater the partial pressure of CO_2 in comparison to CO, the more readily oxidation and decarburization reactions will occur, thereby accelerating the rate of material deterioration. In superalloys with chromium content exceeding 18%, the dominant oxide that forms is Cr_2O_3. This has been noted in iron-based superalloys like Incoloy 800H and Alloy 800H, as well as in nickel-based superalloys such as Hastelloy X and Alloy 617. The compositions of these alloys are detailed in Tables 1.4 and 1.5.

Table 1.4. The composition of iron-based superalloy [8]

Alloy	C	Cr	Fe	Ni	Mn	Al	Si	Cu	Ti
Incoloy 800H	0.084	21.28	Base	30.96	0.97	0.249	0.320	0.120	0.251

Table 1.5. The composition of iron- and nickel-based superalloy [9]

Alloy	Ni	Fe	Cr	Al	Mo	Mn	Ti	C	Co	W	Cu
Alloy 800 H	31.7	Bal	21.4	0.4	-	1.5	0.4	0.05	-	-	0.75
Hastelloy X	Bal	22	18	-	9	1	-	0.1	1.5	0.6	-
Alloy 617	Bal	0.9	21.6	1.2	9.5	0.1	0.3	0.06	12.5	-	0.5

The Cr_2O_3 oxide layer has protective properties against further oxidation of the alloy. However, Cr_2O_3 can degrade when it reacts with water vapor, as illustrated by the reaction in equation 3 [9].

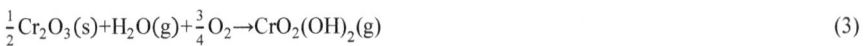

$$\frac{1}{2}Cr_2O_3(s)+H_2O(g)+\frac{3}{4}O_2\rightarrow CrO_2(OH)_2(g) \tag{3}$$

The formation of $CrO_2(OH)_2$ is a volatile process that leads to the reduction of Cr_2O_3 in the oxide layer, causing it to convert into a gaseous form. Alongside the degradation of the oxide layer, internal oxidation has been observed in alloys like Incoloy 800H, Alloy 800H, and Alloy 617. This phenomenon occurs due to the greater stability of aluminum oxide compared to Cr_2O_3. As oxygen is transferred between Cr_2O_3 and aluminum, the aluminum in the alloy undergoes oxidation.

2.4 Radiation resistance: effects of neutron irradiation on alloy properties

In various types of nuclear reactors, including light water reactors (LWRs) and fast reactors, nickel alloys have been utilized successfully without encountering problems related to radiation exposure. However, when these alloys are employed in the unique CANDU reactors, which are proprietary to Atomic Energy of Canada Limited, distinct challenges arise. The operational characteristics of CANDU reactors result in a neutron spectrum that exerts a considerable influence on nickel. This phenomenon gives rise to several issues, including:

- Enhanced Damage Rates: The most common nickel isotope (^{58}Ni) undergoes a readily observable transformation to ^{59}Ni under the influence of neutron bombardment. This new isotope undergoes further nuclear reactions with neutrons, resulting in heightened atomic damage and the generation of helium and hydrogen gas within the alloy.

- The accumulation of helium within the reactor is a further consequence of the aforementioned processes. In contrast to other elements that produce helium (such as boron), nickel is capable of continuous generation due to the ready availability of ^{58}Ni. The buildup of helium can reach significant levels (over 20,000 ppm) in CANDU reactors after extended operation, which can impact the material's integrity and creep performance.

Therefore, while nickel alloys may be suitable for LWRs and fast reactors, their limitations under CANDU reactor conditions due to the unique neutron spectrum and ^{58}Ni transformation must be carefully considered.

The neutron spectrum in CANDU reactors represents a significant concern for nickel alloy components, given its capacity to influence stress relaxation, swelling, and embrittlement. This effect is particularly pronounced in comparison to LWR reactors, due to:

- Higher thermal neutron flux: CANDU reactors have a higher thermal neutron flux, which increases the production of ^{59}Ni from ^{58}Ni.

- Longer residence time: Nickel alloy components reside in CANDU cores for longer periods, leading to higher exposure to damaging radiation.

The combination of these factors results in significant helium production and displacement damage, even in peripheral regions of CANDU reactors. This can lead to issues such as weld cracking and loss of component integrity. In LWRs, the lower thermal neutron flux and shorter residence times of nickel alloy components make helium production less of a concern. However, even small amounts of helium can impact weldability, necessitating careful consideration during maintenance.

Superalloys: Fundamentals and Applications   Materials Research Forum LLC
Materials Research Foundations 178 (2025)143-199   https://doi.org/10.21741/9781644903698-9

3.   Composition and Classification

Superalloys, metallic alloys designed for high-temperature performance, can operate at temperatures exceeding 70% of their absolute melting points. These materials are often referred to as high-performance alloys because of their remarkable structural integrity, superior corrosion durability, excellent surface stability, and greater resistance to thermal creep deformation compared to other alloys [10,11]. Known for their outstanding physical and mechanical properties and durability under harsh conditions, superalloys are critical in industries such as aerospace, automotive, manufacturing, and particularly in nuclear energy. Engineers rely on superalloys for their exceptional performance and durability in demanding environments. Numerous manufacturers have developed a variety of superalloys, each engineered with specific chemical compositions to meet particular performance requirements for their intended applications. The alloys are made in cast or wrought forms. Current superalloys typically contain more than ten elements but are mostly composed of the three base metals: nickel, cobalt, and iron.

There are hundreds of superalloy grades available to date. Thus, the alloys are classified based on three aspects to have a better understanding of the superalloys. The three ways to classify the superalloys are [12]:

- Classify by trademark: MONEL, INCONEL, INCOLOY, HASTELLOY
- Classify by matrix: nickel-based, iron-based, cobalt-based
- Classify by strengthening mechanism: Solid solution strengthened, precipitation strengthened, dispersion strengthened.

Below are explanations of superalloy classifications based on the matrix: nickel-based, iron-based, and cobalt-based superalloys [13].

3.1   Nickel-based superalloys

The first superalloys to be discussed are nickel-based superalloys, celebrated for their exceptional strength, corrosion resistance, and thermal stability, especially under high-temperature conditions. They are commonly used in aerospace, gas turbines, and chemical processing equipment where high strength is required, typically within the temperature range of 1024 to 1371 °C [14]. These superalloys usually contain a high nickel content, ranging from 50% to 70%, and include critical solutes like aluminum and titanium, along with chromium, yttrium, boron, and zirconium as additional alloying elements [15]. Chromium and aluminum contribute to improved oxidation resistance, while yttrium aids in the compliance of the oxide scale to the base material. Boron and zirconium are known to accumulate at grain boundaries, thereby reducing grain boundary energy and enhancing both creep strength and ductility. Although costly, rhenium markedly boosts creep resistance by facilitating rafting and slowing diffusion rates within the alloy [16]. A significant advantage of nickel-based superalloys is their ability to form a protective oxide layer, primarily composed of aluminum oxide, which provides enhanced oxidation and corrosion resistance at elevated temperatures. Nickel-based superalloys play a pivotal role in the construction and functionality of nuclear energy generators for their exceptional high-temperature capabilities, ability to resist corrosion, and resilience under severe environmental conditions. This discussion explores the composition and categories of these alloys, emphasizing their essential elements and the mechanisms that render them ideal for nuclear settings.

The exceptional properties of nickel-based superalloys arise from their unique microstructures. There are two-phase equilibrium microstructures in nickel-based superalloys, comprising γ and γ', formed by the addition of less than 10 atomic percent of aluminum and/or titanium as alloying elements [10]. The matrix phase is designated as γ (gamma) phase, a solid solution rich in nickel, incorporating elements including chromium, cobalt, tungsten, and molybdenum. The γ' (gamma prime) phase, a nickel-aluminum intermetallic complex, which can occupy up to 60-70% of the volume, acts as a precipitate inside the matrix by suppressing dislocation movement and atomic diffusion during creep deformation [16–19]. The γ phase possesses a face-centered cubic (FCC) lattice, wherein all atomic species are dispersed in a random configuration. Conversely, the γ' possesses a primitive cubic lattice, with nickel atoms located at the face centers and aluminum or titanium atoms situated at the cube corners. The γ' phase, which can constitute 60-70% of the volume, serves as a strengthening phase by inhibiting dislocation motion and atomic diffusion during creep deformation. [18,19]. The composition of the alloy and temperature will affect the quantity of γ' in the matrix and alloy microstructure, and the ternary Ni-Al-Ti phase diagram shows this two-phase field created [10].

The lattices of both γ and γ' phases are cubic, with closely matching lattice parameters, and the cell edges of γ' align exactly parallel to those of the γ phase [10]. This similarity in lattice parameters creates what is known as a coherent relationship. The coherency between the γ and γ' phases is vital for the alloys' creep resistance and strength at high temperatures, because:

- Strong γ/γ' interface: coherency between γ and γ' phases creates a strong, ordered interface that resists dislocation movement. A coherent interface hinders dislocation movement, making the material much stronger and more resistant to deformation at high temperatures [10,13].

- Order-strengthening: the γ' precipitates are ordered, meaning they have a specific repetitive pattern of atoms. The ordered structure adds to the material's strength. The coherent interface further enhances this effect by preventing the easy movement of dislocations across the γ/γ' boundary[13,15].

- Enhanced Creep Resistance: At elevated temperatures, creep deformation primarily occurs through dislocation movement. The coherent structure of γ/γ' acts to "pin" or impede these dislocations, markedly reducing the rate of creep and preserving the structural integrity of the alloy under high temperature conditions [13].

If the γ/γ' interface lacks coherence or shows misalignment in the crystal lattice orientation, dislocations may traverse an incoherent interface more readily, resulting in accelerated deformation and early failure under high temperatures. Hence, managing the size, distribution, and coherence of the γ' throughout the manufacturing process to tailor these alloys for particular uses effectively is very important.

Elements that are commonly added to nickel-rich superalloys are sorted based on their affinity for specific microstructures. Elements such as Ni, Co, Fe, Cr, Ru, Mo, Re, and W have a propensity to enrich the γ phase and stabilize it. Conversely, elements that possess larger atomic radii, like Al, Ti, Nb, and Ta, tend to promote the development of ordered precipitate phases like gamma prime (γ'). In contrast, due to their disparate atomic sizes, elements like B, C, and Zr often concentrate along the grain boundaries of the γ phase. Other elements are known as carbides and borides forming, including Cr, Ti, Mo, Ta, W, and Nb [13].

Nickel-based superalloys are categorized into three distinct types:

- Solid Solution Strengthened Alloys: This group includes elements such as Co, Cr, Fe, Mo, W, and Ta. The tendency of certain alloying elements to dissolve homogeneously within the nickel-based superalloy matrix is largely governed by their similarities in atomic size, electronic configuration, and crystallographic structure to that of nickel itself. The capacity of a solute element to enhance the mechanical strength of the solid solution is often directly correlated with the degree of atomic radius mismatch relative to the host nickel atoms. Additions of elements like aluminum, titanium, manganese, niobium, molybdenum, and tungsten, which exhibit the ideal balance of solubility and atomic size differential, are particularly effective at strengthening the alloy through this solution hardening mechanism [20].

- Precipitation/age hardened alloys: aluminum, tantalum, titanium, and niobium are integral in forming γ' and γ" precipitates within the γ matrix [13,21]. Precipitation-hardened alloys are often utilized in high-temperature applications due to their superior strength compared to solid solution strengthened alloys [21].

- Oxide dispersion strengthened (ODS) alloys: characterized by superior radiation resistance and high-temperature mechanical qualities, ODS alloys are a promising candidate material for nuclear reactors and aerospace engines [22].

The unique blend of elevated temperature strength, oxidation resistance, and creep resistance possessed by nickel-rich advanced alloys makes them indispensable for numerous applications, particularly within nuclear energy facilities. In reactor components, these alloys are employed in critical areas such as the drive pipe seats of pressure vessels, the heat transfer systems of steam boilers, cooling systems, and also the reactor vessel [23,24]. Their capability to endure high temperatures and corrosive environments is crucial for maintaining the integrity and prolonged functionality of these vital components. When designated for steam generators and heat exchangers, nickel-based superalloys are chosen because of their exceptional performance in maintaining their properties in an elevated temperature environment and corrosion resistance. This is particularly important in settings where fuel combustion generates heat to operate steam turbines [15]. The nickel-based superalloys are also applied as waste disposal components. Some containers for disposing of radioactive waste are made from nickel-containing stainless steel and high-nickel alloys, ensuring safe containment over long periods [24].

In summary, the specific composition and categorization of nickel-based superalloys are designed to fulfill the demanding needs of nuclear plant operations. The deliberate incorporation of alloying elements like chromium, tungsten, and molybdenum, along with controlled heat treatment techniques, ensures these alloys preserve their structural integrity and functionality in harsh conditions. This renders them indispensable in maintaining the safety and efficiency of nuclear power plants.

3.2 Cobalt-based superalloys

Although less prevalent than nickel-based varieties, cobalt-based superalloys remain essential for certain uses. These superalloys are regarded as high-performance materials crucial in several

Superalloys: Fundamentals and Applications   Materials Research Forum LLC
Materials Research Foundations 178 (2025)143-199   https://doi.org/10.21741/9781644903698-9

demanding environments, such as nuclear power facilities, where their superior high-temperature capabilities, corrosion resistance, and mechanical robustness are essential. Typically, these alloys include chromium, tungsten, and nickel as key alloying elements [25,26]. Cobalt-based superalloys offer outstanding resistance to oxidation at high temperatures and to thermal fatigue, making them suitable for applications such as gas turbine blades, industrial furnaces, and medical implants [25,27]. However, they generally provide less strength and creep resistance than their nickel-based counterparts. The inclusion of chromium contributes to their oxidation resistance, while tungsten and molybdenum enhance their solid solution strengthening and thermal stability [26].

In cobalt-based superalloys, the cobalt quantity ranges from 35% to 70% of the alloy, and the amount provides good high-temperature performance, welding properties, and overall structural integrity [28]. The main alloying elements in cobalt-based superalloys are nickel, chromium, tungsten, molybdenum, niobium, tantalum, and carbon [26]. Nickel is included in amounts between 5% and 25% to stabilize the austenite structure of the alloy and improve corrosion resistance [28]. Chromium in cobalt-based superalloys is a vital alloying element, typically found in greater concentrations than in other superalloys, and it improves high-temperature corrosion resistance by creating a stable oxide film that safeguards the alloy from corrosive conditions [29].

The microstructure of cobalt-based superalloys is crucial for their high-temperature properties. Here are the key aspects:

- γ-γ' microstructure: recent improvements indicate that cobalt-based superalloys can be treated to achieve a γ-γ' microstructure, equivalent to that of nickel-based superalloys. This microstructure comprises a face-centered cubic (FCC) matrix and cuboidal $L1_2$ precipitates, which markedly improve the alloy's strength and creep resistance [30].

- carbide precipitation: Cobalt-based superalloys typically have 0.25%–1.0% carbon, with hardness achieved through the precipitation of incoherent carbides after thermal treatment. The fundamental structures of cobalt-based alloys rely on the carbide morphology inside the cobalt matrix and the grain boundaries to enhance mechanical strength [26].

- solid-solution strengthening: Elements like tungsten, molybdenum, niobium, and tantalum integrate into the cobalt matrix, promoting solid-solution strengthening. This process enhances the alloy's stability and strength under high temperature conditions [28].

Cobalt-based superalloys can be classified based on their manufacturing techniques and microstructural characteristics as casting, powder metallurgy, and single-crystal alloys. Each has a different microstructure, mechanical and physical properties, and manufacturing applications.

Cobalt-based superalloys offer several advantages over other types of superalloys, and the conditions that make these alloys possible for further development are:

- Higher melting points: These alloys generally have higher melting points than nickel-based superalloys, often exceeding 1300°C, which is beneficial for high-temperature applications [28].

- Improved corrosion resistance: The higher chromium content in cobalt-based superalloys provides better high-temperature corrosion resistance, making them more durable in corrosive environments [28].

- Potential for single-crystal production: Recent studies have shown that cobalt-based superalloys can be processed into large single crystals without the formation of surface defects, which is a significant advancement over nickel-based superalloys [31].

Cobalt-based superalloys are employed in different parts of nuclear power plants because of their distinctive characteristics:

- Turbine blades and components: These alloys are used in the manufacture of turbine blades and other components in thermal and nuclear power plants, where high-temperature resistance and corrosion resistance are critical [28,32].

- Valves and pumps: Cobalt-based alloys are commonly utilized in areas susceptible to wear and galling, such as valves and pumps. Their excellent wear resistance enhances the lifespan of components and diminishes the need for frequent maintenance [33,34].

- Heat exchangers and cooling systems: The high-temperature corrosion resistance and durability of cobalt-based superalloys make them suitable for use in heat exchangers and cooling systems within nuclear reactors.

In summary, cobalt-based superalloys' composition, microstructure, and production methods are specifically designed to fulfill the demanding conditions of the nuclear power environment. With the ability of cobalt-based superalloys to maintain mechanical strength and oxidation resistance at elevated temperatures, and ongoing advancements in research, these alloys are planned to continue playing a crucial role in the functionality and efficiency of nuclear power facilities.

3.3   Iron-based superalloys

Iron-based superalloys, including those in the A-286 and Incoloy 800 series, are typically utilized in applications that demand lower temperatures compared to nickel and cobalt-based superalloys, yet they still provide robust strength and corrosion resistance [35]. These alloys are more cost-effective and offer superior weldability compared to their nickel and cobalt alternatives, while also delivering commendable oxidation resistance and mechanical strength at elevated temperatures. Thus, these iron-rich superalloys are ideal for components in nuclear plants that do not necessitate the extremely high-temperature capabilities of other superalloys. Iron-based superalloys are commonly employed in parts such as discs, shafts, and certain steam turbine components, which operate at relatively lower temperatures [34].

Iron-based superalloys are characterized by their iron content, which typically ranges between 15% and 60% of the alloy composition. The base metal, iron, provides the fundamental structure and mechanical properties of the alloy. Nickel, chromium, and molybdenum are essential alloying elements that are commonly added to iron-based superalloys [21]. Nickel and chromium are added to improve corrosion resistance and high-temperature corrosion properties, while molybdenum is added for solid-solution strengthening and enhances the alloy's mechanical properties at elevated temperature. Other alloying elements that might be added to iron-based superalloys are titanium, aluminum, cobalt, boron, and zirconium [36].

Superalloys: Fundamentals and Applications   Materials Research Forum LLC
Materials Research Foundations 178 (2025)143-199   https://doi.org/10.21741/9781644903698-9

The microstructure of iron-based superalloys distinctly differs from that of nickel-based and cobalt-based superalloys. In certain iron-based superalloys that incorporate niobium as an alloying element, such as IN718 and IN706, the γ" phase is more predominant than the γ' phase. Superalloys reinforced by the γ" phase may form an orthorhombic δ phase when overaged. This phase typically does not align well with γ, and thus fails to contribute to the strength of the alloy, even in significant quantities [4].

The microstructure of iron-based superalloys is critical for their performance:

- Austenitic structure: The addition of nickel stabilizes the austenitic structure, which is essential for maintaining high-temperature properties. The austenitic phase provides good ductility and resistance to corrosion [37].

- Precipitation strengthening: The introduction of titanium and aluminum leads to the creation of γ' and γ" phases, which bolster the alloy's strength. Nonetheless, the proportion of these precipitates in iron-based superalloys is usually smaller than in nickel-based superalloys, generally not exceeding 20% [34,37].

Iron-based superalloys can be classified based on their manufacturing techniques and microstructural characteristics:

- Conventional casting: This method is widely used but results in alloys with lower mechanical properties compared to advanced techniques.

- Powder metallurgy: This technique fabricates alloys that exhibit a finer dispersion of carbides and reduced grain size, which contribute to enhanced mechanical properties. The application of hot isostatic pressing (HIP) can improve these properties further by eliminating possible points of failure [38].

Although iron-based superalloys do not perform as well at elevated temperature conditions as other type of superalloys, their affordability and satisfactory performance make them suitable for certain roles within nuclear power plants:

- Components under low stress: Iron-based superalloys are appropriate for components like enclosures or containers that do not endure significant stress. These alloys maintain solid mechanical properties at temperatures below 900°C and exhibit durability in high-temperature settings [37].

- Steam turbines and heat exchangers: These alloys can be used in steam turbines and heat exchangers where the operating temperatures are relatively lower. Their corrosion resistance and mechanical strength make them viable alternatives to more expensive superalloys [39].

Iron-based superalloys, like other materials, have both benefits and limitations concerning cost, properties, and methods of enhancement. As a more budget-friendly option, iron-rich superalloys are commonly cheaper than their nickel and cobalt-based counterparts, making these alloys a reasonable choice for applications where extremely high-temperature properties are not essential. Mechanically, while iron-based superalloys perform well up to 750°C, they lack in strength at elevated temperature and stability seen in nickel and cobalt-rich superalloys. They offer better ductility but have reduced creep resistance and oxidation resistance at higher temperatures. Iron-based superalloys can undergo a unique grain refining process that finely disperses the precipitation strengthening phase, thereby improving their mechanical properties. However, this process has limitations, including potential instability of the γ' phase at high temperatures.

Iron-based superalloys, though less prevalent than nickel and cobalt-based options in nuclear power applications, provide an economical alternative for certain uses. Their formulation includes substantial quantities of nickel, chromium, and molybdenum, with the judicious addition of titanium and aluminum to promote precipitation strengthening. This makes them suitable for components that do not require the high-temperature capabilities of more advanced superalloys. The categorization of these alloys, based on their production processes and microstructural characteristics, underscores their potential suitability for various nuclear plant applications, especially in scenarios where affordability and moderate high-temperature durability are key considerations.

3.4 Function of Alloying Elements like Chromium, Tungsten, and Molybdenum

The efficacy of superalloys greatly relies on the alloying elements incorporated into the base metal. These elements crucially modify the mechanical, thermal, and chemical characteristics of the alloy, thereby boosting its functionality in environments with high temperatures. Often, superalloys contain more than 10 different alloying elements, predominantly transition metals from the d block in the periodic table, whose phase stability and behavior of each alloying element is affected by its atomic number, which will define its periodic table placement [13].

In the context of nuclear power applications, superalloys depend extensively on the deliberate incorporation of specific alloying elements to attain essential properties such as high-temperature endurance, corrosion resistance, and mechanical robustness. The pivotal roles played by chromium, tungsten, and molybdenum in enhancing these attributes in superalloys will be explored.

3.4.1 Chromium (Cr)

Chromium is a crucial alloying element in superalloys due to its ability to form a protective oxide layer (commonly Cr_2O_3) that enhances corrosion and oxidation resistance, as well as high-temperature strength and wear resistance. These properties are especially vital in nuclear power plants, where components face high temperatures and aggressive environments [19,40]. Chromium serves three primary functions in superalloys:

- Oxidation and corrosion resistance: Chromium is predominantly incorporated to boost oxidation and corrosion resistance. It creates a stable Cr_2O_3 oxide layer that serves as a barrier against corrosive conditions, essential in nuclear settings where components are subjected to harsh environments [13,41].

- Precipitation strengthening: In cobalt-based superalloys, chromium contributes to the formation of carbides such as M_3C_2, M_7C_3, and $M_{23}C_6$, aiding in precipitation strengthening. These carbides typically localize at grain boundaries, improving the alloy's creep and corrosion resistance [28,41].

- Solid solution strengthening: In most superalloys, chromium will dissolve in the γ solid solution, inducing lattice distortion and creating an elastic stress field that enhances the strength of the solid solution. This effect is particularly beneficial for increasing high-temperature strength and stability [42].

Superalloys: Fundamentals and Applications   Materials Research Forum LLC
Materials Research Foundations 178 (2025)143-199   https://doi.org/10.21741/9781644903698-9

Additionally, in nickel-based superalloys, chromium aids in stabilizing the matrix and precipitate gamma prime phases. In cobalt and iron superalloys, it plays a similar role, ensuring the materials can endure high temperatures without substantial degradation [16,43].

3.4.2 Aluminum (Al) and Titanium (Ti)

Aluminum and titanium are essential for the development of γ' (gamma prime) precipitates in nickel-based superalloys. The precipitates fortify the alloy by obstructing dislocation migration at elevated temperatures, thus improving creep resistance [13,16]. Aluminum enhances oxidation resistance by generating a protective Al_2O_3 layer.

Aluminum and titanium are incorporated into iron-based superalloys to initiate precipitation strengthening phases. Compared to nickel-based superalloys, iron-based variants typically require a larger amount of titanium relative to aluminum for effective precipitation strengthening. If the titanium ratio exceeds twice that of aluminum, it may destabilize the γ' precipitates, potentially directing to the formation of the η phase, and will impact the strength, thermal stability, and other properties of the alloys [44,45].

3.4.3 Molybdenum (Mo) and Tungsten (W)

Alloying elements with high melting points, such as molybdenum and tungsten, in nickel-based and cobalt-based superalloys can improve strength at elevated temperatures and solid-solution strengthening [19]. Tungsten notably improves the creep resistance in nickel-based superalloys by slowing the rate of atomic diffusion within the lattice [43]. With its huge atomic radius, tungsten produces substantial expansion of the alloy's crystal, boosting its strength. Tungsten is the most important solid solution strengthening element in cobalt-based superalloys, typically present in concentrations of 7% to 15% [28]. It dissolves in the cobalt matrix, enhancing the alloy's strength and stability at high temperatures. Tungsten's high atomic mass and large atomic radius make it highly effective in hindering dislocation movement.

Molybdenum is another essential alloying element in superalloys, contributing to both solid solution strengthening and precipitation hardening [13,29]. Molybdenum improves corrosion resistance in certain working environments by passivating the superalloy surface. This is particularly beneficial in nuclear applications where the alloy must withstand both high temperatures and corrosive environments. Molybdenum also plays a role in carbide formation, which can help in reducing grain boundary sliding and improving overall creep strength [20].

Molybdenum is used in cobalt-based superalloys, particularly in older formulations, to provide solid solution strengthening [26,28]. It dissolves in the cobalt matrix, enhancing the alloy's mechanical properties at elevated temperatures. In iron-based superalloys, molybdenum is added in quantities of 1% to 6% for similar purposes [37]. Molybdenum in nickel-based superalloys creates carbides and other precipitates that enhance precipitation hardening. These precipitates, typically in the form of MC-type carbides, are located at grain boundaries and diminish the propensity for grain boundary sliding, thus improving creep strength and ductility [15,20]. Molybdenum and tungsten increase the solid solution strengthening effect in almost all types of superalloys by providing additional resistance to deformation under load [16,43].

Superalloys: Fundamentals and Applications   Materials Research Forum LLC
Materials Research Foundations 178 (2025)143-199   https://doi.org/10.21741/9781644903698-9

3.4.4 Cobalt (Co)

Thermal fatigue resistance and elevated temperature strength of superalloys that use nickel or iron as their base metal are gained through the addition of cobalt. Integrating cobalt as an alloying component affects the matrix, microstructure, and mechanical characteristics of superalloys [46].

a. Influence on the matrix and microstructure:

- o In cobalt-based superalloys, cobalt stabilizes the austenite structure as a matrix. This is crucial for corrosion resistance and high-temperature performance [13,46].

- o Cobalt affects the solubility of other elements in the γ matrix of nickel-based superalloys, thereby influencing the volume fraction of γ'. This increase in γ' volume allows for additional reinforcing phases, enhancing the strength of the material [13,46,47].

- o Carbide stabilization. Cobalt stabilizes MC-type carbides, enhancing strength and high-temperature deformation resistance [13].

b. Influence on mechanical properties

- o Cobalt lowers the stacking fault energy in the nickel matrix, which strengthens it and enhances its resistance to plastic deformation under load [13,46,47].

- o In manufacturing, cobalt increases creep strength and cracking under stress. Heat treatment techniques and alloy composition determine the amount of this effect [47].

3.4.5 Rare Earth Elements

Incorporating rare earth elements like yttrium and hafnium significantly boosts the oxidation resistance of superalloys [48]. These elements act as diffusion barriers that hinder the movement of metal atoms and oxygen, thereby enhancing the adhesion of the protective oxide layer [49].

Niobium, hafnium, and tantalum partition in nickel-based superalloys preferentially into the γ' phase, further enhancing its strengthening effect. Niobium, for instance, can form a γ'' phase (Ni_3Nb) in alloys like Inconel 718, which provides additional strengthening at lower temperatures [19,43]. In some modern cobalt-based superalloys, niobium and tantalum can replace tungsten, providing solid-solution strengthening and enhancing oxidant capacity [28]. Rhenium in nickel-based superalloys will enhance creep resistance by reducing diffusion rates.

3.4.6 Other Elements

Other elements like zirconium, boron, and carbon are also used to refine grain boundaries and improve overall mechanical properties in nickel-based and iron-based superalloys [16,19]. Boron and zirconium additions in iron-based superalloys help in grain boundary strengthening and reducing grain boundary energy [37]. The carbon content plays a significant role in carbide precipitation in cobalt-based superalloys, and the number is generally higher compared to other superalloys, typically ranging from 0.25% to 1% [25,28].

Nickel plays a crucial role in cobalt-based superalloys by stabilizing the austenite or gamma (γ) structure, as cobalt exhibits various crystal structures at different temperatures [28]. In iron-based superalloys, nickel is incorporated in substantial quantities, typically between 25% and 45%, to stabilize the alloy's austenitic structure while also improving corrosion resistance and maintaining the alloy's properties across various temperatures [37]. Additionally, iron is often added in amounts ranging from 9% to 20% in cobalt-based superalloys to help stabilize the austenitic structure of the alloy.

4.   Manufacturing Processes

The production of superalloys consists of multiple phases, such as melting, casting, and machining. During the melting phase, techniques like vacuum induction melting (VIM) or electro-slag remelting (ESR) are employed to create high-quality ingots. These ingots are subsequently shaped using methods like investment casting or forging to achieve the required form. The process concludes with the machining and finishing of the components to meet exact specifications [5,12].

4.1   Casting and solidification

Superalloys are extensively utilized across several industries, particularly in nuclear power plants. They are commonly employed in pure jets, turbofan engines, turboprop engines, modern gas turbines, as well as in the manufacturing of vanes, blades, and turbine discs. The superalloy described in this book is produced by the process of vacuum induction melting (VIM). Vacuum induction melting is beneficial in various applications, especially when dealing with intricate metals utilized in nuclear power plants. The subsequent benefits exert a significant impact on the exponential growth of metal production through vacuum induction melting: enhanced adaptability resulting from the use of small quantities in production; rapid program adaptation for various steel and alloy types; effortless functioning; minimal losses of alloying components due to oxidation; attainment of extremely precise compositional tolerances; accurate temperature regulation; minimal environmental harm resulting from dust emissions; elimination of unwanted trace elements with elevated vapor pressures, elimination of dissolved gases, such as hydrogen and nitrogen [50].

Vacuum induction melting is essential in the production of superalloys, which need melting in a vacuum or inert gas environment due to their high reactivity with ambient oxygen and nitrogen. This procedure is ideal for manufacturing high-purity metals in an environment devoid of oxygen. Nevertheless, the utilization of ceramic oxide crucibles may give rise to the presence of exogenous nonmetallic inclusions, which can pose a potential issue. Emerging filter technologies are currently being developed, alongside enhancements in the electron beam treatment of vacuum induction melted (VIM) ingots. There are additional important limitations concerning the structure of VIM, including issues like alloy segregation and the formation of coarse grains with inconsistent dimensions. An uneven microstructure is problematic due to its detrimental effects on the material's properties and its ability to perform under high temperatures. Vacuum arc remelting (VAR) is another technique utilized to enhance workability by using the VIM ingot as a consumable electrode. In VAR processing, the quantity of material that solidifies is considerably smaller compared to the VIM method. Additionally, the Electro Slag Remelt (ESR) process is employed to refine VIM ingots, helping to control grain size and further purify the material. However, the use of ESR material for rotating components in jet engines is still at an early stage of adoption in the United States [51].

Superalloys: Fundamentals and Applications
Materials Research Foundations 178 (2025)143-199

Materials Research Forum LLC
https://doi.org/10.2174/9781644903698-9

The manufacture of superalloys begins with mixing base metal and alloying elements into the furnace. Vacuum Induction Melting (VIM) furnaces are more commonly used in the manufacture of superalloys than Electric Arc Furnaces because the atmosphere conditions in VIM are more controlled, resulting in higher purity of alloys. In the process, oxygen is removed from the molten metal before oxygen-reactive alloying elements such as hafnium, aluminum, and titanium are added. Once the designed composition is achieved, the molten alloy will be molded into ingots or further processed according to the designation and subsequent manufacturing process.

Superalloy products are usually intended for material components in elevated temperature environments, such as turbine blades and vanes, which require intricate geometries. Therefore, one of the suitable casting methods is investment casting. In investment casting, wax will be formed into a prototype object with precise shape and size according to the desired superalloy product. The prototype wax is then coated with high-grade ceramic slurry until the entire wax surface is coated and the ceramic layer dries. Next, the wax will be melted using an autoclave until a hollow ceramic layer or ceramic thin wall remains, which will become the mold of the superalloy product to be made. This ceramic mold is then filled with molten alloy, which can be produced from remelted VIM furnace ingot, and the ceramic shell mold will be broken during the alloy cooling process. The as-cast products will then go through a finishing process (e.g., heat treatment) and some material testing.

Based on the casting technique applied in the manufacture of turbine blades, there are three kinds of grain structure, all of which describe the development of turbine blades manufacturing and casting technology of superalloys. At the beginning of the manufacture of superalloy turbine blades, the casting technique used is conventional casting, which results in an equiaxed polycrystalline grain structure, where the casting product will have many grain boundaries. Grain boundaries, especially in large quantities, will reduce creep resistance in high-temperature applications, so a directionally solidified casting technique was developed. In directionally solidified superalloys, the casting product will have a columnar grain structure, where the number of grain boundaries is still large but less than conventional casting, and the grain interface orientation is aligned to the stress axis, resulting in better creep resistance than equiaxed grain structure. Furthermore, a casting technology was developed that can produce single-crystal products, i.e., there is only one crystal in a product, so there are no grain boundaries. The result of the single-crystal casting technique has the best creep resistance. As mentioned earlier, casting products can be given heat treatment cycles to reduce segregation, regulate the size and distribution of precipitates, apply coatings, and mitigate compositional gradients caused by solidification [52].

4.2   Powder metallurgy techniques

Over the last decade, there have been significant advancements in the field of superalloy powder metallurgy (P/M). The ongoing development and optimization of the P/M superalloys are focusing on efforts to improve the workability of high-performance alloys and capitalizing on near-net-form manufacturing capabilities. Powder metallurgy products, including the ones that undergo VAR/ESR process, have a refined, equiaxed grain structure and minimized segregation. Typically, P/M superalloys demonstrate greater strength than conventional wrought superalloys, though their heat resistance does not match that of high-strength cast superalloys. An additional benefit of this processing method is the production of near-final geometric components, which leads to higher material yield compared to traditional forging techniques. Key technical challenges in near-net-shape P/M processing include producing pure powder and compacted forms and removing reactive defects and carbides at particle boundaries. Extensive hot working can effectively mitigate these problems. High-strength superalloys are particularly susceptible to defects under cyclic and impact load conditions. Currently, innovations in superalloy melt processes are concentrated on producing flawless metal. The primary objective is to develop alloys with no ceramic contaminants, whether they stem from the ceramic crucibles in vacuum induction melting processes or the P/M atomizing process [51].

Powder metallurgy in Ni-based superalloys production is developed to avoid defects in casting products due to the solidification process. High-strength nickel-based superalloys with polycrystalline structure usually use various refractory metals and γ-forming as alloying elements to improve their strength to levels that traditional processing techniques cannot effectively handle. The alloying elements in the alloy could make the alloy more brittle compared to other alloys that have lower alloying elements concentration, hence increasing its susceptibility to cracking during cooling. Under these conditions, production using powder metallurgy techniques is usually chosen.

The first step in powder metallurgy is to turn the VIM ingot alloy into powder through physical or mechanical methods. The physics process is done by atomization of re-melting VIM ingot using high-pressure water or inert gas, and this process is the most common. The process of making the alloy into powder mechanically can be done using a milling machine, such as ball mill or hammer mill. When the molten alloy becomes powder, the solidification process takes place quickly so that macrosegregation can be avoided. Defects in powder metallurgy products will depend on the size of initial defects, so the particle size of the powdered alloy also needs to be considered. Powdered alloys are usually sorted into categories larger than 100μm and 50μm using standard 150 or 270 meshes. The powder size directly influences the initial size of potential cracks in the final product. Although smaller particle sizes are preferred to reduce initial defect sizes, the cost significantly increases due to reduced yields.

The next step after making powdered alloy is consolidation. There are many consolidation methods that can be applied, including mechanical pressing, injection molding, isostatic pressing, and hot extrusion. In mechanical pressing, powder is fed into the die and then pressed uniaxially into a green body, to further proceed to the sintering stage. Mechanical pressing is the most widely used method in the production of powder metallurgy products. In injection molding, the powdered alloy is mixed with binder, usually made of wax or thermoplastic, then granulated. The alloy-binder mix is heated and then injected at elevated temperature into the mold to produce the desired part. Debinder is then done after the part hardens, followed by the sintering process. Isostatic pressing can be performed at room temperature (cold isostatic pressing) and high temperature (hot isostatic pressing). In hot isostatic pressing, pressure is applied in all directions up to 310 MPa while heating it to just below the γ' solvus point for 4 to 5 hours. This pressure compacts the powders into a fully dense superalloy billet, and can be further processed. Powdered alloy that will be fed in hot extrusion method needs to be placed in a metal can, followed by heating, and then extruded through various dies that reduce in diameter. Extrusion temperatures are cautiously regulated to remain beneath the γ' solvus temperature so that plastic deformation and adiabatic heating can occur [52].

Superalloys: Fundamentals and Applications   Materials Research Forum LLC
Materials Research Foundations 178 (2025)143-199   https://doi.org/10.21741/9781644903698-9

4.3   Forging and machining

The transformation of superalloy ingots into functional components typically involves elevated-temperature working processes, specifically forging and cogging. For Ni-based materials, these operations are conducted at temperatures approaching 1000°C. These techniques aim to enhance microstructural characteristics while achieving desired component geometries and ensuring uniform material properties. Initial homogenized materials typically display extensive grain dimensions exceeding 10mm with columnar morphology. The processing objective is to develop uniform, equiaxed structures with dimensions ranging from 5 to 50 μm (corresponding to ASTM 6-12), based on specific application needs. This structural refinement occurs through recrystallization mechanisms, both during processing and in subsequent stages. Success depends heavily on careful regulation of processing variables, including deformation rates, temperature conditions, and applied forces. For smaller components like sheets or billets up to 13 cm, direct processing methods suffice. However, larger turbine components require the preliminary processing of substantial ingots (36–60 cm in diameter) through open-die forging to produce intermediate-sized billets (15–40 cm), thereby ensuring complete structural transformation.

Powder-based processing offers distinct benefits in microstructural control. Rapid solidification during powder formation minimizes compositional variations, eliminating the need for homogenization treatments. The resulting consolidated materials feature refined, uniform structures suitable for direct component fabrication. Traditional forging approaches prove less effective for powder-processed nickel alloys due to their restricted plasticity, attributed to high refractory element content. These materials require specialized isothermal forging techniques, where tooling temperatures match workpiece temperatures. Processing occurs just below the γ' dissolution point, maintaining fine structural characteristics while enabling superplastic deformation under controlled conditions [52].

While nickel-based superalloys offer exceptional performance characteristics, including strength, corrosion resistance, and thermal stability, these same properties create significant manufacturing challenges. Despite technological advancements in processing equipment and quality control, fabrication remains complex due to high processing stresses and strict quality requirements. The single-crystal variant CMSX-4 demonstrates particular advantages in high-temperature applications, owing to its enhanced mechanical properties and grain-boundary-free structure. Electrochemical processing methods have emerged as effective fabrication techniques, offering stress-free surfaces, superior finish quality, and the capability for complex geometries [53].

4.4   Surface treatments and coatings for enhanced performance

Inconel 718, a nickel-based superalloy, is extensively used in gas turbine components for its outstanding mechanical properties, excellent ability to resist oxidation, and high corrosion durability. To counteract damage in harsh operating conditions, various surface treatments are applied, such as shot peening, cavitation peening, deep cold rolling, and laser shock peening (LSP). These methods aim to increase the longevity of essential parts [54].

Gas turbine parts often undergo shot peening to boost their fatigue resistance, thus extending their service life. The process involves propelling high-speed particles, known as shot, at the surface, inducing surface work hardening and compressive residual stresses that help prevent the initiation and growth of cracks. Cylindrical Inconel 718 specimens subjected to high-cycle fatigue (HCF) testing have shown that compressive residual stresses and surface hardening can lead to crack initiation beneath the surface, potentially increasing fatigue life by up to twenty times under specific conditions. However, the dimpling effect caused by shot peening may act as stress concentrators, diminishing the effectiveness of this treatment. Research by Bagherifard and Guagliano has shown a direct relationship between the intensity of shot peening and surface roughness on low-alloy steel, noting that in low-cycle fatigue (LCF) scenarios, the fatigue performance of shot-peened specimens, especially those peened at an intensity of 8 A, was reduced compared to untreated samples [55].

Laser shock peening (LSP), a newer surface treatment method, significantly improves the fatigue resistance of metals by generating a deeper layer of residual compressive stress and causing ultrahigh strain rate plastic deformation [56]. Studies by Correa et al. found that laser-peened Al2024-T351 samples exhibited a fatigue life approximately 3.5 times longer than those without peening [57].

This chapter will also explore the field of coating alongside surface treatment methods previously discussed. Traditional protective coatings like aluminide and MCrAlY are widely adopted for their superior resistance to high temperatures and oxidation. Nevertheless, these coatings encounter substantial obstacles regarding the interdiffusion effects from the varying chemical compositions found in advanced alloys. Such interdiffusion can weaken the coating's oxidation protection and adversely affect superalloys' mechanical characteristics. Conversely, coatings with nanocrystalline microstructure represent an innovative development in high-temperature coating layers, offering enhanced oxidation resistance for their specific composition and structural attributes. These coatings share a similar chemical composition with the superalloy, thereby minimizing interdiffusion issues. Produced through magnetron sputtering, nanocrystalline coatings feature a columnar nanostructure with a dense array of grain interfaces, which effectively lowers the minimum amount of aluminum necessary to form a protective oxide layer on the coating. Recent developments have seen these nanocrystalline coatings applied to the 2nd generation single-crystal N5 nickel-rich superalloy [58]. Research by Wang and colleagues has shown that at temperatures of 1000 and 1100°C, the oxidation rates of these nanocrystalline coatings are significantly lower, by less than 50%, compared to those of NiCrAlY coatings on the same N5 superalloy [59].

5. Microstructure and Strengthening Mechanisms

The microstructure of superalloys showcases the advanced engineering required to create materials that can withstand extreme conditions. By understanding and controlling the various phases within the microstructure, scientists and engineers can customize superalloys for specific performance requirements. The ongoing advancement of new alloy compositions, processing techniques, and coating technologies will further enhance superalloy performance for more challenging environments and applications. Cutting-edge characterization techniques and computational tools are used to analyze and optimize the microstructure, guiding the development of new alloy compositions and processing techniques. The exceptional properties of superalloys are due to their meticulously engineered complex microstructure, providing strength, stability, and resistance to degradation. This chapter delves into the intricate microstructure of superalloys, focusing on the key phases and their roles.

Researchers design advanced metal composites by carefully coordinating the relationship between dissolved elements in the primary metal structure and distributed reinforcing particles.

Superalloys: Fundamentals and Applications   Materials Research Forum LLC
Materials Research Foundations 178 (2025) 143-199   https://doi.org/10.21741/9781644903698-9

This strategic arrangement maximizes durability through two mechanisms: the strengthening effect of dissolved elements and the reinforcement provided by scattered hard particles throughout the material. Intermetallic compounds are favored in advanced applications owing to their robust bonding and elevated melting points. Nevertheless, their reduced ductility at lower temperatures presents an obstacle. Progress has been made using dual-phase or multiphase materials and composites. The analysis then extends to cover all key and additional strengthening mechanisms identified in superalloys.

5.1   Grain structure and boundaries

Superalloys are fascinating materials with diverse grain structures that directly impact their performance. These grain structures are influenced by processing methods and alloy composition, leading to the emergence of three primary types [60]:

- Equiaxed Grains: These grains, formed during conventional casting, boast uniform size and excellent mechanical properties. However, due to grain boundary sliding, they may be limited in high-temperature applications.

- Columnar Grains: These elongated grains, typically developed during directional solidification, offer enhanced creep resistance and high-temperature strength. Their alignment in a specific direction significantly improves mechanical properties, making them ideal for components subjected to unidirectional stress.

- Single Crystals: By eliminating grain boundaries, single-crystal superalloys deliver exceptional high-temperature performance, warding off grain boundary sliding and reducing the likelihood of creep and fatigue failures. They are particularly prized in turbine blades for their superior mechanical properties and resistance to thermal fatigue.

Grain boundaries, also called grain interfaces, which form where two grains intersect, play a crucial role in defining the mechanical properties of superalloys. Notably, Low-Angle Grain Boundaries (LAGBs), characterized by a misorientation angle of less than 15°, and High-Angle Grain Boundaries (HAGBs) with an angle over 15°, are significant. Additionally, Special Grain Boundaries such as coincident site lattice (CSL) boundaries contribute positively to mechanical enhancements. These boundaries reinforce superalloys through mechanisms like the Hall-Petch Effect, carbide precipitation, and oxide dispersion. Notably, oxide dispersion-strengthened (ODS) superalloys are distinguished by their superior characteristics. These alloys are designed with finely distributed oxide particles in their entire microstructure, which serve to obstruct dislocation movement and boost resistance to high temperature and creep. The resilience of oxides such as yttria (Y_2O_3) at elevated temperatures promotes enduring reinforcement by preserving the dispersion and minimizing particle coarsening. Methods to engineer grain boundaries include thermomechanical processing, strategic alloying, and specific heat treatments to enhance mechanical attributes. Furthermore, the addition of elements like boron and zirconium markedly enhances grain boundary cohesion, thereby mitigating the potential for sliding and cracking. These elements are prone to accumulating at grain boundaries, thereby increasing their robustness and enhancing their capacity to withstand high-temperature conditions. The integration of rare earth elements such as lanthanum and cerium also boosts the oxidation resistance of the grain boundaries, paving the way for advancements in superalloy technology [61].

When incorporating new elements into materials for enhancement, it's essential to understand the potential phases that might form. Superalloys typically consist of various phases: the matrix gamma (γ), precipitates such as the intermetallic compounds gamma prime (γ') and gamma double prime (γ''), carbide phases, and second-phase particles like oxides, nitrides, and borides, along with Topologically Close-Packed Phases (TCP). Among the precipitates, three primary types are identified: geometrically close-packed (GCP), topologically close-packed (TCP), and carbides. Mechanical properties are bolstered by GCP phases, while TCP phases might detract from them. The gamma (γ) matrix serves as the foundational phase in superalloys, crucial for maintaining structural integrity and mechanical performance at elevated temperatures. Its face-centered cubic (FCC) solid solution structure promotes high atomic density, thereby enhancing the material's strength and density as depicted in Figure 1(a). The γ matrix is capable of incorporating a variety of alloying elements, which further improves its mechanical characteristics and environmental resistance. This continuous matrix predominantly contains high percentages of Co in Ni-based superalloys and Fe-Ni in Co-based superalloys, along with elements like Cr, Mo, and W, all contributing to the ductility and structural arrangement of the precipitates [62].

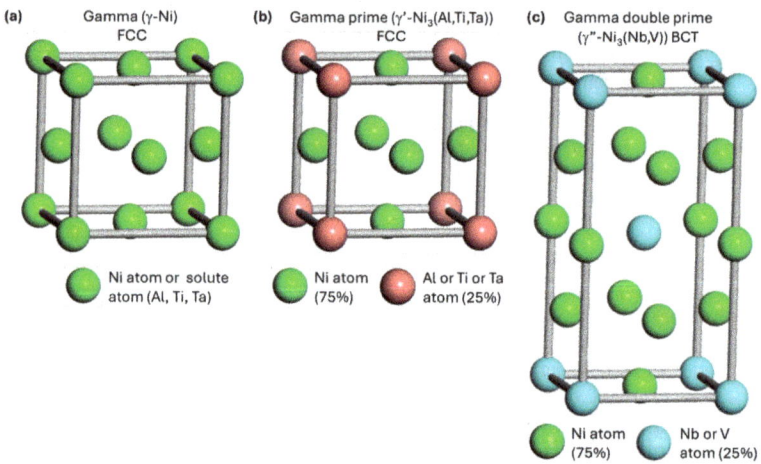

Figure 1. Crystal structure of (a) FCC gamma γ-Ni with solute atoms randomly substitute Ni atoms, (b) FCC gamma prime γ'-Ni$_3$(Al,Ti,Ta) with ordered Al or Ti or Ta atoms at the cube corners, (c) BCT gamma double prime γ''Ni$_3$(Nb,V) with ordered Nb or V atoms at the cube corners and a center of BCT.

The gamma prime (γ') phase in superalloys consists of intermetallic compounds such as Ni$_3$Al or Ni$_3$(Al,Ti,Ta), distributed within the γ matrix. This phase significantly boosts the alloy's resistance to high-temperature deformation and strengthens it (Figure 1(b)). The material's ability to strengthen depends on multiple interconnected characteristics, ranging from the energy levels within gamma prime defects to the dimensional compatibility stresses, as well as the proportional amount, overall toughness, and dimensional measurements of the reinforcing components. The γ'

Superalloys: Fundamentals and Applications   Materials Research Forum LLC
Materials Research Foundations 178 (2025)143-199   https://doi.org/10.21741/9781644903698-9

phase allows for atomic substitutions or excess vacancies, making it non-stoichiometric. Elements like Nb, Hf, and Ta tend to concentrate within the γ' phase. By manipulating the size, morphology, and distribution of γ' precipitates through specific heat treatments, the mechanical properties can be tailored to enhance performance in various applications, increasing the superalloys' versatility and efficacy.

At relatively lower temperatures, the presence of the γ'' phase as precipitates, mainly composed of Ni3Nb or Ni3V, is pivotal in strengthening nickel superalloys. It forms small, coherent discs with an ordered BCT structure (Figure 1(c)), particularly evident in alloys like Inconel 718. Despite its effectiveness, the γ'' phase becomes unstable above 650°C, restricting its applicability in higher temperature environments. Its strengthening mechanism involves lattice mismatch and order hardening, forming anisotropic discs. However, due to thermal instability above 650°C, its use is limited to lower temperature applications.

Carbides are often located along the crystalline border regions and dispersed throughout the internal grain structure, and have an important role in the alloy's microstructure. The alloys typically contain 0.05–0.2% carbon that will impact carbide formation, such as TiC, TaC, or HfC. Over time and along heat treatments, these carbides can transition into lower carbides like $M_{23}C_6$ and M_6C, primarily at grain boundaries. Such transformations substantially bolster the alloy's strength in elevated temperature environments and resistance ability of creep by providing resistance to grain boundary sliding. The formation of carbides and their sequence—HfC, TaC, NbC, and TiC—is influenced by the alloy's formulation and the cooling rate during solidification. In alloys with medium to high levels of Cr, the development of $M_{23}C_6$ and M_6C carbides enhances rupture strength. Notably, M_6C carbides maintain stability at higher temperatures compared to $M_{23}C_6$, playing a crucial role in grain size control during processing. Additionally, the existence of carbides at grain interfaces significantly boosts the alloy's ability to resist oxidation and corrosion at elevated temperatures [60].

During the vacuum-melting production of advanced superalloys, it's challenging to completely remove gaseous impurities such as oxygen and nitrogen. These gases tend to react with reactive elements within the alloy, forming stable second-phase particles. While oxide precipitates are infrequent, nitrogen sometimes results in the formation of nitrides. Adding small quantities of boron and zirconium has proven to significantly improve the stress-rupture properties of these alloys by preventing dislocation movement at grain boundaries. It's essential to keep boron concentrations below its solubility limit to prevent the excessive formation of borides and to maintain optimal properties of the superalloy.

Prolonged exposure to high temperatures will degrade the alloy's mechanical properties, such as the development of detrimental phases like sigma (σ) and mu (μ). Controlling the growth of TCP phases is vital, requiring precise management of alloy compositions and heat-treating methods. Engineers must strive to promote the formation of geometrically close-packed (GCP) phases while avoiding TCP phases. Modifications in structural attributes such as grain size are also strategic in mitigating the negative impacts of TCP phases. Utilizing advanced diagnostic tools such as transmission electron microscopy (TEM) and atom probe tomography (APT) is crucial for a detailed understanding of TCP phase development and distribution. The delta δ phase, which has an orthorhombic (Ni3Nb) structure and forms needle-like shapes, although not coherent with γ, does not inherently weaken the structure. The η phase (Ni3X), which forms after extended high-temperature exposure, plays a role in managing grain boundaries. The Sigma (σ)

phase, emerging at high temperatures, reduces both creep strength and ductility, while Laves phases are electron compounds that negatively influence ductility and yield strength. The mu (μ) phases, including Fe_2Nb, Co_2Ti, and Fe_2Ti, feature hexagonal structures and are usually seen in rounded or plate-like morphologies, categorized as TCP phases [63].

5.2   Strengthening via precipitation

Precipitation strengthening, also referred to as age hardening or particle hardening, is an effective heat treatment technique that significantly boosts the yield strength of various materials. Commonly applied to superalloys, this method exploits variations in solid solubility with temperature to generate fine particles that obstruct dislocation movement, thereby enhancing high-temperature strength. In the heat treatment process, elements such as Nb, Al, and Ti coalesce, generating tiny, coherent precipitates within the matrix of the superalloy. As the temperature increases in Ni-based superalloys, structured $Ni_3(Al,Ti)$ precipitates (γ' phase) form, markedly elevating strength at operational temperatures. These structured precipitates vary in composition based on the amounts of aluminum and titanium in the alloy, with typical phases including gamma prime (γ') and gamma double prime (γ") [60].

The primary precipitating elements in Ni-rich superalloys, besides Al, include Ti and Ta. High-performance single-crystal alloys with nickel as their main element demonstrate distinctive microscopic arrangements where strengthening particles (γ') occupy between 60% and 75% of the total volume. These particles develop into cuboidal formations measuring 0.2 to 0.6 μm when examined at standard room conditions. Achieving these specific dimensional and geometric characteristics requires careful thermal processing, which involves initial solution treatment followed by two distinct stages of temperature-controlled aging. The volumetric presence of these strengthening components is adjusted by modulating the quantities of elements that form precipitates. When aluminum levels are insufficient, metallurgists often incorporate titanium and tantalum as supplementary elements. However, excessive titanium content can compromise the alloy's performance in several ways: it increases structural incompatibility between the FCC matrix and the particles that strengthen the alloys, speeds up the growth of precipitate structures during operation, and promotes the unwanted development of η-Ni_3Ti phase [64].

The precipitation strengthening process begins with a solution treatment, heating the alloy to a high temperature to dissolve the precipitating elements, followed by rapid cooling (quenching) to form a supersaturated solid solution. Aging at a lower temperature then allows the precipitates to develop, substantially improving the alloy's yield strength and creep resistance by hindering dislocation movement. This preservation of a continuous lattice structure with the matrix effectively restricts dislocation activity. Additionally, the variance in shear modulus between the precipitates and the matrix leads to modulus hardening. The effectiveness of this strengthening mechanism depends on the precipitates' size, shape, and distribution within the alloy. Fine and evenly distributed precipitates are optimal for enhancing strength, while over-aging, where precipitates become overly large, can reduce the alloy's overall strength.

Antiphase boundaries (APBs) are crucial in the precipitation strengthening process of superalloys. These boundaries form during alloy ordering, creating shifts in atomic arrangements that disrupt the regularity of the crystal lattice. This disruption, also known as order strengthening, substantially improves the strength and creep resistance of superalloys, making them highly suited for high-temperature uses like turbine blades. The presence of APBs hinders dislocation movement within the crystal, thereby enhancing the material's resistance to shear stress. This characteristic is integral to the exceptional performance of superalloys under demanding conditions.

Understanding antiphase boundary energy (APBE) is vital for appreciating the disruptive impact of these boundaries on crystal structures. APBE is the heightened energy resulting from atomic misalignment, which deviates from the ideal lattice structure. As dislocations pass through the particle, they disrupt the balance of lattice planes, leading to an energy increase termed APBE. This heightened energy state has a critical function in enhancing the efficacy of precipitation hardening, complicating the machining of materials, and facilitating Orowan bowing around precipitates. Precipitates act as dislocation anchors, aiding in the unpinning process through bowing, which leaves dislocation loops around the precipitates. Factors like atom type, misalignment degree, and temperature influence the level of APBE [64]. Techniques such as transmission electron microscopy (TEM) and atom probe tomography (APT) allow for the experimental measurement of APBE, providing critical insights into the atomic behavior of materials.

Calculating antiphase boundary energy (APBE) through computational methods like density functional theory (DFT) and molecular dynamics (MD) simulations offers crucial knowledge about atomic-level interactions and energy dynamics at APBs. These boundaries significantly impact material mechanical properties by restricting dislocation movement, thereby increasing strength and hardness. APBs are particularly valuable in superalloys for their contributions to strength at elevated temperature and creep resistance, fundamentals for utilization in the aerospace and power generation sectors. Ongoing research into APB energy, enhanced by advanced experimental and computational techniques, is critical for deepening our understanding of the mechanisms of APB formation and their effects on material properties. This knowledge is pivotal in guiding the innovation of new materials tailored for specific high-performance applications.

5.3 Dispersion strengthening

Dispersion strengthening is a method that involves evenly distributing fine, stable particles within a metal matrix. These particles hinder the movement of dislocations, which primarily cause metal deformation. By impeding dislocation motion, the material's strength and high-temperature performance are significantly improved. The dispersed particles, such as oxides, carbides, or nitrides, act as obstacles to dislocation motion. They are introduced during the alloy's production process and do not significantly dissolve or coarsen at high temperatures, maintaining their strengthening effect over time. Common examples of dispersoids used in dispersion strengthening include yttria (Y_2O_3) in oxide dispersion-strengthened (ODS) superalloys, titanium carbide (TiC), tantalum carbide (TaC), and aluminum nitride (AlN).

Oxide dispersion strengthening (ODS) entails the deliberate dispersal of small oxide particles throughout the matrix of a material. This technique is particularly beneficial for utilizations in elevated-temperature turbine vanes and heat exchanger pipes. The embedded oxide particles obstruct dislocation movements in the material's lattice structure, effectively reducing creep. Due to the oxide particles' incoherency, dislocations must navigate through climbing. The oxides' thermal stability is crucial, maintaining the strengthening effect even under high temperatures, which makes ODS superalloys ideal for challenging environments such as turbine engines and combustion chambers. The addition of oxide dispersoids not only improves the performance of superalloys but also extends their service life, delivering considerable advantages for the aerospace and power generation sectors [64].

Dispersion strengthening leverages sophisticated processing methods like Mechanical Alloying, Powder Metallurgy, and In-Situ Precipitation. Mechanical Alloying involves a process where powder particles are continuously welded, broken down, and re-welded, resulting in a fine and uniform distribution of particles. Powder metallurgy entails the blending, compacting, and sintering of fine powders from the base alloy along with dispersants to produce a solid material characterized by evenly distributed particles. On the other hand, In-Situ Precipitation generates dispersoids directly within the alloy matrix either during solidification or through subsequent heat treatments. The benefits of dispersion strengthening are substantial, offering better strength at elevated temperature, also improved the ability to resist creep, oxidation, and corrosion. However, this sophisticated method presents challenges, including intricate processing requirements, elevated production costs, and difficulties in achieving consistent particle distribution.

5.4 Work hardening and phase transformations

When a metal undergoes deformation, it initiates a remarkable process known as work hardening, or strain hardening, which enhances the metal's strength and hardness through plastic deformation. This occurs as dislocations within the crystal structure multiply and interact, significantly increasing the material's toughness and durability. Work hardening mechanisms, including Dislocation Multiplication, Dislocation Interaction, and Strain Hardening Exponent, play a crucial role in this transformation. As plastic deformation takes place, the density of dislocations intensifies, leading to interactions and entanglements that impede further movement, creating internal stress fields. This impediment restrains the motion of other dislocations, thereby elevating the material's strength. The degree of work hardening is typically characterized by the strain hardening exponent (n), which depends on the material and its microstructure.

By deforming the alloy at room temperature (Cold Working), we can significantly increase the density of dislocations and enhance work hardening. This can be achieved through techniques such as rolling, forging, and drawing. Additionally, deforming the alloy at elevated temperatures (Hot Working) can also induce work hardening, though the effects may be somewhat less pronounced due to dynamic recovery processes. Post-deformation heat treatments offer the opportunity to optimize the balance between strength and ductility by controlling the dislocation structure. The advantages of work hardening are substantial, including improved mechanical strength, enhanced fatigue resistance, and better wear resistance. However, it's important to note that work hardening can also lead to increased brittleness and reduced ductility, potentially limiting the material's ability to undergo further deformation without cracking.

The strain-induced face-centered cubic (FCC) to hexagonal close-packed (HCP) phase transformation is an incredible process where mechanical stress triggers a remarkable change in the crystal structure of a material. This transformation is like a metamorphosis, where the typically ductile and deformable FCC structure morphs into the harder and less ductile HCP structure. The movement of dislocations within the crystal lattice sets off this transformation, as dislocations under applied stress rearrange the atomic planes, fundamentally altering the crystal structure. Picture this as a dance of atomic planes, with the glide of partial dislocations on

Superalloys: Fundamentals and Applications
Materials Research Foundations 178 (2025)143-199

Materials Research Forum LLC
https://doi.org/10.21741/9781644903698-9

specific crystallographic planes leading to the formation of the HCP phase. In some cases, the transformation may proceed through intermediate phases, like the fascinating ε-martensite phase in certain steels. The resulting HCP phase is like a superhero version of the FCC phase, being harder and stronger, creating a "secondary hardening" effect that boosts the overall hardness of the material. While this transformation increases hardness, it typically reduces the material's ductility, making it less capable of further plastic deformation without fracturing. Understanding and mastering this transformation can lead to the creation of materials with tailored properties for specific applications.

The concept of stacking fault energy (SFE) is fundamental in assessing how materials deform. SFE quantifies the energy needed to form a stacking fault within a crystal structure, which alters the typical atomic arrangement. Alloys with lower SFE tend to develop stacking faults more readily during deformation, which can serve as initiation points for phase transitions, such as from FCC to HCP or toward martensitic transformations. This understanding is crucial for developing materials with superior mechanical properties by fine-tuning the SFE through specific alloying and processing strategies to enhance both strength and ductility for optimal material performance. Additionally, the temperature sensitivity of SFE and its impact on martensitic transformations are critical for understanding material behavior under various environmental conditions, essential for creating materials that perform excellently across a broad temperature range.

Several methods are available for determining SFE, each offering distinct insights and accuracy levels, as illustrated in Figure 2. Among experimental approaches, Transmission Electron Microscopy (TEM) and X-ray Diffraction (XRD) are prevalent. TEM provides a direct visualization of dislocations and stacking faults at high resolution, offering a clear view of the crystal structure. Conversely, XRD employs line-profile analysis and weak-beam techniques to assess SFE by analyzing the diffraction patterns produced by the material. In the realm of computational analysis, ab-initio calculations, particularly those using density functional theory (DFT), are highly valued for their precision. These calculations assess SFE by delving into the electronic structure of materials, providing a theoretical perspective that enhances experimental findings [65]. Molecular Dynamics (MD) simulations also offer valuable insights into SFE by utilizing semi-empirical potentials that draw from both experimental results and quantum mechanical theories, achieving a good balance of computational speed and accuracy. Thermodynamic models contribute to SFE estimation as well, considering the material's chemical composition and phase stability through equilibrium calculations that forecast SFE variations. These models integrate various thermodynamic variables to anticipate how compositional or temperature changes might influence SFE. Each technique has its own strengths and weaknesses. Employing a combination of these methods can yield a more thorough understanding of SFE, facilitating improved predictions of how materials will react under diverse conditions [65].

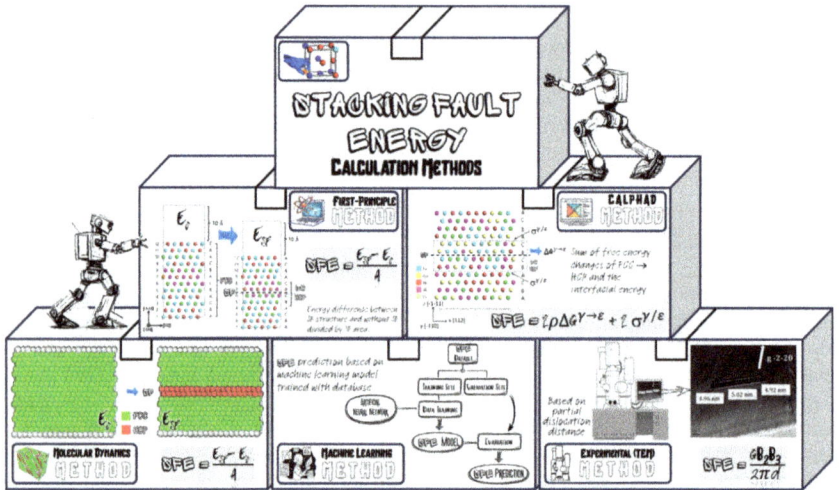

Figure 2. Illustration of various Stacking Fault Energy (SFE) calculation methods.

Machine learning (ML) has become a powerful tool for predicting SFE, offering significant advantages over traditional methods. By leveraging large datasets and advanced algorithms, ML models can predict SFE with high accuracy and efficiency. For example, combining ab-initio calculations with machine learning has accelerated the prediction process by approximately 80 times compared to direct density functional theory (DFT) calculations. This approach not only speeds up the prediction but also reduces the computational cost significantly. These models are trained on extensive datasets that include experimental and theoretical SFE, allowing them to learn complex relationships between chemical composition, temperature, and SFE. One notable example is the use of ML to predict SFE in FCC high-entropy alloys [66]. Researchers have developed models that can accurately predict SFE across various compositions, helping optimize these materials' mechanical properties. Overall, integrating machine learning in SFE prediction represents a significant advancement, enabling faster and more efficient alloy design and optimization.

6.   Applications in Nuclear Plants

Generally, nickel superalloys used in nuclear generating station applications are categorized based on their alloying elements, which include combinations such as Ni and Ni-Cu; Ni-Cr-Fe and Ni-Cr-Fe-Mo; Ni-Mo-Fe and Ni-Mo-Cr-Fe, among other variations of nickel superalloys. Originally, they were adopted as a substitute for austenitic stainless steel alloys, primarily due to their enhanced resistance to corrosion, particularly from chloride ions. Furthermore, nickel exhibits greater nobility than copper, iron, and zinc, hence improving its ability to resist corrosion compared to these elements that were once utilized as alloying materials in nuclear power plants [67].

a. Ni and Ni-Cu Alloys

In this variant, two distinct alloys, designated Alloy 200 and Alloy 201, are identified as suitable for use in base evaporation pipes. In addition to Alloy 200 and Alloy 201, Alloy 400 is also a viable option. This alloy exhibits resistance to oxidative environments, rendering it suitable for use in salt and seawater environments. A variant of Alloy 400 that undergoes precipitate strengthening, designated Alloy K-500, is employed extensively in pump shafts, impellers, and springs.

In nuclear power facilities, Alloy 400 is often used for constructing pipes and casings in boiler feed water heat exchangers, while Alloy 500 is typically used in the manufacture of rotating shafts and impellers for marine environment pumps. Additionally, Alloy 400 is commonly found in steam generator (SG) pipes within CANDU reactors (PHWR type).

b. Ni-Cr-Fe and Ni-Cr-Fe-Mo Alloys

The X-750 alloy, which incorporates titanium, aluminum, and niobium, develops a precipitation-strengthened γ' phase, leading to the creation of $Ni_3(Ti, Al, Nb)$ when subjected to external stimuli. This alloy is widely used in the manufacturing of screws and coil springs. Additionally, it is used to produce thin wires and springs. Similarly, alloys exhibiting comparable crawling resistance to Alloy X-750, such as Alloy A286, can also be utilized as bolting material. Additionally, Alloy 718 can be employed as a spring material for water-cooled reactors, given its crawling resistance up to 400°C.

The ability of Alloy 600 to resist cracking triggered by chloride ions in an applied stress and corrosive environment renders the alloy appropriate for application in steam generator pipes. Furthermore, both alloys mentioned can undergo heat treatment to develop resistance to primary water stress corrosion cracking (PWSCC), occurring in environments involving high-purity, high-temperature water.

After undergoing a heat treatment process, Alloy 600 continues to be a suitable material for constructing steam generator piping and nozzles for the control rod drive mechanism (CRDM). While Alloy X-750 is not used for support pins and flexure pins in PWRs, it is suitable for use in bolting applications. Additionally, Alloy 800 serves as an alternative material for steam generator piping in several PWR reactors in Germany and the PHWR-CANDU reactor.

c. Ni-Mo-Fe and Ni-Cr-Mo-Fe Alloys

Alloy 625 may be suitable for supercritical light water-cooled reactors. It also shows promise for use in components within the reactor core and control rods. Moreover, Alloy C-22 is considered another viable option for materials used in the storage of high-level radioactive waste.

d. Other Alloys

XR alloy has proven to be an excellent choice for structural material in Generation IV reactors, as demonstrated by its effective performance in tests conducted on Japan's High Temperature Engineering HTGR Test Reactor (HTTR).

6.1   Components in nuclear reactors

The steam generator (SG) pipes at Chooz-A/SENA (France, PWR) were initially made from 306 stainless steel but have since been upgraded with Inconel 600, which is known for its superior resistance to chloride-induced corrosion. Initially, Inconel 600 was chosen for steam generator pipes, but it was later replaced by Inconel 690 due to Inconel 600's vulnerability to various forms of corrosion, including stress corrosion cracking (SCC). Additionally, Inconel 600 is used in the lower cap of the primary circuit pressurizer at Calvert Cliffs-2 (USA, PWR). Along with its use in steam generators, Inconel 600 also features in the control rod adapter that penetrates the vessel. Moreover, Inconel 182 is employed as a buttering and welding material in the control rod adapter for joining the IN 600 bimetallic structure to stainless steel. Within the vessel, a secondary support mechanism is in place to mitigate radioactive materials' emission in the event of core barrel failure. In this setup, Inconel X-750 is utilized for screw components. Beyond its application in screws, Inconel X-750 is also used for guide tube spindles, which are crucial for rod injection and fuel transportation in nuclear reactors [11].

The active core component, which serves as the site for nuclear fission reactions, is constructed from a combination of zirconium-based alloy fuel rods (Zircoloy) with interstitial barriers (grids) utilizing Inconel 718 material. However, Inconel 718 can also be substituted with Zircoloy-4 due to its enhanced neutron transparency, which mitigates the risk of neutron absorption [11].

The European Pressurized Reactor (EPR) employs two distinct reactor monitoring systems: aeroball instrumentation for point scanning and cobalt collectors for continuous scanning. In this cobalt collector, a collector part made from Inconel 600 is utilized in the chamber of the fission mobile (CFM). In addition to Inconel, cobalt-based superalloys, including Stellite, are also widely used in PWR. These are employed as part of the shell for the primary circuit pump (Novovoronezh-1), as coating materials for stress-bearing areas in the primary circuit, including the shaft and sealing, primary circuit valves, and material for pins on the upper core plate. They are also used as coatings on turbine blades to reduce corrosion and erosion [11].

Fourth-generation nuclear power plants (NPPs) encompass various designs, each with unique features. Gas-cooled fast reactors (GFRs) operate using a fast-neutron spectrum and utilize helium gas for cooling, alongside a closed fuel cycle. Graphite moderation and helium gas cooling are used to characterize very-high-temperature reactors (VHTRs), employing a once-through uranium fuel cycle. By utilizing water cooling while remaining below the critical point of water, the supercritical-water-cooled reactors (SCWRs) operate at elevated temperatures and pressures. Sodium-cooled fast reactors (SFRs) also utilize a fast-neutron spectrum, employing sodium as a refrigerant within a closed fuel cycle to enhance the efficiency of actinides and the conversion of fertile uranium. A eutectic lead-bismuth alloy is incorporated in lead-cooled fast reactors (LFRs) for cooling in a fast-spectrum reactor, facilitating efficient change of fertile uranium and actinides within a closed fuel cycle. Lastly, molten salt reactors (MSRs) utilize a molten salt fuel mixture in epithermal spectrum designs, featuring recycling processes for actinides to generate fission energy effectively [12].

Given the operational temperatures and potential failure modes in these reactor types, nickel-based alloys are considered suitable structural materials, particularly for GFR, MSR, and VHTR configurations. They are also considered for secondary applications in SCWR variants. Beyond structural applications, nickel-based alloys are utilized in heat exchangers due to their capability to endure temperatures up to 750°C, making them appropriate for reactors with helium and molten salt cooling systems. Inconel 718 is a commonly used nickel-based alloy for heat exchangers, capable of operating between 1200 to 1250°C. Other suitable materials for intermediate heat exchangers in VHTR reactors include Inconel 617 and Haynes 230 [12].

6.2   Turbine components in nuclear power generation systems

Superalloys and stainless steels have been extensively utilized in the construction of gas turbines and steam turbines, respectively, including those employed in nuclear power plants. In gas turbines, superalloys are employed in the construction of bucket and nozzle components, as well as combustion components. In contrast, stainless steel is utilized in the fabrication of compressor blade components. In steam turbines, martensitic stainless steel (12Cr steel) remains the predominant material used for turbine components, including turbine blades, rotors, casings, and bolts.

Fundamentally, the working mechanisms of turbine drive systems in nuclear generating stations and steam power stations are quite similar. The primary difference lies in the type of fuel used, with the turbine drive fluid generally being steam. Consequently, the materials employed in turbine components are either identical or very similar. Given that steam turbine operations typically occur within a temperature range of 500-650°C, the materials employed tend to be iron-based alloys, with the martensitic stainless steel group (12-13% Cr) being a notable exception. Martensitic stainless steels exhibit a favorable combination of strength, toughness, and corrosion resistance. For the utilization of advanced ultra-supercritical steam with operating temperatures exceeding 700°C, nickel-based superalloys that exhibit superior high-temperature capabilities can be employed instead of martensitic stainless steels. Moreover, austenitic stainless steels are being developed as a more cost-effective alternative with optimal creep strength and corrosion resistance among stainless steel grades. Austenitic stainless steel (Type A286, Incoloy A286) is the most prevalent material utilized for steam turbine components. Austenitic stainless steel contains 13.5-20% Cr and 24-27% Ni. It can be strengthened with $Ni_3(Al,Ti)$ precipitates, which categorizes this type of steel as precipitation-strengthened steel.

6.3   Superalloy applications in safety and support structures

Several Ni-based alloys have been used in various nuclear power plant components, such as safety structures and supports. Alloy 400, for instance, has been widely used in the tubes and shells of boiler feed water heat exchangers, while Alloy 500 has been a common choice for marine pump components. Alloy 400 was also chosen for steam generator (SG) tubes in multiple CANDU reactors. Alloy X-750 has been favored for bolting and coil spring applications in both light and heavy water reactors due to its proven effectiveness in jet engines and outstanding creep resistance. Similarly, Alloy A286 has been chosen for bolting for similar reasons. Alloy 718, known for its use in jet engines, its ability to resist creep development, also superior yield strength in hot environments, which can reach 400°C, has been selected for use in springs and bellows.

Alloy 800, which has iron as its base metal, is valued for its resistance to cracking failure due to combined stress and corrosion in hot water and caustic conditions. It is utilized in steam generator tubes in some German pressurized water reactors (PWRs) and CANDU-type reactors. Alloy 625, a nickel-chromium-molybdenum-iron alloy, is considered a robust alternative to austenitic stainless steels because of its high strength, corrosion resistance, and resistance to

SCC, making it suitable for reactor core and control rod components in water-cooled and supercritical water-cooled reactors. Its property of having a wide range of allowable stress to be applied in hot service conditions also makes it a viable option for advanced high-temperature reactors.

Alloy C-22 has undergone evaluation for its ability to resist corrosion in very alkaline and chloride-rich solutions, and it is being considered for containers for high-level radioactive waste disposal. Alloy XR, used as a structural material in the Japanese HTTR reactor, shows excellent resistance to oxidation in high-temperature helium gas environments. Its composition, including chromium and manganese, forms a protective spinel layer that enhances its stability and longevity in low-oxidizing conditions. Alloy XR is deemed suitable for Generation IV reactors. Additionally, a nickel-chromium-tungsten alloy is under investigation for analogous uses as an innovative and updated version of Alloy XR.

7.   Performance and Reliability

7.1   Creep and fatigue behavior under nuclear plant conditions

Advanced Ni-base superalloys have been created to achieve high-efficiency technology for nuclear plant settings. These alloys exhibit remarkable creep resistance due to their strengthening microstructure, which includes cuboidal γ' precipitates. When Ni-base superalloys are utilized as structural components, the combination of high temperatures and external tensile loads can lead to the formation of creep fractures. Furthermore, the configuration of the structural component can induce notch effects. This investigation involved carrying out in-situ observational creep crack growth experiments using IN100, alongside Finite Element Method (FEM) studies to explore how microstructural reinforcement influences crack branching behaviors. The findings revealed that the creep fracture growth behavior of creep-susceptible materials like IN100 is significantly influenced by the grain distribution at the notch tip, leading to a stress concentration effect. This factor is responsible for the observed differences in creep crack growth behaviors. The study examined the mechanical aspects of creep deformation, creep damage formulations, and creep crack growth behaviors through two-dimensional elastic-plastic creep finite element analyses and FE-SEM/EBSD studies. The microstructural reinforcement in IN100 is linked to branch cracking, which arises from the strengthened microstructural configuration and grain distribution [68].

Fatigue failure can occur at high temperatures, even below the melting point. This failure mode can manifest with minimal plastic deformation and will exhibit the distinctive characteristics associated with fatigue failure. Nevertheless, it is well noted that elevated temperatures lead to a decrease in both fatigue and static strength qualities. As previously mentioned, the application of a load, especially at higher temperatures relative to the material's melting point, will result in time-dependent plastic deformation, also known as creep. The creep-rupture strength diminishes as the temperature rises. Typically, alloys that exhibit resistance to creep also demonstrate resistance to fatigue [69].

Nickel-based superalloys are predominantly utilized as the principal material for turbine discs in gas turbines due to their superior high-temperature performance. During operation, turbine discs are subjected to both mechanical strains and high temperatures, making them susceptible to low-cycle fatigue (LCF) degradation. This vulnerability can lead to a decrease in service life and disrupt their normal operations. The cyclic deformation and accumulation of fatigue damage in nickel-based superalloys are significantly affected by the microstructure and the modes of deformation observed during testing. The main reason for fatigue damage and the onset of fractures in nickel-based superalloys, particularly those with few metallurgical defects, is the buildup of dislocations that slide permanently. The microstructure in most nickel-based superalloys consists of a complex mix, featuring unevenly distributed precipitates, varied grain sizes, and distinct grain boundary characteristics. Under cyclic fatigue loading, such an uneven microstructure may either produce dislocations or act as a local stress concentrator [70].

7.2 Longevity and lifecycle assessments

Evaluating the durability and lifecycle of superalloys in nuclear plant applications necessitates a detailed approach that considers a variety of factors, including manufacturing processes, material properties, operating conditions, environmental impacts, and system reliability. These factors are integrated into the design, operation, and maintenance of nuclear power facilities. The process of lifetime and aging management for a nuclear power plant begins in the design stage, where components likely to face harsh operational conditions are planned to be easily accessible and replaceable. Selecting the most appropriate material and determining the best thermal and fabrication treatments can be complex and not always straightforward. Specific nuclear environments or thermal histories, possibly in combination with other factors, may lead to the unexpected development of new phases, alteration of existing phases, and diffusion of elements, which all influence the lifespan of a component. Facilitating the replacement of parts has the potential to substantially lower downtime and enhance the availability of the nuclear facility over its operational life. The advancement of materials and the application of improved nondestructive testing and monitoring techniques, especially those that can be performed online, will help utility companies set realistic and effective schedules. This, in turn, aids in organizing timely component replacements and preventative maintenance to manage component degradation without excessive caution [71].

Environmental assessment of nuclear generation facilities reveals that initial production phases generate the most substantial ecological impact. These manufacturing activities, encompassing the entire spectrum from raw material extraction to facility construction, account for approximately 30% to 60% of total environmental effects. Within this category, nuclear fuel production chain activities - including extraction, processing, refinement, and fuel assembly - constitute 25% to 50% of the overall environmental burden. The production of building materials, however, shows comparatively lower environmental consequences. Significant ecological impacts also emerge during facility operation and eventual dismantling. Daily operations create 10% to 20% of environmental effects, primarily due to dependence on fossil fuel-generated electricity. The dismantling phase exhibits even greater environmental implications than operational activities, primarily because of the extensive power requirements for handling and disposing of radioactive components. To achieve optimal environmental stewardship in nuclear power development, particular attention must be directed toward optimizing both fuel production processes and facility decommissioning procedures throughout the project's lifespan [72].

Managing long-term functioning and longevity extension of existing nuclear power plants is crucial for guaranteeing energy resources and transitioning to a low-carbon future. The present objective is to prolong the operational lifespan of generating reactors to exceed 60 years.

Advanced approaches for predicting the life cycle are essential to ensure the prolonged, secure, and dependable functioning of a system. The sustainability of long-term operations is shaped by multiple aspects, including aging management, asset management (which encompasses decisions regarding repairs, replacements, modifications, or shutdowns), economic optimization, technological advancements, operational infrastructure, maintenance, development, as well as regulatory and public acceptance. The following sections delve into predicting the lifecycle of metallic materials. Stress corrosion cracking (SCC) emerges as the primary corrosion mechanism responsible for the structural degradation of metallic parts in nuclear power facilities. This type of cracking stems from the synergistic effects of corrosion, tensile stresses, and susceptibility of the material to damage. The degradation process can be categorized into two distinct stages: the initiation phase and the propagation phase, both influencing the material's durability. Risk-informed inspection represents a progressive approach that improves the accuracy of in-service inspections [73].

7.3   Case studies of superalloy performance in existing nuclear plants

Numerous case studies have illustrated the utility of superalloys in nuclear power plants. Take Inconel 625, a nickel-based superalloy, for instance, which is highly esteemed in both civilian and defense nuclear reactor sectors due to its superior high-temperature attributes like corrosion resistance and the ability to withstand deformation under continuous stress. These qualities render it ideal for critical internal components of nuclear reactors, particularly steam generator tubes. Traditionally, Inconel 625 has been shaped through methods such as forging or casting. Recently, however, there has been a shift towards considering powder metallurgy and hot isostatic pressing (PM-HIP) as a more economical option to create components nearing their final shapes. This shift necessitates a microstructure analysis to verify the appropriateness of PM-HIP for reactor component use [74].

Similarly, Inconel X750 is another superalloy known for its high-temperature resilience and capability to be strengthened through aging. It offers exceptional corrosion resistance along with robust mechanical strength and ductility, making it suitable for nuclear reactor applications [50]. The operational physical properties of reactors may change due to neutron-induced displacement damage. This kind of damage can displace atoms from their lattice positions, resulting in dislocation loops, tangles, voids, and precipitates [75].

Furthermore, the nickel-based superalloy C276 has been chosen as a primary material for the atomic power generation industry due to its compatibility in mechanical characteristics and substantial ability to resist the corrosion process [76]. Current design concepts for Sodium Fast Reactors (SFR) are being developed for Generation IV (Gen IV) nuclear reactors. Key challenges for these reactors include maintaining dimensional stability under severe thermal stress, ensuring compatibility with heat-transfer agents such as sodium coolants, achieving long-term stability and reliability, and streamlining manufacturing processes.

8.   Technological Advancements and Innovations

8.1   Recent developments in alloy composition

The alloying elements in nickel-based superalloys have a considerable impact on their characteristics. Gaining insight into how these elements influence the physical and mechanical attributes of nickel superalloys can assist in the creation of novel and specialized versions of these materials. The subsequent observations are offered for evaluation.

- The incorporation of refractory metals such as tungsten (W), rhenium (Re), and molybdenum (Mo) significantly boosts the high-temperature capabilities and creep resistance of superalloys. Solution strengthened mechanism and formation of stable carbide phase are significantly affected by these alloying elements. Additionally, ruthenium has been introduced to improve phase stability along with resistance to oxidation and corrosion. An increased concentration of Ru has been noted to slow dislocation movement by reinforcing the γ matrix and altering the γ' phase, resulting in reduced dimensions of the initial γ' phase. Consequently, this leads to prolonged durations of the primary and secondary creep stages. However, despite the extended creep life due to higher Ru levels, it was observed that this does not prevent the formation of the rare TCP phase [77].

- Adding grain refiners like CrMoNb and CoFeNb can significantly improve the grain size and morphology of the K4169 alloy. These refiners can lead to smaller equiaxed grains, a higher fraction of equiaxed grains, and diminished elements' dispersion, including chromium, niobium, iron, and molybdenum. Moreover, grain refinement can decrease the temperature gradient in the melt, which in turn enhances melt undercooling, the cooling speed, nucleation supercooling, and the stability of grain-refining particles.

- The addition of yttrium (Y) to superalloy K417 can facilitate the purification of the alloy by reducing the content of nitrogen (N) and oxygen (O). Yttrium (Y) forms a new phase, a $Ni_{3.55}Co_{0.5}Cr_{0.2}Al_{0.75}Y$ phase with a hexagonal crystal structure. The addition of further Y does not result in the formation of additional Y-rich secondary phases. Yttrium enhances the solubility of nitrogen (N) in the alloy, reducing its concentration. The degree of high vacuum also affects the concentration of N and O. The addition of Y can reduce the concentration of O, but the concentration of N may increase slightly due to its solubility in Y [78].

- Boron (B) and zirconium (Zr) have been found to strengthen grain boundaries, reducing the risk of intergranular fracture and improving overall mechanical properties, especially at high temperatures. This makes the alloys more robust and reliable under operating conditions [79].

- The addition of rare-earth elements (REs) in a small amount has demonstrated a substantial enhancement in the microstructure of superalloys, assists in the purification of grain juncture, and increases their hot oxidation resistance. Additionally, the inclusion of REs has been found to enhance the creep resistance of these alloys. Owing to their high reactivity, rare earth elements can react with sulfur (S) and oxygen (O) to form a protective oxide layer that also improves the density of the alloy. However, there is a notable loss of REs during the melting and casting processes due to their volatility and the likelihood of reactions with ceramics.

- Hafnium (Hf) is critical in promoting the formation of specific defects called freckles in alloys. It has a strong preference for the solid phase and is absent from the dendritic cores within freckles. Hafnium's distribution is consistent between freckles and the solid-liquid (S-L) phases, indicating its tendency to segregate into the solid phase during solidification. Freckle development is impacted by Hf, yet its levels must exceed 1-2% to be effective. The impact of Hf is more noticeable in alloys that contain high levels of cobalt or chromium, which enhance the solubility of Hf in the alloy [80].

8.2   Advanced manufacturing technologies (e.g., 3D printing of superalloys)

Nickel-based superalloys are primarily produced using casting methods, especially investment casting. However, researchers are exploring alternative manufacturing methods, including additive manufacturing (AM) and powder metallurgy, for future applications. Additive manufacturing (AM) methods, such as selective laser melting (SLM) and electron beam melting (EBM), have revolutionized the fabrication of superalloy parts. These processes offer meticulous control over the alloy's composition, microstructure, and phase distribution, facilitating the production of gradient alloys. These gradient alloys feature variable properties across the component, enabling tailored optimization of different sections to meet specific needs, such as improved wear resistance on external surfaces and greater toughness at the core.

A comparative analysis was conducted on the microstructures of Inconel 718 superalloy components fabricated through suction casting and additive manufacturing (AM). The investigation revealed that the initial microstructures of both as-built and as-cast specimens displayed unique characteristics that affected the uniformity of niobium (Nb) distribution within the matrix, which is crucial for grain morphology during the homogenization phase. In the suction-cast samples, there was a notable improvement in Nb uniformity, while the AM samples showed a decrease in this attribute when subjected to isothermal heat treatment at 1180°C. Additionally, the phase transformations occurring during the homogenization process were different between the suction-cast and AM samples, owing to their varying initial solidification microstructures.

The microstructural evolution varied between different processing methods. Brief heat treatment of suction-cast specimens revealed the presence of Laves_C14, which subsequently disappeared during prolonged homogenization processes. In contrast, additively manufactured materials exhibited rapid dissolution of this phase, taking merely 20 minutes, followed by NbC precipitation. The formation of these strengthening structures was heavily influenced by Nb content, where microsegregation effects triggered the initial development of the Laves_C14 structure.

The initial Nb uniformity in AM alloys was advantageous because of the solute trapping effect during rapid solidification. However, these alloys showed limited grain growth compared to the suction-cast alloys, which exhibited significant grain growth and required longer homogenization periods. AM alloys also demonstrated finer typical grain sizes and a continuous recrystallization process. The grain refinement in AM alloys was further supported by the Zener pinning effect of NbC carbides. Additionally, the presence of dislocation arrays and loops in AM alloys could impact the hardening characteristics of the alloy. Carbides that were not coherent with the matrix served as sources for dislocations. These dislocations, generated from carbide growth and local strains during homogenization, accumulated at the grain boundaries, potentially facilitating local strain relief and preventing fractures, thus contributing to the hardening of the alloy.

8.3 Innovations in thermal and environmental barrier coatings

The use of thermal or anti-erosive/corrosive coatings can help protect superalloys, but it's crucial to understand that these coatings might also affect the physical and mechanical characteristics of the base nickel superalloy. This section summarizes the results from various research studies exploring how coatings impact the material properties, oxidation resistance, and overall durability of nickel superalloys.

One study examined how thermal barrier coatings (TBCs) affect the low-cycle fatigue (LCF) performance of nickel superalloys that were manufactured using investment casting. TBCs are designed to prolong the operational lifetime of superalloy parts by protecting them from high temperatures. Nonetheless, it's important to recognize that TBCs can alter mechanical properties, including LCF behavior. The research showed that both coated and uncoated samples experienced rafting during LCF tests. Crack initiation in uncoated samples typically began near defects, whereas in coated samples, cracks started at the TBC/bond coat or bond coat/substrate interfaces. Moreover, it was found that the frequency of fatigue cracks propagating through the TBC was higher in coated samples compared to uncoated ones.

The findings indicated that the spreads and allocation of plastic strain across the gauge length of the coated samples were more even compared to uncoated samples. When subjected to the same stress amplitudes, coated samples showed higher levels of plastic strain. Fatigue life curves revealed that both coated and uncoated superalloys had similar fatigue lives at the highest amplitudes, but at higher cycle counts, coated samples displayed better fatigue properties under increased stress amplitudes. Ultimately, while TBCs did not significantly improve fatigue life, they proved beneficial for shielding alloys from extremely high-temperature conditions. Researchers conducted an investigation examining how thermal cycling affected TBC specimens that utilized single-crystal substrates. These specimens featured a distinctive layered structure: a base of Amdry 995 (Co-based bond coat), followed by thin ODS Co-based flash coatings, and finished with a porous YSZ top layer applied through atmospheric plasma spraying techniques.

Compared to samples with single-layer bond coats, those using thin flash coats enhanced with oxide dispersions exhibited a doubling in lifetime. Exceptional resistance to thermal cycling was demonstrated by specimens featuring the dual-layer bond coat with ODS. These samples maintained their integrity through more than 4,000 thermal cycles, withstanding temperature variations between 1080°C and 1400°C. The eventual failure of these test specimens occurred due to bond coat oxidation processes, which determined their maximum service duration. Enhanced durability during thermal cycles was achieved through the implementation of the ODS flash coating. This improvement manifested in several ways: it restricted yttrium aluminate formation within the TGO layer, facilitated better oxygen movement throughout the alumina scale structure, and strengthened the bonding capabilities of the uppermost coating layer. These factors collectively reduced cracking in the TGO, aided stress relaxation, and helped prevent topcoat failure.

9.   Challenges and Solutions

9.1   Addressing radiation damage and swelling

The materials employed in nuclear energy systems, whether fission or fusion, can be a limiting factor in prolonging lifetimes due to harsh in-service circumstances [81]. Superalloys face several critical challenges when used in nuclear plant applications, including radiation damage, and swelling, high-temperature corrosion and wear, and supply/cost issues related to rare alloying elements. Radiation damage and swelling pose significant threats to the structural integrity of superalloy components in nuclear reactors. These issues can compromise the structural integrity and performance of reactor components, potentially leading to safety concerns and reduced operational efficiency.

Swelling associated with cavity formation and growth is one of the most damaging radiation-induced degradation modes [82]. To address this, researchers are using ion irradiation as an accelerated method to study and predict swelling behavior. Advanced modeling techniques are also being developed to better understand cavity nucleation and growth processes across different radiation dose rates. For high-nickel superalloys, the transmutation of 58Ni to 59Ni can significantly enhance both atomic displacement and helium production rates, especially in thermal neutron environments [82]. Careful alloy design and composition control are needed to mitigate these effects.

Superalloys in nuclear plants face persistent challenges with high-temperature corrosion and wear. Techniques such as electron beam melting have been effective in enhancing the microstructure and corrosion resistance of superalloys like Inconel 617 [83]. This method refines the microstructure and ensures a more uniform distribution of alloying elements, thus boosting corrosion resistance. For improving wear resistance, methods like boronizing and boro-aluminizing are being investigated, which have been shown to significantly increase hardness and wear resistance at both ambient and high temperatures [84]. Additionally, the procurement and cost of rare alloying elements such as rhenium and ruthenium remain significant issues, with modern nickel superalloys incorporating 3-6% of these costly elements [85]. In response, strategies are underway to refine alloy compositions to lessen dependency on these rare materials, enhance manufacturing techniques to decrease waste, and develop improved recycling methods to reclaim valuable elements from superalloy waste.

9.1.1   Radiation damage

Radiation damage is radiation impacts denotes any macroscopic alterations in the physical and chemical properties of a material subjected to radiation, preceded by microscopic changes collectively [86]. One of the primary challenges is the displacement of atoms in the crystal lattice due to neutron irradiation. In light water reactors (LWRs), fuel cladding typically experiences about 20 displacements per atom (dpa) at a burnup of 40 GWd/tU [87]. However, future fast reactor systems may subject materials to 150-200 dpa, significantly increasing the potential for radiation damage [87]. Solutions to mitigate radiation damage include:

- Developing advanced alloys with enhanced radiation resistance. For decades, academics, labs, and international organizations have sought ways and concepts to accelerate radiation damage in structural materials for nuclear energy use to advance the development of radiation-resistant materials [82].

Materials Research Forum LLC
https://doi.org/10.21741/9781644903698-9

- Incorporating nano-scale oxide dispersoids to improve high-temperature stability and creep resistance.

- Optimizing microstructures to create features like grain boundaries that can capture and hold migrating radiation defects [87].

9.1.2 Swelling

Swelling is another critical issue, particularly for austenitic alloys used in nuclear reactors. The swelling rate depends on various factors, including dislocation structure, damage rate, temperature, and helium generation rate [88]. To address swelling challenges:

- Researchers are exploring the use of oxide dispersion strengthened (ODS) superalloys, which exhibit superior resistance to swelling. ODS materials exhibit superior swelling resistance (up to doses of 200 dpa), diminish fracture embrittlement at low temperatures, and mitigate creep rupture at elevated temperatures [81].

- Refining alloy compositions to minimize helium production is vital, as helium significantly influences cavity stabilization and swelling at elevated temperatures [88]. The atoms of He and H are generated by fast neutrons through neutron–alpha particle (n, α) and neutron–proton (n, p) reactions. These atoms can form gas bubbles that lead to swelling and create voids within the material [87].

- Developing alloys with specific grain boundary properties to reduce helium bubble accumulation is crucial because the swelling from phase changes in the alloy is more significant than that caused by helium bubble formation, due to the limited penetration depth of helium ion implantation [88,89].

9.1.3 High Temperature Performance

Future nuclear reactor designs aim to operate at higher temperatures (500-1000°C) and with more corrosive coolants, presenting additional challenges for superalloys [87].
Solutions for this challenge include:

- Investigating novel alloying elements to enhance high-temperature stability and corrosion resistance [90].

- Developing advanced processing techniques to create optimized microstructures for extreme conditions.

- Exploring the potential of bulk-nanocrystalline oxide nuclear fuels to increase fission gas retention, plasticity, and radiation tolerance [91].

9.1.4 Computational and experimental approaches

To better understand and predict material behavior under extreme conditions:

- Developing advanced modeling techniques to simulate long-term material performance.

- Utilizing accelerated testing methods to bridge the gap between laboratory experiments and real-world applications.

- Utilizing collaborative research approaches to study the effects of radiation damage on the materials' properties under elevated temperatures and radiation exposure [87].

By addressing these challenges through innovative materials design, advanced processing techniques, and comprehensive research efforts, the nuclear industry can develop superalloys capable of withstanding the extreme conditions in next-generation nuclear plants, ensuring safer and more efficient operations.

9.2 Managing high-temperature corrosion and wear

High-temperature corrosion and wear are significant challenges for superalloys used in nuclear plants. These issues can compromise the structural integrity and performance of reactor components, potentially leading to safety concerns and reduced operational efficiency.

9.2.1 High-temperature corrosion

Superalloys in nuclear facilities must withstand severe conditions such as elevated temperatures, high pressures, and corrosive environments [92]. These factors pose considerable challenges in preserving the material's integrity over extended durations. Strategies to combat high-temperature corrosion include:

- Alloying with corrosion-resistant elements: Inconel alloys, composed mainly of nickel, chromium, and iron, exhibit excellent corrosion resistance. The addition of elements like molybdenum and tungsten can further enhance this property [92].

- Surface modification: Techniques such as electron beam surface melting have shown promise in improving corrosion resistance. This process leads to microstructural refinement and homogenization of alloying elements, enhancing the material's resistance to corrosive environments [83].

Protective oxide layer formation: Alloys that form stable, adherent oxide layers (like Cr_2O_3 or Al_2O_3) on their surface can provide superior protection against corrosion. For instance, Cr_2O_3-forming alloys have shown better resistance to hot corrosion at 900°C compared to Al_2O_3-forming alloys [93].

9.2.2 Wear Resistance

Wear is another critical issue, particularly in components subject to mechanical stress and friction. Approaches to improve wear resistance include:

- Microstructural optimization: Developing alloys with optimized microstructures can enhance wear resistance. This may involve controlling grain size and orientation, and introducing dispersed strengthening precipitates [87].

- Surface engineering: Methods such as laser surface melting are known to enhance the hardness and wear resistance of structural materials in conditions that mimic those found in nuclear settings [87].

- Advanced coatings: Application of wear-resistant coatings can significantly extend the life of components subject to mechanical wear [94].

Superalloys: Fundamentals and Applications   Materials Research Forum LLC
Materials Research Foundations 178 (2025)143-199   https://doi.org/10.21741/9781644903698-9

9.2.3   Challenges in extreme conditions

Future nuclear reactor designs aim to operate at even higher temperatures (500-1000°C) and with more corrosive coolants, presenting additional challenges for superalloys [92]. To address these challenges:

- Novel alloy development: Researchers are exploring new alloying elements and compositions to create superalloys capable of withstanding these extreme conditions.

- Innovative fabrication methods: Methods like additive manufacturing are being investigated to produce parts featuring complex geometries and customized microstructures.

- Comprehensive testing: Accelerated testing methods and advanced modeling techniques are being developed to predict long-term material behavior under these extreme conditions.

By addressing these challenges through innovative materials design, advanced processing techniques, and comprehensive research efforts, the nuclear industry can develop superalloys capable of withstanding the harsh conditions in next-generation nuclear plants, ensuring safer and more efficient operations.

9.3   Overcoming supply and cost issues related to rare alloying elements

Rare earth elements (REE) are essential for contemporary technologies and civilization, ranking among the most crucial elements. The development and production of superalloys for nuclear plants face significant challenges related to the supply and cost of rare alloying elements. These issues can impact the economic viability and sustainability of superalloy production for nuclear applications.

9.3.1   Supply risk of rare elements

A major challenge in the use of superalloys is the supply risk related to specific scarce elements used in their composition [95]. A semi-quantitative evaluation of supply risks for elements in nickel-based superalloys shows that rhenium, molybdenum, and cobalt are among those with the highest supply risks [96]. These elements are essential for improving superalloy performance, yet their scarce availability and geopolitical influences may create vulnerabilities in the supply chain. Strategies to mitigate these supply risks include:

- Diversifying supply sources to reduce dependence on a single region or supplier. Supply chain diversification will bring more agility to the supply availability.

- Developing alternative alloy compositions that use more readily available elements.

- Implementing recycling programs to recover rare elements from superalloy scraps. Rare earth elements are vital, yet only 1% are recycled. Increasing REE recycling will help overcome some of these elements' criticality difficulties [95].

9.3.2   Cost considerations

The significant expense associated with rare alloying elements presents a notable challenge. For example, incorporating rhenium into second-generation single crystal alloys has expanded the range of supply risk values due to the impact of its cost share [97]. Strategies to address these cost concerns include:

- Optimizing alloy compositions to reduce the content of expensive elements without compromising performance.

- Developing more efficient manufacturing processes to minimize material waste.

- Exploring alternative, less expensive elements that can provide similar performance benefits.

9.3.3 Recycling and recovery

Recycling superalloy scraps and the recovery of rare elements is a crucial strategy for addressing both supply and cost challenges. Innovative recycling methods are being developed to efficiently recover valuable elements:

- Ultrasonic leaching methods have shown promise in recovering rare metals from superalloy scraps. This technique can achieve high leaching percentages for elements like rhenium, nickel, cobalt, aluminum, and chromium [97].

- Two-stage separation processes have been developed to recover rare metals from leaching solutions, allowing for more efficient recycling of valuable elements [97].

- Hydrometallurgical technologies are being explored for the production of compounds like manganese(II) perrhenate dihydrate from recycled materials, offering a pathway to recover both manganese and rhenium [98].

9.3.4 Technological advancements

To address these issues, continuous research and development activities are concentrated on:

- Developing new alloy compositions that maintain or improve performance while reducing reliance on scarce elements.

- Improving manufacturing techniques, such as additive manufacturing, to optimize material usage and reduce waste.

- Enhancing recycling technologies to increase the recovery rates of rare elements from superalloy scraps.

- Exploring the potential of computational modeling to predict alloy performance and optimize compositions, potentially reducing the need for extensive experimental testing.

By tackling supply and cost challenges with inventive approaches and technological progress, the nuclear sector can secure the ongoing evolution and utilization of high-performance superalloys in nuclear facilities. These initiatives will enhance the continued reliability and economic feasibility of nuclear power generation.

Superalloys: Fundamentals and Applications   Materials Research Forum LLC
Materials Research Foundations 178 (2025)143-199   https://doi.org/10.21741/9781644903698-9

10.   Future trends and prospects

10.1   Potential of superalloys in next-generation nuclear reactors

The future of superalloys in next-generation nuclear reactors is marked by significant potential and ongoing innovation. These advanced materials are being developed to meet the stringent requirements of Generation IV and advanced small modular reactors (SMRs), which demand exceptional performance under extreme conditions. Laves phase alloys, in particular, are gaining attention for their high-temperature strength, phase stability, and resistance to oxidation and corrosion, making them ideal for ultra-high-temperature structural applications. As the industry shifts towards alternative coolants like molten salt and liquid metals, superalloys are being tailored to exhibit enhanced thermal stability and compatibility with these new environments.

Advanced manufacturing methods, particularly additive manufacturing (AM), are transforming the way superalloys are produced, enabling the construction of intricate shapes and custom-designed microstructures. These advancements contribute to superior mechanical characteristics and improved performance in nuclear settings. There is an increasing focus on developing high-entropy alloys (HEAs) and other sophisticated compositions known for their exceptional strength, ductility, and resistance to radiation damage. High-throughput computational modeling and advanced analytical methods are expediting the creation of new alloy systems by enhancing understanding of the connections between structure and properties.

Environmental sustainability is now a pivotal factor in the development of superalloys. Initiatives aimed at minimizing environmental impacts are enhancing manufacturing efficiencies and promoting the use of recycled materials. The importance of lifecycle assessments is growing, ensuring that these materials are both effective and eco-friendly. Efforts to prolong the lifespan of superalloys, enhance their recyclability, and develop more environmentally conscious compositions are advancing, aiming to improve the overall lifecycle management of superalloys. As ongoing research continues to expand the capabilities of these materials, even stronger, more efficient, and more sustainable superalloys are projected to emerge, playing a vital contribution to the future of nuclear reactors and thereby boosting the safety, efficiency, and environmental friendliness of nuclear energy.

10.2   Emerging research and new alloy systems

The prospects for superalloys in nuclear facilities are set for considerable progress, fueled by current research initiatives and the creation of innovative alloy systems. Notably, the Laves phase alloys are gaining interest because of their superior high-temperature capabilities, positioning them as suitable materials for ultra-high-temperature structural uses in future nuclear reactors. Despite their potential, researchers are actively working to overcome the challenge of room-temperature brittleness through alloy design, doping, and crystal structure modulation. Advanced analytical techniques are enabling a deeper understanding of these alloys, paving the way for optimized properties and enhanced performance.

High-entropy alloys (HEAs) represent another exciting frontier in superalloy development for nuclear applications. These multi-principal element alloys offer unique properties such as superior strength, ductility, and resistance to radiation damage, making them highly suitable for next-generation nuclear reactors. The design and optimization of HEAs are being facilitated by high-throughput computational simulations and experimental techniques, accelerating their development and potential integration into nuclear systems.

Additive manufacturing (AM) is reshaping the fabrication of superalloys, enabling the creation of complex designs and custom-tailored microstructures previously unattainable. Innovations like AMALLOY-HT highlight AM's ability to generate materials that excel under harsh conditions. This method not only speeds up the prototype testing and development of new alloy systems but also supports eco-friendly manufacturing practices by reducing waste and conserving energy.

The shift towards novel cooling and fuel systems in contemporary nuclear reactors is driving the need for advanced superalloy formulations. Alloys intended for use with molten salts, liquid metals, or gas coolants must exhibit exceptional thermal stability, corrosion resistance, and compatibility with these innovative systems. This necessity is fostering research into alloy compositions that are robust and reliable in these challenging environments, thus pushing the boundaries of nuclear material science.

Furthermore, sustainability is becoming an essential element in the development of superalloys for nuclear use. Efforts are concentrated on diminishing the environmental effects of superalloy production by optimizing manufacturing processes and recycling materials. The importance of lifecycle assessments is on the rise, verifying that these materials are efficient and sustainable. Actions are being taken to extend the service life of superalloys, enhance their recyclability, and develop more eco-friendly alloy compositions, aligning nuclear practices with broader ecological goals.

Ongoing research continues to navigate the complexities of these sophisticated materials, with expectations set on the development of even more robust, efficient, and sustainable superalloys essential for future nuclear reactors. The persistent application of advanced analytical and manufacturing techniques is crucial for enhancing these alloys' characteristics and ensuring their applicability. The future of superalloys in nuclear facilities is characterized by an amalgamation of state-of-the-art material science, progressive manufacturing strategies, and a strong commitment to sustainability, promising significant progress in reactor design, safety, and efficiency.

10.3   Sustainability in the production and use of superalloys

The future of superalloys in nuclear plants is increasingly intertwined with sustainability considerations, as the industry evolves to meet environmental challenges and economic demands. A key trend in sustainable superalloy production is the adoption of more efficient manufacturing processes, particularly additive manufacturing (AM). This advanced technique, exemplified by the development of high-strength aluminum alloys like AMALLOY-HT, offers significant sustainability benefits by reducing material waste and energy consumption compared to traditional methods. The ability to create complex geometries and tailored microstructures with minimal waste not only enhances environmental sustainability but also reduces economic costs associated with material usage and energy expenditure.

Recycling and closed-loop systems represent another crucial trend in the sustainable use of superalloys. The implementation of advanced recycling technologies to recover valuable elements from used superalloys supports a more circular economy approach. This closed-loop system optimizes the lifecycle of superalloys, significantly reducing the demand for raw materials, lowering energy consumption, and minimizing waste. Such practices are essential in reducing the environmental footprint of superalloy production and use in nuclear applications.

The shift towards alternative coolants and fuel systems in next-generation nuclear reactors also has profound implications for the sustainability of superalloys. Coolants such as molten salt, liquid metals, and gases offer improved safety and efficiency, potentially leading to more efficient energy production and reduced fuel consumption. Superalloys designed for these new coolant systems must be compatible and resistant to specific chemical and thermal conditions, thereby enhancing the overall sustainability of reactor systems. This trend aligns with the broader goal of developing more efficient and environmentally friendly nuclear power generation.

Efforts to minimize material consumption and prolong the operational life of superalloys are central to sustainability initiatives. Modern nuclear reactor designs, such as small modular reactors (SMRs) and Generation IV reactors, utilize modular construction and incorporate passive safety mechanisms that enhance operational longevity and decrease waste. Investigating high-entropy alloys and other cutting-edge materials for their exceptional qualities could also lengthen the lifespan of superalloys in nuclear settings, thereby curtailing the frequency of replacements and lessening environmental burdens.

Lifecycle assessments (LCAs) are increasingly pivotal in assessing and enhancing the sustainability of superalloys. By analyzing every phase of a material's lifecycle—from the extraction of raw materials through to recycling or disposal at the end of its life—LCAs pinpoint opportunities to improve sustainability. This holistic approach, enhanced by advanced computational modeling and simulation tools, facilitates the more effective design and optimization of superalloys, aiming to reduce energy consumption and material waste.

As the nuclear sector progresses, its emphasis on sustainability continues to influence the development and utilization of superalloys. Through the adoption of advanced technologies, the establishment of closed-loop recycling practices, and the optimization of material use and lifespan, the industry is advancing towards a more environmentally sustainable and economically feasible future. These trends not only mitigate current environmental issues but also set the stage for more sustainable methods in the production and application of superalloys in nuclear facilities, ensuring their enduring importance in the quest for clean and efficient energy solutions.

11. Conclusions

11.1 Summary of the role and future of superalloys in nuclear energy technology

Superalloys are fundamental to the advancement of nuclear energy technology, deeply integrated with the sector's progress. These materials are crucial in nuclear facilities for their ability to maintain high-temperature strength, resist corrosion, and preserve mechanical properties in severe conditions. They are widely utilized in critical reactor components such as turbine blades, heat exchangers, and structural elements, largely because of the superior heat tolerance and creep resistance provided by nickel-based superalloys. Looking forward, the innovation of new superalloy systems designed for the rigorous environments of emerging nuclear reactors is a key development. High-entropy alloys (HEAs) are currently under exploration for their exceptional strength, ductility, and resistance to radiation, making them suitable for the extreme conditions of cutting-edge reactor designs. Additionally, the use of pioneering manufacturing technologies like additive manufacturing (AM) is reshaping superalloy production, enabling the creation of intricate designs and tailored microstructures that reduce waste and conserve energy.

Ongoing research is concentrated on improving superalloy compositions, including oxide dispersion-strengthened (ODS) alloys and Laves phase alloys, noted for their resistance to high-temperature creep and robustness against neutron damage. Advanced analytical and computational methodologies are crucial in elucidating the structural and property relationships within these alloys, supporting their refinement for nuclear applications. Environmental sustainability is increasingly significant in the lifecycle of superalloys. Efforts are being made to reduce the ecological footprint of these materials through more efficient production methods and the incorporation of recycled materials. Lifecycle assessments are critical in verifying the eco-friendliness of superalloys, involving strategies to prolong their lifecycle, improve their recyclability, and create alloy compositions that minimize reliance on rare elements. The progression of superalloys is intimately connected with the evolution of advanced nuclear reactors, including small modular reactors (SMRs) and Generation IV reactors, which operate at elevated temperatures and utilize diverse coolants such as liquid metals and molten salts. These applications demand superalloys with enhanced thermal stability and corrosion resistance that are compatible with new coolant environments.

In conclusion, superalloys remain vital for the continuous development and safety of nuclear reactors. As the industry progresses towards innovative designs and alternative coolants, the demand for superalloys that can perform under harsh conditions will escalate. Continued research into novel alloys, advanced manufacturing methods, and sustainability considerations is essential for meeting these challenges. These endeavors will help propel the nuclear industry toward a more sustainable, efficient, and secure future, ensuring that nuclear energy remains a crucial element of the worldwide energy framework.

11.2   Reflection on ongoing research and development needs

The conclusion of this chapter highlights the critical need for continuous research and development to progress and sustain superalloys in the nuclear industry. Superalloys, primarily composed of nickel, cobalt, and iron, are indispensable in the extreme environments of nuclear reactors due to their ability to maintain mechanical integrity under high temperatures, resist oxidation and corrosion, and withstand irradiation effects. However, with the evolving specifications of next-generation nuclear reactors, such as Generation IV and advanced small modular reactors (SMRs), further enhancements in superalloy technologies are required to meet the challenges posed by increased operational temperatures and the introduction of novel coolant systems. A significant research focus is on developing new and improved alloy formulations, with high-entropy alloys (HEAs) and oxide dispersion-strengthened (ODS) alloys showing considerable promise due to their superior strength, ductility, and resistance to radiation damage, making them well-suited for demanding nuclear environments. Concurrently, the implementation of innovative manufacturing techniques like additive manufacturing (AM) is revolutionizing superalloy fabrication. AM facilitates the creation of complex structures and tailored microstructures, potentially boosting mechanical properties while reducing material waste and energy consumption. However, challenges such as porosity and cracking in AM components need to be addressed to fully capitalize on this technology. A thorough knowledge of superalloy metallurgy, particularly phase transformations and microstructural stability, is essential for optimizing their effectiveness. Advanced analytical and computational techniques are applied to explore these phenomena and predict the behavior of superalloys under varying conditions, which is essential for developing materials that can withstand the rigorous demands of future nuclear reactors.

Sustainability has become a crucial factor in the production and usage of superalloys. Initiatives are underway to minimize their environmental impact by optimizing manufacturing processes, leveraging recycled materials, and performing extensive lifecycle evaluations. These measures are designed to enhance the environmental sustainability of superalloys from their inception through to their disposal, aligning with wider environmental targets. Moreover, as the nuclear industry explores alternative cooling and fuel systems, such as molten salts, liquid metals, and gases, there is an increasing demand for superalloys that are resistant to corrosion and capable of maintaining their structural integrity in these challenging environments. Research is also focused on formulating alloys that are chemically suited for specialized roles, like the breeding blankets in fusion reactors. Furthermore, advancements in nanotechnology are being explored to develop self-healing alloys, which could dramatically increase the service life of nuclear reactors and reduce maintenance costs. These progressive approaches aim to significantly bolster the extended service life and reliability of superalloys in nuclear applications.

In conclusion, the ongoing progress of superalloys in nuclear facilities is underpinned by essential and continuous research and development efforts. Meeting these challenges is crucial for maintaining superalloys as a leading material technology, which supports the advancement of safer, more efficient, and sustainable nuclear power generation. As the nuclear sector advances, the adoption of cutting-edge analytical methods, nanotechnology, and progressive manufacturing techniques will be essential to address the material challenges and to fully harness the capabilities of superalloys in this industry. The sustained development of these materials will significantly influence the future of nuclear energy and aid in the global shift toward cleaner, more sustainable energy solutions.

References

[1] J.T. Busby, Overview of structural materials in water-cooled fission reactors, in: G.R. Odette, S.J. Zinkle (Eds.), Structural Alloys for Nuclear Energy Applications, Elsiever, 2019. https://doi.org/10.1016/B978-0-12-397046-6.00001-0.

[2] R. Mitra, R.J.H. Wanhill, Structural Intermetallics, in: N. Eswara Prasad, R.J.H. Wanhill (Eds.), Aerospace Materials and Material Technologies, Springer, Singapore, 2017. https://doi.org/https://doi.org/10.1007/978-981-10-2134-3.

[3] M. Griffiths, Ni-based alloys for reactor internals and steam generator applications, in: G.R. Odette, S.J. Zinkle (Eds.), Structural Alloys for Nuclear Energy Applications, Elsevier, 2019. https://doi.org/10.1016/B978-0-12-397046-6.00009-5.

[4] A. Strasser, F.P. Ford, High Strength Ni-alloys in Fuel Assemblies, ANT International, Molnlycke, 2012.

[5] R. Cozar, A. Pineau, Morphology of y' and y" precipitates and thermal stability of inconel 718 type alloys, Metall. Trans. 4 (1973) 47–59. https://doi.org/10.1007/BF02649604.

[6] R.C. Read, The Superalloys Fundamentals and Application, Cambridge, UK (2006).

[7] R.G. Davies, N.S. Stoloff, On the yield stress of aged Ni-Al alloys, Trans. Metall. Soc. AIME. 233 (1965) 714.

[8] W. Zheng, H. Zhang, B. Du, H. Li, H. Yin, X. He, T. Ma, High-temperature corrosion behavior of incoloy 800h alloy in the impure helium environment, Sci. Technol. Nucl. Install. 2022 (2022) 8098585. https://doi.org/https://doi.org/10.1155/2022/8098585.

[9] C.J. Tsai, T.-K. Yeh, M.Y. Wang, High temperature oxidation behavior of nickel and iron based superalloys in helium containing trace impurities, Corros. Sci. Technol. 18 (2019) 8–15. https://api.semanticscholar.org/CorpusID:139444008.

[10] H.K.D.H. Bhadeshia, Nickel based superalloys. https://www.phase-trans.msm.cam.ac.uk/2003/Superalloys/superalloys.html, 2003 (accessed October 7, 2024).

[11] I.G. Akande, O.O. Oluwole, O.S.I. Fayomi, O.A. Odunlami, Overview of mechanical, microstructural, oxidation properties and high-temperature applications of superalloys, Mater. Today Proc. 43 (2021) 2222–2231. https://doi.org/10.1016/j.matpr.2020.12.523.

[12] AEETHER, Clearly Understand Superalloys through 3 Classification Methods, (2021). https://www.aeether.com/AEETHER/media/media-32/media.html (accessed October 9, 2024).

[13] R.C. Reed, The Superalloys, Cambridge University Press, 2006. https://doi.org/10.1017/CBO9780511541285.

[14] M.J. Donachie, S.J. Donachie, Superalloys, ASM International, 2002. https://doi.org/10.31399/asm.tb.stg2.9781627082679.

[15] T.M. Pollock, S. Tin, Nickel-Based Superalloys for Advanced Turbine Engines: Chemistry, Microstructure and Properties, American Institute of Aeronautics, (2006) 361–374. https://doi.org/10.2514/1.18239.

[16] C. Chen, Q. Wang, C. Dong, Y. Zhang, H. Dong, Composition rules of Ni-base single crystal superalloys and its influence on creep properties via a cluster formula approach, Sci. Rep. 10 (2020) 21621. https://doi.org/10.1038/s41598-020-78690-8.

[17] K.A. Rao, Nickel based superalloys – properties and their applications, Int.J. Manag. Technol. Eng. 8 (2018).

[18] U. Nwachukwu, A. Obaied, O.M. Horst, M.A. Ali, I. Steinbach, I. Roslyakova, Microstructure property classification of nickel-based superalloys using deep learning, Model. Simul. Mat. Sci. Eng. 30 (2022) 025009. https://doi.org/10.1088/1361-651X/ac3217.

[19] Total Materia, Nickel-Based Superalloys: Part-Two. https://www.totalmateria.com/en-us/articles/nickel-based-superalloys-2/ (2010) (accessed October 9, 2024).

[20] J.N. DuPont, J.C. Lippold, S.D. Kiser, Welding Metallurgy and Weldability of Nickel-Base Alloys, Wiley, 2009. https://doi.org/10.1002/9780470500262.

[21] J.R. Davis, Heat Resistant Materials - ASM Specialty Handbook, ASM International, Ohio, 1997.

[22] L. Tan, G. Wang, Y. Guo, Q. Fang, Z. Liu, X. Xiao, W. He, Z. Qin, Y. Zhang, F. Liu, L. Huang, Additively manufactured oxide dispersion strengthened nickel-based superalloy with superior high temperature properties, Virtual. Phys. Prototyp. 15 (2020) 555–569. https://doi.org/10.1080/17452759.2020.1848283.

[23] M. Griffiths, Ni-based alloys for reactor internals and steam generator applications, in: Structural Alloys for Nuclear Energy Applications, Elsevier, 2019, pp. 349–409. https://doi.org/10.1016/B978-0-12-397046-6.00009-5.

[24] Nickel Institute, Nickel Alloys in Energy and Power, https://nickelinstitute.org/en/nickel-applications/nickel-alloys-in-energy-and-power/ 2024 (accessed October 13, 2024).

[25]   D. Coutsouradis, A. Davin, M. Lamberigts, Cobalt-based superalloys for applications in gas turbines, Mater. Sci. Eng. 88 (1987) 11–19. https://doi.org/10.1016/0025-5416(87)90061-9.

[26]   D. Klarstrom, P. Crook, S. Mridha, Cobalt alloys and designation system, in: Reference Module in Materials Science and Materials Engineering, Elsevier, 2018. https://doi.org/10.1016/B978-0-12-803581-8.02558-3.

[27]   M.G. Raucci, A. Gloria, R. De Santis, L. Ambrosio, K.E. Tanner, Introduction to biomaterials for spinal surgery, in: Biomaterials for Spinal Surgery, Elsevier, 2012: pp. 1–38. https://doi.org/10.1533/9780857096197.1.

[28]   AEETHER, Introduction of Cobalt-based Superalloys - The Choice for Extreme Environments, https://www.aeether.com/AEETHER/media/alloy-knowledge/index.html/ 2024 (accessed October 10, 2024)

[29]   Z. Zhang, Q. Ding, X. Wei, Z. Zhang, H. Bei, Effects of alloying elements on the microstructure of precipitation strengthened Co-based superalloys, J. Alloys Compds. 989 (2024) 174401. https://doi.org/10.1016/j.jallcom.2024.174401.

[30]   F.L. Reyes Tirado, J. Perrin Toinin, D.C. Dunand, $\gamma+\gamma'$ microstructures in the Co-Ta-V and Co-Nb-V ternary systems, Acta. Mater. 151 (2018) 137–148. https://doi.org/10.1016/j.actamat.2018.03.057.

[31]   J. Coakley, E.A. Lass, D. Ma, M. Frost, H.J. Stone, D.N. Seidman, D.C. Dunand, Lattice parameter misfit evolution during creep of a cobalt-based superalloy single crystal with cuboidal and rafted gamma-prime microstructures, Acta. Mater. 136 (2017) 118–125. https://doi.org/10.1016/j.actamat.2017.06.025.

[32]   Cobalt Institute, Superalloys, https://www.cobaltinstitute.org/essential-cobalt-2/cobalt-innovations/superalloys/ 2024 (accessed October 8, 2024).

[33]   R. Moat, Cobalt-free Hard-facing for Reactor Systems, https://gtr.ukri.org/projects?ref=EP%2FT016728%2F1#/tabOverview/ 2024 (accessed October 5, 2024).

[34]   R. Zhang, C. Zhang, Z. Wang, J. Liu, Evolution of recrystallization texture in a286 iron-based superalloy thin plates rolled via various routes, Metals. (Basel) 13 (2023) 1527. https://doi.org/10.3390/met13091527.

[35]   A. Günen, M. Keddam, S. Alkan, A. Erdoğan, M. Çetin, Microstructural characterization, boriding kinetics and tribo-wear behavior of borided Fe-based A286 superalloy, Mater. Charact. 186 (2022) 111778. https://doi.org/10.1016/j.matchar.2022.111778.

[36]   A.P. Mouritz, Superalloys for gas turbine engines, in: Introduction to Aerospace Materials, Elsevier, 2012: pp. 251–267. https://doi.org/10.1533/9780857095152.251.

[37]   AEETHER, What is Iron-based superalloy? Which characteristics does It have?, (2022). https://www.aeether.com/AEETHER/media/media-71/media.html (accessed October 11, 2024).

[38]   X. Zhou, H. Dong, Y. Wang, M. Yuan, Microstructure characteristics and mechanical performance of Fe-Cr-Ni-Al-Ti superalloy fabricated by powder metallurgy, J. Alloys Compds. 918 (2022) 165612. https://doi.org/10.1016/j.jallcom.2022.165612.

[39]   Teknovak, What is a Super Alloy? Where is it used?, https://www.teknovak.com/en/what-is-a-super-alloy-where-is-it-used/ 2022 (accessed October 5, 2024).

[40]   ICDACr, Superalloys, https://www.icdacr.com/applications/superalloys/2024 (accessed October   6,2024).

[41]   AEETHER, Clearly Understand the 4 Roles of Chromium in Superalloys,
https://www.aeether.com/AEETHER/media/media-6/media.html/2021   (accessed October 6,
2024).

[42]   R. Sowa, S. Arabasz, M. Parlinska-Wojtan, Classification and microstructural stability of
high generation single crystal Nickel-based superalloys, Zastita Materijala. 57 (2016) 274–
281. https://doi.org/10.5937/ZasMat1602274S.

[43]   The Alloys Network, Nickel Based Superalloys,   https://nickel-alloys.net/article/nickel-
based-superalloys.html/2024 (accessed October 9, 2024).

[44]   M. Seifollahi, S. Kheirandish, S.H. Razavi, S.M. Abbasi, P. Sahrapour, Effect of η Phase
on mechanical properties of the iron-based superalloy using shear punch testing, ISIJ Int. 53
(2013) 311–316. https://doi.org/10.2355/isijinternational.53.311.

[45]   AEETHER, What is Iron-based Superalloy? Which Characteristics does It
have?https://www.aeether.com/AEETHER/media/media-71/media.html/2022 (accessed
October   7,   2024).

[46]   J.K. Tien, R.N. Jarrett, Effects of cobalt in nickel-base superalloys, in: High Temperature
Alloys for Gas Turbines, Springer Netherlands, Dordrecht, 1982, pp. 423–446.
https://doi.org/10.1007/978-94-009-7907-9_17.

[47]   AEETHER, Clearly Understand the Role of Cobalt in All 3 Types of Superalloys,
https://www.aeether.com/AEETHER/media/media-37/media.html/2021 (accessed October 7, 2024).

[48]   C. Hu, S. Hou, Effects of rare earth elements on properties of ni-base superalloy powders
and coatings, Coatings. 7 (2017) 30. https://doi.org/10.3390/coatings7020030.

[49]   X.L. Pan, H.Y. Yu, G.F. Tu, W.R. Sun, Z.Q. Hu, Effect of rare earth metals on
solidification behaviour in nickel based superalloy, Mater. Sci. Technol. 28 (2012) 560–564.
https://doi.org/10.1179/1743284711Y.0000000069.

[50]   K. Herfurth, S. Scharf, Casting, Springer Handbooks, (2021) 325–356.
https://doi.org/10.1007/978-3-030-47035-7_10.

[51]   J.K. Tien, V.C. Nardone, The U.S. superalloys industry — status and outlook, JOM, 36
(1984) 52–57. https://doi.org/10.1007/BF03338563.

[52]   T.M. Pollock, S. Tin, Nickel-Based Superalloys for Advanced Turbine Engines:
Chemistry, Microstructure, and Properties, American Institute of Aeronautics, 2006, 361–374.
https://doi.org/10.2514/1.18239.

[53]   E. Kaya, B. Akyüz, Effects of cutting parameters on machinability characteristics of Ni-
based superalloys: A review, Open Eng. 7 (2017) 330–342. https://doi.org/10.1515/eng-2017-
0037.

[54]   Y. Tang, M.Z. Ge, Y. Zhang, T. Wang, W. Zhou, Improvement of fatigue life of GH3039
superalloy by laser shock peening, Materials. 13 (2020) 3849.
https://doi.org/10.3390/ma13173849.

[55]   T. Klotz, M. Lévesque, M. Brochu, Effects of rolled edges on the fatigue life of shot
peened Inconel 718, J. Mater. Process. Technol. 263 (2019) 276–284.
https://doi.org/10.1016/j.jmatprotec.2018.08.019.

[56]   P. Peyre, R. Fabbro, P. Merrien, H.P. Lieurade, Laser shock processing of aluminium
alloys, Application to high cycle fatigue behaviour, Mater. Sci. Eng. A. 210 (1996) 102–113.
https://doi.org/10.1016/0921-5093(95)10084-9.

[57]   C. Correa, L. Ruiz De Lara, M. Díaz, J.A. Porro, A. García-Beltrán, J.L. Ocaña, Influence
of pulse sequence and edge material effect on fatigue life of Al2024-T351 specimens treated

by laser shock processing, Int. J. Fatigue. 70 (2015) 196–204.
https://doi.org/10.1016/j.ijfatigue.2014.09.015.

[58] B. Meng, S. Yang, J. Zhao, J. Wang, M. Chen, F. Wang, The impact of various superalloys on the oxidation performance of nanocrystalline coatings at high temperatures, Coatings. 13 (2023) 1770. https://doi.org/10.3390/coatings13101770.

[59] J. Wang, M. Chen, S. Zhu, F. Wang, Ta effect on oxidation of a nickel-based single-crystal superalloy and its sputtered nanocrystalline coating at 900-1100 °C, Appl. Surf. Sci. 345 (2015) 194–203. https://doi.org/10.1016/j.apsusc.2015.03.157.

[60] A. Behera, Advanced Materials: An Introduction to Modern Materials Science, Springer, International Publishing, 2021. https://doi.org/10.1007/978-3-030-80359-9.

[61] R.C. Reed, Superalloys - Fundamentals and Applications, Cambridge University Press, 2021, pp.291.

[62] M.J. Donachie, S. James Donachie, Superalloys : A Technical Guide, Second ed., ASM International, 2002.

[63] Superalloys, InTech, http://dx.doi.org/10.5772/59358/2015 (accessed October 14, 2024).

[64] H. Long, S. Mao, Y. Liu, Z. Zhang, X. Han, Microstructural and compositional design of Ni-based single crystalline superalloys — A review, J. Alloys Compds. 743 (2018) 203–220. https://doi.org/10.1016/j.jallcom.2018.01.224.

[65] Y. Du, R. Schmid-Fetzer, J. Wang, S. Liu, J. Wang, Z. Jin, Computational Design of Engineering Materials, Cambridge University Press, 2023. https://doi.org/10.1017/9781108643764.

[66] T.L. Achmad, F.F. Baskara, Design of high entropy superalloy fenicralcu using computational thermodynamic and machine learning: effect of alloying compositions and temperatures on the stacking fault energy, in: E3S Web of Conferences, EDP Sciences, 2024. https://doi.org/10.1051/e3sconf/202454303010.

[67] T. Yonezawa, 2.08 - Nickel Alloys: Properties and Characteristics, Elsevier Inc., 2012. https://doi.org/10.1016/B978-0-08-056033-5.00016-1.

[68] Y. Nagumo, A.T. Yokobori, R. Sugiura, T. Matsuzaki, H. Takeuchi, Y. Ito, Behavior of branch cracking and the microstructural strengthening mechanism of polycrystalline Ni-base superalloy, IN100 under creep condition, Mater Trans 52 (2011) 1876–1884. https://doi.org/10.2320/matertrans.M2011074.

[69] L.K. Murty, I. Charit, An Introduction to Nuclear Materials, 2015, pp. 398.

[70] P. Zhang, Q. Zhu, G. Chen, H. Qin, C. Wang, Fatigue behavior and life prediction model of a nickel-base superalloy under different strain conditions, Mater. Trans. 57 (2016) 25–32. https://doi.org/10.2320/matertrans.M2015316.

[71] P. Tipping, Lifetime and ageing management of nuclear power plants: A brief overview of some light water reactor component ageing degradation problems and ways of mitigation, Int. J. Press. Vessels Pip. 66 (1996) 17–25. https://doi.org/10.1016/0308-0161(95)00082-8.

[72] L. Wang, Y. Wang, H. Du, J. Zuo, R. Yi Man Li, Z. Zhou, F. Bi, M.P. Garvlehn, A comparative life-cycle assessment of hydro, nuclear and wind power: a China study, Appl. Energy. 249 (2019) 37–45. https://doi.org/10.1016/j.apenergy.2019.04.099.

[73] I. Aho-Mantila, O. Cronvall, U. Ehrnstén, H. Keinänen, R. Rintamaa, A. Saarenheimo, K. Simola, E. Vesikari, Lifetime prediction techniques for nuclear power plant systems, Nucl. Corros. Sci. Eng. (2012) 449–470. https://doi.org/10.1533/9780857095343.4.449.

[74]   E. Getto, B. Tobie, E. Bautista, A. Bullens, D.W. Gandy, J.P. Wharry, Grain Evolution in Thermally Aged Cast and Hot Isostatic Pressed Inconel 625, Microscopy and Microanalysis 24 (2018) 666–667. https://doi.org/10.1017/s1431927618003823.

[75]   O. Woo, C. Judge, H. Nordin, D. Finlayson, C. Andrei, The microstructure of unirradiated and neutron irradiated inconel X750, microscopy and microanalysis 17 (2011) 1852–1853. https://doi.org/10.1017/s1431927611010130.

[76]   Y. Lu, J. Liu, X. Li, L. Jiang, Z. Li, G. Wu, X. Zhou, Effect of long time thermal exposure on microstructure and mechanical properties of C276 superalloy, 8th Pacific Rim International Congress on Advanced Materials and Processing, PRICM 8 (2013) 345–352. https://doi.org/10.1007/978-3-319-48764-9_42.

[77]   S.K. Selvaraj, G. Sundaramali, S. Jithin Dev, R. Srii Swathish, R. Karthikeyan, K.E. Vijay Vishaal, V. Paramasivam, Recent advancements in the field of ni-based superalloys, Adv. Mater. Sci. Eng. 2021 (2021) 1-60. https://doi.org/10.1155/2021/9723450.

[78]   W.D. Bian, H.R. Zhang, M. Gao, Q.L. Li, J.P. Li, T.X. Tao, H. Zhang, Influence of yttrium and vacuum degree on the purification of K417 superalloy, Vacuum. 152 (2018) 57–64. https://doi.org/10.1016/J.VACUUM.2018.02.031.

[79]   E. Cortes, A. Bedolla-Jacuinde, M. Rainforth, I. Mejia, A. Ruiz, N. Ortiz, J. Zuno, Role of titanium, carbon, boron, and zirconium in carbide and porosity formation during equiaxed solidification of nickel-based superalloys, J. Mater. Eng. Perform. 28 (2019) 4171–4186. https://doi.org/10.1007/s11665-019-04179-9.

[80]   Y. Amouyal, D.N. Seidman, The role of hafnium in the formation of misoriented defects in Ni-based superalloys: An atom-probe tomographic study, Acta. Mater. 59 (2011) 3321–3333. https://doi.org/10.1016/J.ACTAMAT.2011.02.006.

[81]   M. Samaras, Nuclear energy systems: advanced alloys and cladding, in: K.H.J. Buschow, R.W. Cahn, M.C. Flemings, B. Ilschner, E.J. Kramer, P. Veyssiere, S. Mahajan (Eds.), Encyclopedia of Materials: Science and Technology, Elsevier, 2011, pp. 1–7. https://doi.org/10.1016/B978-0-08-043152-9.02263-6.

[82]   S. Taller, G. VanCoevering, B.D. Wirth, G.S. Was, Predicting structural material degradation in advanced nuclear reactors with ion irradiation, Sci. Rep. 11 (2021) 2949. https://doi.org/10.1038/s41598-021-82512-w.

[83]   S. Basak, S.K. Sharma, K.K. Sahu, S. Gollapudi, J.D. Majumdar, Surface modification of structural material for nuclear applications by electron beam melting: Enhancement of microstructural and corrosion properties of Inconel 617, SN Appl. Sci. 1 (2019) 708. https://doi.org/10.1007/s42452-019-0744-5.

[84]   A. Günen, Ö. Ergin, A comparative study on characterization and high-temperature wear behaviors of thermochemical coatings applied to cobalt-based haynes 25 superalloys, Coatings.13 (2023) 1272. https://doi.org/10.3390/coatings13071272.

[85]   A. Kollová, K. Pauerová, Superalloys - characterization, usage and recycling, Manuf. Technol. 22 (2022) 550–557. https://doi.org/10.21062/mft.2022.070.

[86]   T. Wiss, A. Benedetti, E. De Bona, Radiation effects in UO2, in: R.J.M. Konings (Ed.), Comprehensive Nuclear Materials, Elsevier, 2020: pp. 125–148. https://doi.org/10.1016/B978-0-12-803581-8.12047-8.

[87]   Development of radiation resistant reactor core structural materials, in: International Atomic Energy Agency, 2007. https://api.semanticscholar.org/CorpusID:111380797.

Superalloys: Fundamentals and Applications　　　　　　　　Materials Research Forum LLC
Materials Research Foundations 178 (2025)143-199　　　https://doi.org/10.21741/9781644903698-9

[88]　M. Griffiths, S. Xu, J.E. Ramos Nervi, Swelling and he-embrittlement of austenitic stainless steels and ni-alloys in nuclear reactors, Metals. (Basel) 12 (2022) 1692. https://doi.org/10.3390/met12101692.

[89]　F. Zhang, L. Boatner, Y. Zhang, D. Chen, Y. Wang, L. Wang, Swelling and helium bubble morphology in a cryogenically treated fecrni alloy with martensitic transformation and reversion after helium implantation, Materials. 12 (2019) 2821. https://doi.org/10.3390/ma12172821.

[90]　T. Wenga, W. Gwenzi, I.A. Jamro, W. Ma, High-temperature corrosion-resistant alloy for waste-to-energy plants: Alloy designing, fabrication, and possible corrosion-resistance mechanism, Heliyon. 10 (2024) e30177. https://doi.org/10.1016/j.heliyon.2024.e30177.

[91]　S.J. Zinkle, G.S. Was, Materials challenges in nuclear energy, Acta. Mater. 61 (2013) 735–758. https://doi.org/10.1016/j.actamat.2012.11.004.

[92]　AEETHER, Application of Inconel Alloy in Nuclear Power Industry. https://www.aeether.com/AEETHER/media/media-74/media.html/2023 (accessed October 6, 2024).

[93]　T.K. Tsao, A.C. Yeh, C.M. Kuo, H. Murakami, High temperature oxidation and corrosion properties of high entropy superalloys, Entropy. 18 (2016) 62. https://doi.org/10.3390/e18020062.

[94]　L. Zhu, P. Xue, Q. Lan, G. Meng, Y. Ren, Z. Yang, P. Xu, Z. Liu, Recent research and development status of laser cladding: A review, Opt. Laser. Technol. 138 (2021) 106915. https://doi.org/10.1016/j.optlastec.2021.106915.

[95]　S.M. Jowitt, T.T. Werner, Z. Weng, G.M. Mudd, Recycling of the rare earth elements, Curr. Opin. Green Sustain. Chem. 13 (2018) 1–7. https://doi.org/10.1016/j.cogsc.2018.02.008.

[96]　C. Helbig, A.M. Bradshaw, A. Thorenz, A. Tuma, Supply risk considerations for the elements in nickel-based superalloys, Resources. 9 (2020) 106. https://doi.org/10.3390/resources9090106.

[97]　L. Wang, S. Lu, J. Fan, Y. Ma, J. Zhang, S. Wang, X. Pei, Y. Sun, G. Lv, T. Zhang, Recovery of rare metals from superalloy scraps by an ultrasonic leaching method with a two-stage separation process, Separations. 9 (2022) 184. https://doi.org/10.3390/separations9070184.

[98]　K. Leszczyńska-Sejda, A. Palmowski, M. Ochmański, G. Benke, A. Grzybek, S. Orda, K. Goc, J. Malarz, D. Kopyto, Recycling of rhenium from superalloys and manganese from spent batteries to produce manganese(II) perrhenate dihydrate, Recycling. 9 (2024) 36. https://doi.org/10.3390/recycling9030036.

Superalloys: Fundamentals and Applications   Materials Research Forum LLC
Materials Research Foundations 178 (2025) 200-214   https://doi.org/10.21741/9781644903698-10

Chapter 10

Opportunities and Challenges of Superalloys for Coating

Anupama Rajput[1*], B K Bhuyan[1], Prachika Rajput[2]

[1]SET, Manav Rachna International Institute of Research and Studies

[2]Netaji Subhas University of Technology

* anupamarajputchem@gmail.com

Abstract

Superalloys, also referred to as high-performance alloys, have an interesting past that stems from the requirement for materials that can endure harsh environments. In many high-temperature applications, such as gas turbines, power plants, and jet engines, they are indispensable. Therefore, underlying one of the applications i.e. super alloys are often used in coatings to protect components like air turbines from extreme temperatures and harsh environments. Nickel is the most used metal which is used as a combinant in the formulation of super alloys owing to its high temperature withstand behaviour. The further modification and types include cobalt based superalloys; copper-nickel based super alloys, etc. This chapter discusses the overview of types of super alloys, types of respective coatings and methods of formulation that impact the various essential properties like high temperature stability, redox resistance and creep resistance.

Keywords

Superalloys, Coatings, Nickel Based Superalloys, Creep Resistance, High Temperature Resistance

Contents

1. Introduction

In many different industrial applications, superalloys have been widely used [1, 2]. These materials are recognized for their exceptional properties compared to many other alloys, including remarkable corrosion resistance, exceptional mechanical strength, excellent surface stability, and notable creep resistance [3, 4]. Due to its capacity to function at a significant portion of its melting point, the superalloy finds extensive use, especially in high-temperature environments [5, 6]. Face-centered cubic austenitic is the crystal structure that most distinguishes super alloys, which are mostly created by chemical advances [7,8]. Usually, they form solitary crystals after casting. At very low temperatures, grain boundaries can improve strength, but they also reduce creep resistance. The techniques of age hardening and reinforcing solid solutions by precipitating secondary phases, including carbides and gamma prime (γ'), are applied to produce super alloys with extraordinary strength at high temperatures. [9,10] Further the super alloys are divided into four generation based on modification and evolvement as shown in Fig.1 [11]

High-temperature materials may be made stronger by single crystal air foils, which can also optimize the balance of properties by varying the orientation of the crystals. When it comes to single crystal materials, the structure lacks grain boundaries, resulting in a single crystal with an airfoil-shaped orientation. Eliminating all grain boundaries and adding chemicals to reinforce them can significantly raise the alloy's melting point and boost its strength at high temperatures. Since 1995, gas turbine engines underline the use of single crystal alloys in their fabrication. The new super alloys, along with advanced protective coatings, will boost the growth potential of gas turbine engines in the future [12].

Hot corrosion significantly accelerates degradation, hence greatly diminishing the lifespan of superalloys. CM 247 LC exhibited significant corrosion among the super alloys, demonstrating its great susceptibility to heat corrosion. Sulphur diffusion and the formation of metal sulphides, especially those of chromium and nickel, are shown to be the important contributors. Nickel-based alloys perform worse in the development of sulphide phases inside super alloys than cobalt- and iron-based alloys, which are especially good at undermining these materials' corrosion resistance [13].

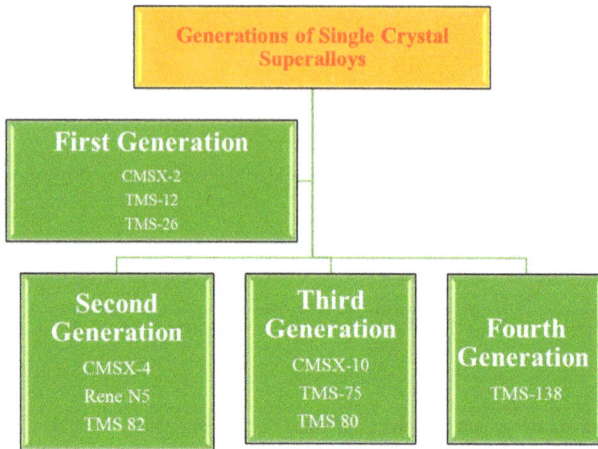

Figure1. Generation of Super alloys

2.   Types of superalloy

2.1   Nickel-based superalloys

Usually used for manufacturing components requiring temperatures above 500 °C, nickel-based super alloys exhibit remarkable resistance to corrosion and high temperatures. Numerous design components for aviation, maritime, nuclear reactor, and chemical fields have been used extensively to run equipment facilities due to their exceptional mechanical and chemical qualities [14, 15]. The mechanical qualities of nickel-based super alloys must be further improved to suit the industrial operational needs resulting from the rise in temperature and component loading, thus preventing the constant material failure. Numerous approaches have been explored by researchers to boost the super alloys based on nickel's mechanical strength for industrial usage. Nowotnik [14] talked about the impact of intermetallic precipitates on the mechanical properties of super alloys based on nickel. It has been illustrated that fine particles dispersed in the matrix contribute to the resistance against dislocations which will eventually produce improved strengthening kinetics in these classes of super alloys.

Due to their superior creep resistance and yield strength, nickel-based super alloys are widely used in aviation and land-based gas turbines. The inclusion of 'Co' in nickel-based superalloys greatly lowers the stacking fault energy, which is necessary to enhance the creep and stress rupture characteristics. It also decreases the solubility of Al and Ti in the matrix and raises the concentration of the γ' phase [15]. When significant creep and/or fatigue resistance are required, superalloys based on nickel are often used at temperatures higher than 500°C. Usually made up of numerous alloying elements, which may include lighter elements such as carbon or boron, aluminium, as well as heavier refractory metals that include tantalum, rhenium, or tungsten[16], these alloys typically consist of Though they were intended for use in turbine engines, the alloys are also increasingly used in other fields like fuel cells, nuclear power plants, ultra-supercritical power plants, and diesel engines fed from fossil fuels. Applications are primarily noted at 750 °C, given that the properties of ferritic steels considerably deteriorate at this temperature [17].

2.2 Cobalt based superalloys

In order to achieve enough resistance to oxidation and effective resistance to hot corrosion, these systems need to absorb significant increases in chromium (more than 20 weight percent). The fabrication of cobalt super alloys, intended to improve oxidation and hot corrosion resistance, has gained significant momentum recently, particularly with the introduction of overlay coating techniques (Fig.2) and comprehensive research aimed at clarifying hot corrosion mechanisms and the influence of alloying elements [18]. Super alloys based on cobalt show remarkable resistance to heat corrosion in harsh settings. These alloys result in the continuous production of compact protective oxide layers. The oxides exhibit enhanced thermodynamic stability, lower growth rates, and better adhesion to the outer layer compared to those built on nickel-based super alloys [19]. These alloys also have better corrosion-wear resistance, weldability, and thermal fatigue resistance, in addition to remarkable creep rupture qualities at temperatures above 1000°C. Cobalt-based superalloys primarily consist of components including cobalt, chromium, tungsten, nickel, and aluminum. Cobalt-based super alloys are extensively utilized in aviation, aerospace, energy, medical, and various other sectors. They are employed in aviation and aerospace for the fabrication of components like turbine engines and gas turbines. Cobalt-based superalloys are utilized in the energy sector for the production of turbine blades and various components in thermal and nuclear power facilities. Medical devices, including artificial bones and joints, are frequently constructed from these superalloys [20].

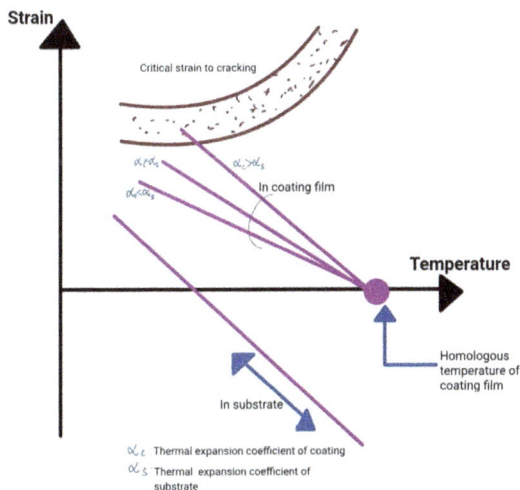

Figure 2.   Coating film of critical strain

Superalloys: Fundamentals and Applications   Materials Research Forum LLC
Materials Research Foundations 178 (2025) 200-214   https://doi.org/10.21741/9781644903698-10

The melting temperature of cobalt-based super alloys typically exceeds 1300 °C. The initial melting temperature of the majority of nickel-based super alloys is below 1280°C. A substantial quantity of refractory metal is frequently used into cobalt-based superalloys to enhance solid solution strengthening. The elevated melting point renders cobalt-based superalloys extensively utilized in the fabrication of guide vanes for aircraft engines. It prevents defects such as combustion or distortion of the guiding vane at elevated temperatures. Cobalt-based super alloys often possess a greater chromium concentration than other super alloys, resulting in enhanced high-temperature corrosion resistance. Cobalt-based super alloys can generate a stable oxide layer at elevated temperatures. It can preserve the integrity and stability of its material under the combined effects of elevated temperature and corrosive substances. This attribute of the cobalt-based super alloy enables it to function reliably in engine blades [21].

2.3   Copper based superalloy

Copper alloys, particularly Cu Ag Zn alloys, exhibit remarkable thermal conductivity, exceptional corrosion and wear resistance, impressive fluidity along with wettability, as well as considerable ductility. Copper alloys and superalloys based on nickel have the necessary material characteristics for sealing. This study utilized a CuAgZn copper alloy, which was manufactured by casting and had 50% Cu, 25% Ag, and 25% Zn by weight. By smelting copper, silver, and zinc in a 2:1:1 weight ratio, the CuAgZn alloy was created [22-23].

3.   Different types of coating

Super alloy coatings work better in demanding and demanding environments since they are particularly made to show a extensive range of advanced material features. The aforementioned characteristics are directly related to the careful alloy composition and specific techniques used during the whole production process. Because super alloy coatings can maintain their mechanical properties and structural integrity at elevated temperatures, which frequently cause conventional materials to degrade, they provide remarkable high-temperature stability [24]. In certain applications, such as gas turbines, where parts are subjected to excessive temperatures, this capability is essential. By carefully selecting alloying elements that retain their mechanical qualities and strength even under extreme heat, coatings may be made to remain stable at high temperatures [25]. Developing unique coating compositions to address specific difficulties is necessary for a super alloy coating with highly specialized properties. This is the process of creating coatings that precisely combine alloying elements to achieve desired material properties, including increased thermal conductivity, increased wear resistance, or resistance to specific corrosive substances [26].

3.1   Protective coating

In order to preserve materials and components, protective coatings are essential because they operate as a vital barrier against the damaging effects of oxidation and corrosion in a variety of environmental settings. These coatings' principal function is to serve as a barrier of protection, shielding surfaces from potentially damaging forces that might compromise their structural soundness and functionality. In hostile environments, where corrosion and oxidation present more challenges, the value of protective coatings is more apparent [27]. Several significant elements emphasize the necessity of applying protective coatings in these situations: Protective coatings

Superalloys: Fundamentals and Applications   Materials Research Forum LLC
Materials Research Foundations 178 (2025) 200-214   https://doi.org/10.21741/9781644903698-10

provide materials an extra line of protection against direct contact to corrosive chemicals and high temperatures, increasing their longevity.

3.2   Diffusion coating

Diffusion coatings can be applied during the Ti-Al, Al-Cr, Al-Si, and Pt-Al processes, although they are typically applied during the diffusion aluminizing process. Coatings consisting of a particular powder mixture of chemical activator, aluminium powder or alloy, and neutral filler (such as Al2O3) are applied to the coated components before they are placed in specialized containers. Once inside the furnace, the closed container is heated to a temperature of between 700°C and 1050°C by the chemical activator, which forms the transporting vapor source. This kind of diffusion aluminizing can take up to twenty hours, and oxidation protection from a powder combination must be applied precisely. Enhancement of metallic components that offer surface protection by diffusion from an external reservoir forms the basis of diffusion coatings. The concentration differential at high temperatures causes elemental diffusion into the substrate and outward diffusion of alloy elements. The inside or outer diffusion process is more evident depending on whether process is quicker. These processes eventually manage to change the composition in both cases by enriching the surface with elements from the diffusion reservoir. A few other processing techniques include hot-dipping in a metal melt, galvanic and metal foil coatings, and slurry-reaction coatings. Each of them employs a similar process that entails heating a reservoir of the metal to encourage diffusion after it has been applied to the surface [28].The chemical vapor deposition process was developed from the previously published diffusion aluminide coating deposition techniques. Turbine blades are placed in the retort and supplied with the gas AlCl3+H2 atmosphere produced in the outer reactor as part of the CVD process. When HCl is processed in the heated generator containing aluminum, gas AlCl3 is produced. Once the AlCl3+H2 vapors are heated to a temperature of around 1000°C, they are placed inside the retort. To heat the CVD retort carrying its charge to the processing temperature, a bell type furnace, soaking pit furnace, or elevator furnace is often utilized. A specialized gas neutralizing system processes the reaction gasses that exit the retort. The outside and inner surfaces of the turbine blades, particularly the cooling channels, may be coated concurrently using this type of aluminizing, which can present difficulties with other coating techniques. Furthermore, this method allows for the regulation of cooling rate, a crucial feature considering the high temperatures at which many super alloys employed in foundries operate [29].

3.3   Thermal barrier coatings

Advanced protective coatings called thermal barrier coatings (TBCs) are used in high-temperature applications, such gas turbines and jet engines, to shield metallic components from severe temperatures. Through the reduction of heat load and provision of oxidation resistance, they enhance the longevity and functionality of these components. It offers resistance to oxidation and corrosion and guarantees strong adherence between the topcoat and substrate [30-32]. For thermal barrier coatings, ZrO_2 x Y_2O_3 normally makes up the outer ceramic zone, while MCrAlY (where M = Ni, Co, Fe) makes up the bond coat. The resistance to oxidation and hot corrosion improves when the temperature in the interlayer below the bond coat decreases because ceramic materials have a limited capacity to transport heat. The methods most often utilized to create thermal barrier coatings are APS (Air Plasma Spray), LPPS (Low Pressure Plasma Spray), and EB-PVD (Electron Beam Physical Vapour Deposition) [30]. Usually, thermal barrier coatings (TBCs) are applied to the combustion chamber and vanes sections instead of covering the blades using EB PVD technology.

The resistance to cyclic temperature changes of the yttria stabilized zirconia coating requires a precisely controlled porosity and microcracks [33, 34]. The MCrAlY bond coat, which is about 100 μm thick, and $ZrO_2X_8Y_2O_3$, which forms a ceramic zone about 300 μm thick, make up the microstructure of the sample plasma-sprayed heat barrier coating. Increasingly sophisticated technological equipment was used to manufacture thermal barrier coatings, particularly for gas turbine spinning blades where significantly greater mechanical stresses occur [35]. Thermal barrier coating durability is known to be significantly impacted by the production of TGO (Thermally Grown Oxide) between MCrAlY and ceramic. At the interface where the bond coat and topcoat meet during operation, a layer develops. Although this layer is crucial to the coating's longevity, if it gets too thick, it might fail. In power generation, aerospace, automotive, and industrial applications, thermal barrier coatings are crucial for safeguarding components in high-temperature settings and greatly increasing their performance, longevity, and efficiency [36-38].

3.4 Slurry coating

Slurry-aluminizing is a method where aluminium-particle-containing slurry is used to protect the substrate. Usually, a binder such a solvent or polymer is used with aluminium powder to create the slurry. To guarantee that the aluminized coating adheres well, the substrate must first be cleansed and prepared. Subsequently, it is applied to the substrate's surface by brushing, spraying, or dipping. After applying the slurry to the coated substrate and allowing it to dry, the aluminum particles are fused together in a furnace. This procedure creates a uniform and long-lasting aluminized coating by bonding the aluminium particles to the substrate (Fig 3). There are some advantages to the slurry approach as compared to alternative techniques [39].

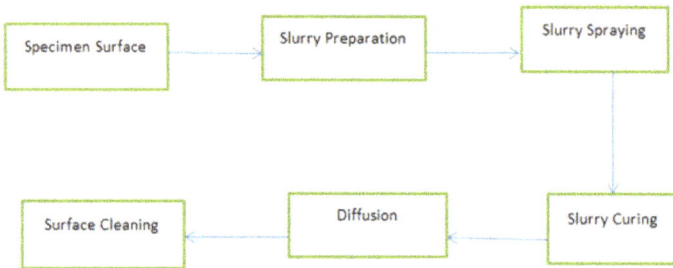

Figure 3 Slurry-Aluminizing Process

Superalloys: Fundamentals and Applications   Materials Research Forum LLC
Materials Research Foundations 178 (2025) 200-214 https://doi.org/10.21741/9781644903698-10

In this approach, large components may be accurately protected, and the heat cycle during the coating preparation process will be quicker. Diffusion aluminide coatings on engine jet components are applied using the slurry technique because of these benefits. A smooth coating with a surface roughness Ra < 4.5 m may be achieved by slurry aluminizing on a nickel-based substrate, as demonstrated by works by Li et al. [40]. Slurry-aluminized coatings perform as designed even in the face of severe service circumstances. Bortoluci Ormastroni et al. [41] report that the CMSX-4 Plus alloy with a NiAl coating outperformed the AM1, CMSX-4, and Rene N5 alloys in terms of fatigue life under the same applied alternating stress (180 MPa), at a high temperature (1000°C), and under completely reversed circumstances (R=1). Efficient ultrafast slurry-aluminizing techniques for nickel purification are described in detail in a few recent papers [42, 43]. Pillai et al. [44] added iron to the NiAl coating in order to reduce production costs and maintain the coating's resistance to cyclic oxidation in air with 10% H2O at 900 °C for a day. Galetz et al. [45] used a modified NiGeAl-aluminized coating to enhance the 602CA alloy's high-temperature performance at 1200°C. Hatami et al. [46] successfully used the slurry technique to deposit a silico-aluminide layer composed of a-(Ni,Co) Alphaseon Hastelloy-X/NiCoCrAlY after heat treating the material in argon. It was demonstrated that the silico-aluminide (slurry)/NiCoCrAlY (HVOF) coating exhibited greater resistance to high-temperature oxidation at 1000°C in contrast to the NiCoCrAlY coating. These astounding findings demonstrate that an extra layer of aluminized protection may be provided to avoid corrosion even in anal ready coated nickel-based super alloys. Mahmoudi et al.'s approach [47] for applying a Ni/Cr/Ti-Alcoating on Inconel 738 involves plasma paste aluminizing.

3.5 Thermal spraying

A collection of techniques known as thermal spraying are used to apply coatings on surfaces to improve characteristics including wear resistance, corrosion resistance, and thermal insulation. This process involves the projection of molten or semi-molten materials onto a substrate, forming a coating. An outline of the main categories of thermal spraying methods is provided here (Fig.4). Melts the coating material with an oxy-fuel gas flame and uses a gas stream to spray the substance onto the substrate. Common materials: metals, alloys, ceramics, and plastics. The coating substance is melted using a high-temperature plasma jet. Extremely high melting point materials are among the many materials for which it is appropriate [48]. Arc spraying is the process of melting wire material by creating an electric arc between two consumable wires, which is subsequently atomized and sprayed onto the substrate [49-50]. Applications: Anti-corrosion coatings, often. By burning fuel and oxygen together, oxy-fuel spraying creates a fast-moving gas stream that melts and drives the coating material. By rapidly accelerating powder particles in a gas stream, cold spraying creates a coating on the substrate without using a lot of heat. Thermal spraying is a significant technique with numerous applications since it is extensively utilized in many sectors to increase the performance and durability of components [51, 52].

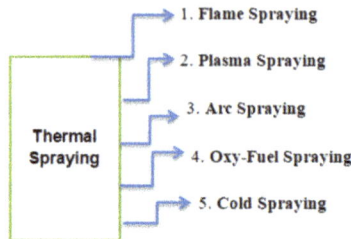

Figure 4. Process of thermal Spraying

3.6   Plasma spraying

The thermal spraying method known as "plasma spraying" uses a high-temperature plasma jet to melt and propel coating components onto a substrate. A few materials that can be coated with this method are metals, alloys, ceramics, and composites. A copper nozzle and tungsten electrode are struck by an electric arc, producing a high-temperature plasma jet that can reach up to 15,000°C. To transfer the coating material and maintain plasma stability, an inert gas like hydrogen, nitrogen, or argon is used. The powdered coating material is fed into the plasma jet either radially or axially [53–55]. The extreme heat of the plasma jet melts the powder particles. The molten or semi-molten particles are driven toward the substrate by means of the plasma jet. As soon as they contact the surface, particles flatten and solidify to form a coating. Usually, the substrate is cleaned and roughened before coating in order to guarantee effective coating adherence. There are numerous innovative industrial and engineering applications that benefit greatly from plasma spraying since it is a very adaptable and efficient way to apply high-performance coatings [56-58].

3.7   Gas phase coatings

The method of forming a solid substance that is deposited on a substrate by chemically interacting with gaseous precursors is known as Chemical Vapor Deposition (CVD). Gas phase coatings are methods of applying thin films or coatings to a substrate using vapor-phase processes. The procedure normally happens in a reaction chamber at high temperatures. It boosts chemical processes at lower temperatures by using plasma and employs high temperatures to initiate chemical reactions. The creation of high-performance coatings with particular qualities suited to a variety of applications is made possible by gas phase coating processes, which are crucial in the advanced manufacturing and technology sectors [59-60].

4.   Opportunities and challenges

Superalloys are a class of high-performance materials designed to endure extreme conditions, making them ideal for use in coatings where mechanical strength, corrosion resistance, and thermal stability are critical. These alloys are primarily used in industries like aerospace, power generation, and nuclear applications. Despite their benefits, superalloys also face challenges when applied as coatings. Below is an overview of the opportunities and challenges:

Superalloys: Fundamentals and Applications   Materials Research Forum LLC
Materials Research Foundations 178 (2025) 200-214   https://doi.org/10.21741/9781644903698-10

4.1   Opportunities

Superalloys, particularly nickel-based ones, can withstand high temperatures without significant degradation, making them ideal for coating turbine blades, engine parts, and other components exposed to extreme heat. They maintain mechanical strength at temperatures above 1000°C, which is crucial in high-performance environments like jet engines or power turbines. Superalloys form protective oxide layers (such as chromia or alumina) that shield the underlying material from oxidation and corrosion in harsh environments, enhancing the lifespan of the components they coat. This makes them excellent candidates for protective coatings in corrosive environments like chemical processing or marine applications. Moreover, superalloy coatings provide excellent resistance to mechanical wear and tear, which is important in applications where parts undergo high friction, such as rotating machinery. Coatings from cobalt-based superalloys, for example, exhibit excellent wear resistance in extreme conditions, which extends the service life of components. Further, newer technologies like additive manufacturing and laser cladding allow for the precise application of superalloy coatings, which reduces material waste and increases the efficiency of coating processes. These advancements make it possible to create complex geometries and tailor the coating properties for specific needs, improving overall performance.

4.2   Challenges

Superalloys are often coated onto surfaces using sophisticated techniques including thermal spraying, laser cladding, physical vapor deposition (PVD), chemical vapor deposition (CVD), and thermal spraying. These processes are complex, costly, and require precise control over parameters like temperature, pressure, and atmosphere. Achieving uniform coatings with the desired thickness and adhesion is a technical challenge that demands expertise. Further, the high cost of superalloys, due to their composition (containing elements like nickel, cobalt, chromium, and rare earth elements), is one of the major barriers to widespread adoption in coatings. Specialized equipment and processes needed to apply these coatings, like electron beam or plasma spraying, further increase the cost. Once damaged, it can be challenging to repair superalloy coatings without completely reapplying the coating, which can be costly and time-consuming. Techniques to repair superalloy coatings without compromising performance are still being developed but remain limited in scope. Moreover, maintaining the desired microstructure during the coating process can be difficult, especially under high temperatures. The mechanical characteristics of coatings can be impacted by problems like as phase transitions, grain development, or the precipitation of unwanted phases (such carbides or sigma phase). For superalloy coatings to remain stable over time, heat treatment and process control are required.

Conclusion

Superalloys offer tremendous opportunities as coating materials, especially in high-temperature, high-stress environments where conventional materials fail. Their excellent wear, corrosion, and oxidation resistance make them attractive for various industries. However, their application poses challenges, particularly in terms of cost, thermal mismatch, and the complexity of deposition techniques. Ongoing research and technological advancements in coating methods and materials are likely to address some of these challenges, broadening the scope of superalloy coatings in the future. The study of the microstructures of super alloys is crucial for developing and optimizing coatings for high-temperature applications. Advanced analytical techniques provide detailed insights into the microstructural features that influence the performance of super alloys, guiding the design and processing of coatings to meet the demanding requirements of industries such as aerospace and power generation.

References

[1] D. Zhu, X. Zhang, H. Ding, Tool wear characteristics in machining of nickel based superalloys, Int. J. Mach. Tools Manuf. 64 (2013) 60-77. https://doi.org/10.1016/j.ijmachtools.2012.08.001

[2] P.S. Gowthaman, S. Jeyakumar, A Review on machining of high temperature aeronautics super-alloys using WEDM, Mater. Today. Proc. 18 (2019) 4782 4791. https://doi.org/10.1016/j.matpr.2019.07.466

[3] J. Xu, H. Gruber, D. Deng, R.L. Peng, J.J. Moverare, Short-term creep behaviour of an additive manufactured non-wieldable nickel-base super alloy evaluated by slow strain rate testing, Acta. Mater. 179 (2019) 142-157. https://doi.org/10.1016/j.actamat.2019.08.034

[4] S. Gao, J. Hou, F. Yang, Y. Guo, L. Zhou, Effect of Ta on microstructural evolution and mechanical properties of a solid-solution strengthening cast Ni based alloy during long-term thermal exposure at 700° C, J. Alloy. Compd. 729 (2017) 903-913. https://doi.org/10.1016/j.jallcom.2017.09.194

[5] D.J. Crudden, A. Mottura, N. Warnken, B. Raeisinia, R.C. Reed, Modeling of the influence of alloy composition on flow stress in high-strength nickel-based super alloys, Acta. Mater. 75 (2014) 356-370. https://doi.org/10.1016/j.actamat.2014.04.075

[6] H.T. Mallikarjuna, N.L. Richards, W.F. Caley, Effect of alloying elements and microstructure on the cyclic oxidation performance of three nickel-based super alloys, Mater. 4 (2018) 487-499. https://doi.org/10.1016/j.mtla.2018.11.004

[7] E. Chauvet, C. Tassin, J.J. Blandin, R. Dendievel, G. Martin, Producing Ni-base super alloys single crystal by selective electron beam melting, Scr. Mater, 152 (2018) 15-19. https://doi.org/10.1016/j.scriptamat.2018.03.041

[8] M. Pröbstle, S. Neumeier, J. Hopfenmüller, L.P. Freund, T. Niendorf, D. Schwarze, M. Göken, Superior creep strength of a nickel-based super alloy produced by selective laser melting, Mater. Sci. Eng. A. 674 (2016) 299-307. https://doi.org/10.1016/j.msea.2016.07.061

[9] R.K. Gupta, V. Anil Kumar, U.V. Gururaja, B. Shivaram, Y. Maruti Prasad, P. Ramkumar, Processing and characterization of Inconel 625 nickel base super alloy, Mater. Sci. Forum. 830 (2015). https://doi.org/10.4028/www.scientific.net/MSF.830-831.38

[10] N. Nadammal, S. Cabeza, T. Mishurova, T. Thiede, A. Kromm, C. Seyfert, L. Farahbod, Effect of hatch length on the development of microstructure, texture and residual stresses in selective laser melted superalloy Inconel 718, Mater. Des. 134 (2017) 139-150. https://doi.org/10.1016/j.matdes.2017.08.049

[11] W.J. Molloy, Investment cast superalloys, Adv. Mater. Process. 138 (1990) 23-30.

[12] D.D. Hass, A.J. Slifka, H.N.G. Wadley, Low thermal conductivity vapour deposited zirconia microstructures, Acta. Mater. 49 (2001) 973-983. https://doi.org/10.1016/S1359-6454(00)00403-1

[13] S. Kumar, B. Satapathy, D. Pradhan, G.S. Mahobia, Effect of surface modification on the hot corrosion resistance of Inconel 718 at 700 C, Mater. Res. Express 6 (2019) 086549. https://doi.org/10.1088/2053-1591/ab1dc7

[14] A. Nowotnik, Reference module in materials science and materials engineering, in: Ni-Based Superalloys, 2016. https://doi.org/10.1016/B978-0-12-803581-8.02574-1

[15] O. Lypchanskyi, T. Sleboda, K. Zyguła, A. Łukaszek-Sołek, M. Wojtaszek, Evaluation of hot workability of nickel-based superalloy using activation energy map and processing maps, Materials. 13 (2020) 3629. https://doi.org/10.3390/ma13163629

[16] K. Horke, A. Meyer, R.F. Singer, Metal injection molding (MIM) of nickel-base super alloys, in: D.F. Heaney (Eds.), Handbook of Metal Injection Molding, Second ed., Woodhead Publishing, 2019, pp. 575-608. https://doi.org/10.1016/B978-0-08-102152-1.00028-3

[17] R.C. Reed, C.M.F. Rae, Physical Metallurgy of the Nickel-Based Superalloys, Fifth ed., Elsevier, Oxford, 2014, pp. 2215-2290. https://doi.org/10.1016/B978-0-444-53770-6.00022-8

[18] D. Coutsouradis, A. Davin, M. Lamberigts, Cobalt-based superalloys for applications in gas turbines, Mater. Sci. Eng. 88 1987 11-19. https://doi.org/10.1016/0025-5416(87)90061-9

[19] S.K. Makineni, B. Nithin, K. Chattopadhyay, Synthesis of a new tungsten-free γ-γ′ cobalt-based superalloy by tuning alloying additions, Acta. Mater. 85 (2015) 85-94. https://doi.org/10.1016/j.actamat.2014.11.016

[20] M. Cartón-Cordero, M. Campos, L.P. Freund, M. Kolb, S. Neumeier, M. Göken, J.M. Torralba, Microstructure and compression strength of Co-based superalloys hardened by γ′ and carbide precipitation, Mater. Sci. Eng. A. 734 (2018) 437-444, https://doi.org/10.1016/j.msea.2018.08.007. https://doi.org/10.1016/j.msea.2018.08.007

[21] Z. Zhang, Q. Ding, X. Wei, Z. Zhang, H. Bei, Effects of alloying elements on the microstructure of precipitation strengthened Co-based superalloys, J. Alloys Compd. 989 (2024) 174401. https://doi.org/10.1016/j.jallcom.2024.174401

[22] Y. Xiao, L. Lang, W. Xu, D. Zhang, Diffusion bonding of copper alloy and nickel-based superalloy via hot isostatic pressing, J. Mater. Res. Technol. 19 (2022) 1789-1797. https://doi.org/10.1016/j.jmrt.2022.05.152

[23] B. Wierzba, W. Skibiński, The interdiffusion in copper-nickel alloys, J. Alloys Compd. 687, 2016, 104-108. https://doi.org/10.1016/j.jallcom.2016.06.085

[24] H. Shahbaz. R.B. Vakilifard, A. Nair, C. Liberati, C. Moreau, R.S. Lima, High entropy alloy (HEA) bond coats for thermal barrier coatings (TBCs): A review, ITSC. 2023 659-666. https://doi.org/10.31399/asm.cp.itsc2023p0659

[25] M.C. Shekar, A. Vaddula, S.D. Yerramsetti, K.M. Buddaraju, Additive Manufacturing of titanium and nickel-based super alloys: A review, Mater. Today Proc. (2023).

[26] A. Gupta, A. Pattnayak, N.V. Abhijith, D. Kumar, V. Chaudhry, S. Mohan, Development of alumina-based hybrid composite coatings for high temperature erosive and corrosive environments, Ceram. Int. 49 (2023) 862-874. https://doi.org/10.1016/j.ceramint.2022.09.059

[27] P. Soumy, R. Garg, K. Kumar Saxena, V.K. Srivastav, H. Vasudev, N. Kumar, Data-driven materials science: application of ML for predicting band gap, Adv. Mater. Process. Technol. (2023) 1-10. https://doi.org/10.1080/2374068X.2023.2171666

[28] M.C. Galetz, X. Montero, H. Murakami, Novel processing in inert atmosphere and in air to manufacture high-activity slurry aluminide coatings modified by Pt and Pt/Ir, Mater Corros. 63 (2012) 921-928. https://doi.org/10.1002/maco.201206773

[29] O. Knotek, F. Loffler, H.J. Scholl, Properties of arc-evaporated CrN and (Cr,Al)n coatings, Surf. Coat. Technol. 45 (1991) 53-58. https://doi.org/10.1016/0257-8972(91)90205-B

[30] S.A.A. Jude, J.T.W. Jappes, M. Adamkhan, Thermal barrier coatings for high-temperature application on superalloy substrates: A review, Mater. Today Proc. 60 (2022) 1670-1675. https://doi.org/10.1016/j.matpr.2021.12.223

[31] Q.Q. Zhou, L. Yang, C. Luo, F.W. Chen, Y.C. Zhou, Y.G. Wei, Thermal barrier coatings failure mechanism during the interfacial oxidation process under the interaction between interface by cohesive zone model and brittle fracture by phase-field, Int. J. Solids Struct. 214-215 (2021) 18-34. https://doi.org/10.1016/j.ijsolstr.2020.12.020

[32] Y. Chen, X. Zhao, Y. Dang, P. Xiao, N. Curry, N. Markocsan, P. Nylen, Characterization and understanding of residual stresses in a NiCoCrAlY bond coat for thermal barrier coating application, Acta. Mater. 94 (2015) 1-14. https://doi.org/10.1016/j.actamat.2015.04.053

[33] J.T. DeMasi-Marcin, D.K. Gupta, Protective coatings in the gas turbine engine, Surf. Coat. Technol. 68-69 (1994) 1-9. https://doi.org/10.1016/0257-8972(94)90129-5

[34] L. Swadzba, G. Moskal, B. Mendala, T. Gancarczyk, Characterization of air plasma sprayed TBC coating during isothermal oxidation at 1100°C, J. Achiev. Mater. Manuf. Eng. 21 (2007) 81-84.

[35] G. Moskal, The porosity assessment of thermal barrier coatings obtained by APS method, J. Achiev. Mater. Manuf. Eng. 20 (2007) 483-486.

[36] G. Walther, et al., Oxidation and corrosion protection of super alloys by coating systems, J. Mater. Eng. Perform. 19 (2010) 725-733.

[37] H. Evans, M. Taylor, Diffusion coatings for super alloys, Surf. Coat. Technol. 188-189 (2004) 170-177. https://doi.org/10.1016/j.surfcoat.2004.08.003

[38] L. Wang, et al., Effects of thermal barrier coatings on the oxidation behavior of nickel-based super alloys, J. Alloys Compd. 452 (2008) 401-406. https://doi.org/10.1016/j.jallcom.2006.11.099

[39] L.A. Dobrzanski, K. Lukaszkowicz, A. Zarychta, Mechanical properties of monolayer coatings deposited by PVD techniques, J. Achiev. Mater. Manuf. Eng. 20 (2007) 423-426.

[40] Z. Li, C. Wang, X. Ding, X. Li, J. Yu, Q. Li, Y. Qu, Effect of slurry thickness on the Quality of aluminized coatings, Materials. 15 (2022) 6758. https://doi.org/10.3390/ma15196758

[41] L.M.B. Ormastroni, T. Kepa, A. Cervellon, P. Villechaise, F. Pedraza, J. Cormier, Very high cycle fatigue rupture mode at high temperatures of Ni-based super alloys coated with a slurry aluminide, Int. J. Fatigue. 180 (2024) 108107. https://doi.org/10.1016/j.ijfatigue.2023.108107

[42] T. Kepa, G. Bonnet, F. Pedraza, Oxidation behavior of ultrafast slurry aluminized nickel, Surf. Coat. Technol. 424 (2021) 127667. https://doi.org/10.1016/j.surfcoat.2021.127667

[43] T. Kepa, F. Pedraza, F. Rouillard, Intermetallic formation of Al-Fe and Al-Ni phases by ultrafast slurry aluminization (flash aluminizing), Surf. Coat. Technol. 397 (2020) 126011. https://doi.org/10.1016/j.surfcoat.2020.126011

[44] R. Pillai, S. Dryepondt, B. Armstrong, M. Lance, G. Muralidharan, Evaluating the efficacy of aluminide coatings to improve oxidation resistance of high performance engine valve alloys, Surf. Coat. Technol. 421 (2021) 127401. https://doi.org/10.1016/j.surfcoat.2021.127401

[45] M.C. Galetz, C. Oskay, S. Madloch, Microstructural degradation and inter diffusion behavior of NiAl and Ge-modified NiAl coatings deposited on alloy 602 CA, Surf. Coat. Technol. 364 (2019) 211-217. https://doi.org/10.1016/j.surfcoat.2019.02.048

[46] E. Hatami, S.M.M. Hadavi, D.S. Doolabi, M. Bahamirian, High-temperature oxidation behavior of a silico-aluminized MCrAlY coating on a Ni-based superalloy, Oxid. Met. 97 (2022) 575-597. https://doi.org/10.1007/s11085-022-10109-3

[47] H. Mahmoudi, S.M.M. Hadavi, Y. Palizdar, Characterization, growth kinetics and formation mechanism of aluminide coating by plasma paste aluminizing on IN738, Vacuum. 184 (2021) 109968. https://doi.org/10.1016/j.vacuum.2020.109968

[48] M.F. Conde, Thermal spraying processes, in: J.R. Davis (Eds.), Handbook of Thermal Spray Technology, ASM Int. (2004) 38-49.

[49] T. Burakowski, T. Wierzchon, Advanced surface engineering techniques, in: T. Burakowski, T. Wierzchon, Surface Engineering of Metals: Principles, Equipment, Technologies, CRC Press, 1999. https://doi.org/10.1201/9781420049923

[50] J.R. Nicholls, D.J. Wortman, Thermal spray coatings, Adv. Mater. Process. 153 (1998) 25-31.

[51] S. Sampath, H. Herman, Thermal spray processing, Mater. Sci. Eng. 139 (1991) 261-288.

[52] R.C. Reed, The Super alloys: Fundamentals and Applications, Cambridge University Press, 2006.

[53] P. Fauchais, A. Vardelle, Plasma spraying process and applications, J. Therm. Spray Technol. 17 (2008) 765-806. https://doi.org/10.1007/s11666-008-9160-x

[54] P. Fauchais, A. Vardelle, B. Dussoubs. Plasma spraying: from conventional to hybrid processes, J. Therm. Spray Technol. 10 (2001) 44-66. https://doi.org/10.1361/105996301770349510

[55] J.R. Davis (Eds.), Plasma Spraying, Handbook of Thermal Spray Technology, ASM International, 2004.

Materials Research Forum LLC

https://doi.org/10.21741/9781644903698-10

[56] P. Fauchais, G. Montavon, A. Vardelle, Plasma spraying: from plasma generation to coating structure, Appl. Phys. A. 83 (2006) 413-422.

[57] J. Mostaghimi, S. Chandra, M.Y. Heberlein, Investigation of plasma spraying process by numerical simulation, J. Therm. Spray Technol. 12 (2003) 102-114.

[58] D.E. Morks, A review on plasma spraying of biomedical coatings, J. Biomater. Appl. 25 (2010) 393-418.

[59] J. Smith, A. Johnson, Gas phase deposition of titanium nitride thin films for optical applications, Thin Solid Films. 500 (2022) 25-31.

[60] T.S. Sidhu, S. Prakash, R.D. Agrawal, Hot corrosion and performance of nickel-based coatings: A review, Surf. Coat. Technol. 198 (2005) 441-446. https://doi.org/10.1016/j.surfcoat.2004.10.056

Superalloys: Fundamentals and Applications
Materials Research Foundations 178 (2025) 215-229

Materials Research Forum LLC
https://doi.org/10.21741/9781644903698-11

Chapter 11

Casting and Forging of Superalloys

S. Ganeshkumar[1], M. Ramesh[2*], J. Maniraj[2]

[1]Department of Mechanical Engineering, Sri Eshwar College of Engineering, Coimbatore, Tamil Nadu, India

[2]Department of Mechanical Engineering, KIT-Kalaignarkarunanidhi Institute of Technology, Coimbatore, Tamil Nadu, India

* mramesh97@gmail.com

Abstract

This chapter explores advancements in casting technologies and the development of superalloys, emphasizing their impact on modern engineering applications. Casting methods, such as sand, die, investment, centrifugal, and continuous casting, have evolved to offer improved precision and efficiency. Superalloys, including nickel-based, cobalt-based, and iron-based variants, exhibit exceptional high-temperature strength and corrosion resistance, crucial for aerospace, power generation, automotive, and chemical industries. Recent innovations, such as additive manufacturing and computational modelling, have further enhanced these materials and processes. Notably, nickel-based superalloys withstand temperatures up to 1,100 °C, and advanced casting techniques achieve dimensional tolerances within ± 0.1%. This chapter highlights the synergy between casting and superalloys, facilitating the production of complex, high-performance components. Future trends focus on sustainability, with efforts to reduce energy consumption and develop recyclable materials. This synergy drives progress in high-stress applications, promising significant advancements in industrial capabilities.

Keywords

Casting Technologies, Superalloys, Aerospace Applications, Additive Manufacturing, Corrosion Resistance

Contents

1.   Introduction

Casting processes have advanced significantly, enabling the production of components with precise dimensions and excellent mechanical properties. Each type of casting offers unique advantages and is suited to different applications. Sand casting involves creating a mould from a sand mixture and pouring molten metal into the cavity. This method is cost-effective and versatile, allowing for the production of large parts weighing up to several tons. Sand casting can accommodate a wide range of metals, including aluminum, cast iron, and steel, with typical mould temperatures ranging from 900 °C to 1400 °C [1]. This process is particularly useful for manufacturing large components such as engine blocks and cylinder heads. Advances in sand casting include the use of improved sand binders and automated moulding machines, which enhance the precision and repeatability of the cast parts. Die casting uses high-pressure to force molten metal into a metal mould (die). This process is ideal for high-volume production and results in parts with excellent surface finish and dimensional accuracy. Commonly used for non-ferrous metals such as aluminum, magnesium, and zinc, die casting operates at mould temperatures ranging from 200 °C to 500 °C [2]. The process allows for the production of complex shapes with thin walls and tight tolerances, making it suitable for automotive components, consumer electronics, and other precision parts.

Innovations in die casting include the development of vacuum die casting and semi-solid metal casting, which reduce porosity and enhance the mechanical properties of the cast parts. Investment casting, also known as lost-wax casting, produces high-precision components by using a wax pattern coated with a ceramic shell. Once the ceramic material hardens, the wax is melted out, and molten metal is poured into the cavity. Investment casting is capable of producing intricate and detailed parts with excellent surface finish and dimensional accuracy. This process is suitable for a wide range of metals, including stainless steel, aluminum, and titanium, with mould temperatures typically ranging from 1000 °C to 1600 °C. It is widely used in the aerospace, medical, and jewellery industries [3]. Recent advancements in investment casting include the use of 3D printed wax patterns and advanced ceramic materials, which improve the efficiency and precision of the process. Centrifugal casting involves pouring molten metal into a rotating mould. The centrifugal

Materials Research Forum LLC
https://doi.org/10.21741/9781644903698-11

force drives the metal into the mould's walls, resulting in a dense and defect-free casting. This process is particularly effective for producing cylindrical parts such as pipes, bushings, and rings. Typical mould temperatures range from 900 °C to 1400 °C, depending on the metal used. Centrifugal casting offers excellent mechanical properties and is capable of producing parts with uniform grain structure and minimal impurities. Recent innovations include the development of horizontal centrifugal casting and advanced control systems, which enhance the quality and consistency of the cast parts. Continuous casting is a process used primarily in the steel industry to produce long lengths of metal with a uniform cross-section. Molten metal is poured into a water-cooled mould, where it solidifies and is continuously withdrawn from the mould [4]. The process operates at temperatures typically around 1500 °C for steel. Continuous casting is highly efficient and capable of producing large volumes of metal with consistent quality. It is widely used for manufacturing steel slabs, billets, and blooms.

Advances in continuous casting include the use of electromagnetic stirring and advanced cooling techniques, which improve the internal quality and surface finish of the cast products. Each casting process offers unique benefits and is suited to different applications, enabling the production of a wide range of components with varying sizes, shapes, and material properties [5]. The choice of casting process depends on factors such as the type of metal, the complexity of the part, the required surface finish, and the production volume. Continuous improvements and innovations in casting technologies are driving the development of more efficient and precise manufacturing methods, meeting the evolving demands of various industries. The integration of advanced materials, automation, and control systems is further enhancing the capabilities and versatility of casting processes, contributing to the production of high-performance components in modern engineering applications [6]. Fig. 1 shows the different types of casting process.

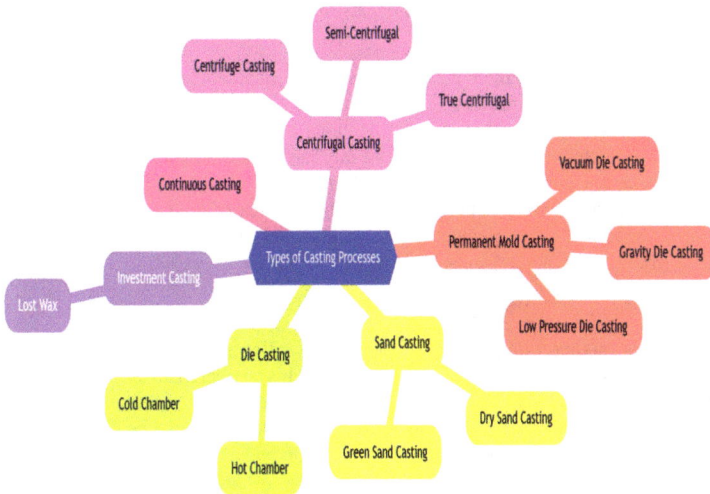

Fig. 1 Types of casting process

2.   Properties and characteristics of superalloys

Superalloys are a class of high-performance materials designed to operate in extreme environments where mechanical stress, high temperatures, and corrosion are prevalent. These alloys are primarily used in the aerospace, power generation, and chemical processing industries due to their exceptional properties. The key characteristics of superalloys include high strength at elevated temperature, creep resistance, oxidation and corrosion resistance, and good surface stability. The following sections delve into the specific properties and the scientific data that underpins their remarkable performance [7]. Fig. 2 shows the prominent properties of superalloys.

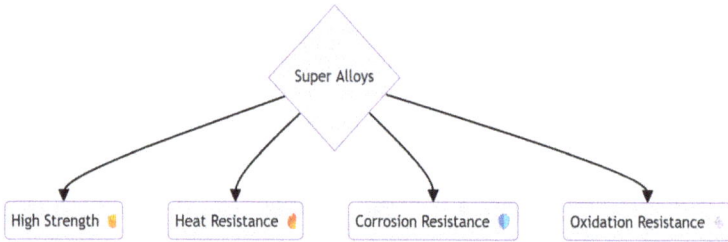

Fig. 2 Properties of Superalloys

2.1   High-temperature strength

Superalloys are renowned for their ability to retain mechanical strength at high temperatures. Nickel-based superalloys, for instance, can maintain their strength at temperatures up to 1100 °C (2012 °F). This property is critical for applications in gas turbines and jet engines, where components are subjected to intense thermal and mechanical stresses. The strength of these alloys is attributed to their unique microstructure, which includes a γ' (gamma prime) phase precipitate that provides significant reinforcement. The γ' phase, typically Ni_3 (Al, Ti), forms a coherent structure with the nickel matrix, enhancing the alloy's ability to resist deformation at elevated temperatures [8].

2.2   Creep resistance

Creep resistance is another vital property of superalloys, referring to their ability to resist gradual deformation under constant stress at high temperatures. This characteristic is particularly important for components such as turbine blades and vanes, which operate under sustained loads. The creep resistance of superalloys is enhanced through solid solution strengthening and precipitation hardening. Elements such as chromium, cobalt, and molybdenum are added to the alloy to form stable precipitates that impede dislocation movement. For instance, the addition of rhenium (up to 6% by weight) has been shown to significantly improve the creep resistance of nickel-based superalloys by stabilizing the γ' phase and delaying the onset of creep deformation [9].

Superalloys: Fundamentals and Applications   Materials Research Forum LLC
Materials Research Foundations 178 (2025) 215-229   https://doi.org/10.21741/9781644903698-11

2.3   Oxidation and corrosion resistance

Superalloys exhibit excellent resistance to oxidation and corrosion, which is essential for maintaining their structural integrity in harsh environments. The high chromium content (15-25% by weight) in nickel-based superalloys forms a stable oxide layer on the surface, protecting the underlying material from further oxidation. Additionally, elements such as aluminum and titanium contribute to the formation of a protective oxide scale. For example, the aluminum content in the alloy can lead to the formation of a dense Al_2O_3 (alumina) layer, which provides a robust barrier against oxidation at high temperatures. Superalloys also resist hot corrosion, a phenomenon caused by the presence of sulphur and other contaminants in the environment. The addition of yttrium and hafnium (0.05-0.1% by weight) enhances the adhesion and stability of the protective oxide layer, further improving the alloy's corrosion resistance [10].

2.4   Surface stability and microstructural stability

The surface stability of superalloys is crucial for maintaining their performance over prolonged periods. This property is linked to the alloy's microstructural stability, which ensures that the desirable phases and precipitates remain intact during high-temperature exposure. Superalloys are designed to minimize the formation of deleterious phases, such as TCP (topologically close-packed) phases, which can embrittle the material. Advanced processing techniques, such as single-crystal growth and directional solidification, are employed to produce superalloy components with optimal microstructures. Single-crystal superalloys, devoid of grain boundaries, exhibit superior creep and fatigue resistance compared to their polycrystalline counterparts. The elimination of grain boundaries reduces the sites for crack initiation and propagation, enhancing the alloy's durability in high-stress applications [11].

2.5   Thermal conductivity and density

Superalloys also have specific thermal conductivity and density values that are tailored to their applications. Nickel-based superalloys typically have thermal conductivities in the range of 10-15 W/m·K at room temperature, which helps in managing the heat distribution within components. The density of these alloys ranges from 7.8 to 8.5 g/cm^3, depending on the composition and processing methods. While the relatively high density may be a drawback in weight-sensitive applications, the unparalleled performance of superalloys justifies their use in critical components where reliability and longevity are paramount [12]. Fig. 3 shows the essential characteristics of superalloys.

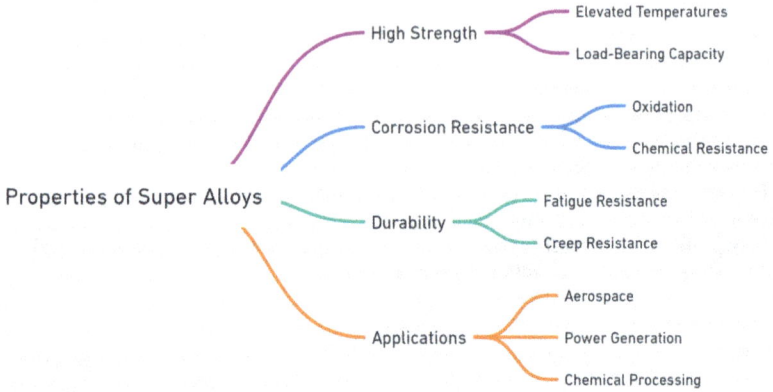

Fig. 3 Characteristics of superalloys

3.   Applications for aerospace engineering

Superalloys play a pivotal role in aerospace engineering, where their unique properties are crucial for the performance, safety, and efficiency of aircraft and spacecraft. The aerospace industry demands materials that can withstand extreme temperatures, mechanical stresses, and corrosive environments. Superalloys, particularly nickel-based variants, are engineered to meet these rigorous requirements, ensuring reliability in high-performance applications such as turbine engines, structural components, and propulsion systems. One of the primary applications of superalloys in aerospace is in gas turbine engines. These engines operate under severe conditions, with temperatures reaching up to 1200 °C (2192 °F) in the combustion section. Nickel-based superalloys, such as Inconel 718 and Rene 80, are commonly used in turbine blades, vanes, and other critical components due to their exceptional high-temperature strength and oxidation resistance. For instance, Inconel 718 retains its tensile strength up to approximately 1000 °C (1832 °F), making it suitable for the high-temperature sections of turbine engines. The alloy's composition, which includes nickel, chromium, and molybdenum, provides excellent creep resistance and maintains mechanical properties under sustained loads.

In addition to turbine engines, superalloys are integral to the production of spacecraft components. For space exploration, materials must endure not only high temperatures but also extreme thermal cycling and radiation [13]. Superalloys like Rene N4 and Haynes 230 are employed in rocket nozzles and spacecraft engines, where they must perform reliably under conditions that include temperatures exceeding 1300 °C (2372 °F). These superalloys are designed to resist thermal fatigue and oxidation, ensuring the durability of components exposed to the harsh environment of space. Superalloys also find applications in aerospace structural components that require both high strength and resistance to environmental degradation. Components such as aircraft landing gear, structural supports, and fasteners are made from superalloys to handle the significant stresses experienced during takeoff, flight, and landing. For example, titanium-based superalloys are used in aircraft frames and engine components due to their high strength-to-weight ratio and resistance to corrosion. Titanium alloys like Ti-6Al-4V, with a composition of approximately 90% titanium, 6% aluminum, and 4% vanadium, offer superior performance with a density of about 4.4 g/cm³, making them ideal for weight-sensitive applications [14]. In addition to their use in engines and structural components, superalloys are critical in the production of advanced propulsion systems, including those used in military aircraft and space exploration vehicles. For instance, superalloys are used in the production of afterburner components, which are subjected to extremely high temperatures and pressures. The advanced superalloy compositions, which include elements such as rhenium and tungsten, enhance the material's ability to withstand the thermal and mechanical stresses encountered in afterburners [15]. Fig. 4 shows the various applications of superalloys in aerospace components. Key properties and characteristics of casting processes and superalloys are compiled and listed in Table 1 [7-15].

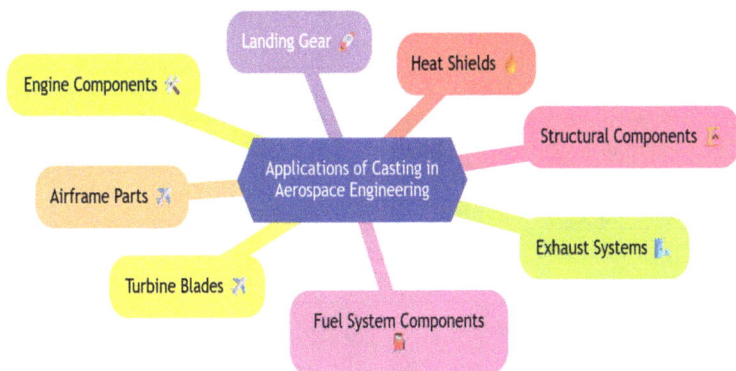

Fig. 4 Applications of casting products in aircraft components

Table 1 Key properties and characteristics of casting processes and superalloys [7-15]

S. No	Category	Properties	Value
1.	Sand casting	Mould temperature	1200 °C
2.	Die casting	Mould temperature	300 °C
3.	Investment casting	Mould temperature	1400 °C
4.	Centrifugal casting	Mould temperature	1200 °C
5.	Continuous casting	Molten metal temperature	1500 °C

6.	Nickel-based superalloy	Maximum operating temperature	1100 °C
7.	Nickel-based superalloy	γ' Phase volume fraction	0.30
8.	Cobalt-based superalloy	Maximum operating temperature	1050 °C
9.	Iron-based superalloy	Maximum operating temperature	1000 °C
10.	Nickel-based superalloy	Rhenium content	6% by weight
11.	Nickel-based superalloy	Thermal conductivity	13 W/m·K
12.	Nickel-based superalloy	Density	8.0 g/cm³
13.	Investment casting	Wax pattern temperature	1600 °C
14.	Die casting	Aluminum content	90%
15.	Centrifugal casting	Casting speed	1500 RPM
16.	Continuous casting	Withdrawal speed	2 m/min
17.	Nickel-based superalloy	Creep resistance improvement (rhenium)	+15%

Recent advancements in superalloy technology have focused on improving the performance and efficiency of aerospace components. Research has led to the development of new superalloy formulations with enhanced properties, such as increased thermal conductivity and reduced density, to optimize the balance between strength and weight. Innovations such as additive manufacturing (3D printing) are also revolutionizing the aerospace industry by enabling the production of complex superalloy components with precise geometries that were previously difficult to achieve using traditional methods [16]. This technology allows for the creation of lightweight, high-strength parts that contribute to the overall efficiency and performance of aerospace systems. In a nutshell, superalloys are indispensable in aerospace engineering due to their ability to withstand extreme conditions and deliver reliable performance. Their applications range from turbine engines and rocket nozzles to structural components and advanced propulsion systems. The continuous development and refinement of superalloys ensure that they meet the evolving demands of the aerospace industry, contributing to the advancement of both commercial and military aviation as well as space exploration. As the aerospace industry pushes the boundaries of technology, superalloys will remain a cornerstone of high-performance materials, enabling the next generation of aerospace innovations [17].

Superalloys: Fundamentals and Applications
Materials Research Foundations 178 (2025) 215-229

Materials Research Forum LLC
https://doi.org/10.21741/9781644903698-11

4. Advances in additive manufacturing

Additive manufacturing (AM), commonly known as 3D printing, has revolutionized the production of complex components by enabling the creation of intricate geometries that were previously unattainable with traditional manufacturing techniques. This technology has found significant applications in casting and superalloys, enhancing their capabilities and expanding their use in modern engineering [18].

4.1 Enhanced design capabilities

One of the most profound advances in additive manufacturing is the ability to produce highly complex and customized designs. Traditional casting methods often face limitations due to mould constraints and complex geometries that are difficult or impossible to achieve. In contrast, additive manufacturing builds components layer by layer directly from digital models, allowing for the creation of intricate structures such as lattice frameworks, internal channels, and customized cooling passages [19]. For example, in aerospace applications, additive manufacturing has enabled the production of optimized turbine blades with internal cooling channels that enhance thermal efficiency and performance. These designs are often impractical with traditional casting methods due to their complexity.

4.2 Material innovations

Additive manufacturing has also driven innovations in material science, particularly in the development and processing of superalloys. Advanced AM techniques, such as selective laser melting (SLM) and electron beam melting (EBM), have been employed to process high-performance alloys that are challenging to manufacture using conventional methods. Nickel-based superalloys, which are renowned for their high-temperature strength and resistance to oxidation, have been successfully processed using these technologies. For instance, Inconel 718, a commonly used nickel-based superalloy, has been fabricated using SLM with impressive results. This alloy, typically used in jet engines and high-stress applications, exhibits a tensile strength of approximately 1200 MPa at 650 °C, and the SLM process has been shown to maintain these properties with minimal degradation [20].

4.3 Precision and efficiency

The precision achieved in additive manufacturing is another significant advancement. The layer-by-layer construction process allows for tight control over the dimensions and tolerances of the produced components. This precision is particularly beneficial for applications requiring high accuracy, such as in the aerospace and medical fields. For example, the production of complex medical implants using titanium alloys has been greatly improved with additive manufacturing. These implants, often required to match patient-specific anatomical features, benefit from the high precision and customization that AM provides. The ability to achieve dimensional tolerances within ±0.1 mm ensures that the implants fit perfectly, reducing the risk of complications and improving patient outcomes [21].

4.4 Reduction of material waste

AM also offers substantial advantages in terms of material efficiency. Traditional casting methods can result in significant material waste, as excess material is often trimmed away after casting. In

Superalloys: Fundamentals and Applications   Materials Research Forum LLC
Materials Research Foundations 178 (2025) 215-229   https://doi.org/10.21741/9781644903698-11

contrast, additive manufacturing builds components directly from the digital model with minimal material waste. For example, conventional casting processes might result in up to 30% of the material being wasted due to the removal of excess metal. Additive manufacturing, however, can achieve near-net-shape production, where the amount of material used is closely aligned with the final component's requirements. This efficiency not only reduces material costs but also contributes to more sustainable manufacturing practices [22].

4.5   Integration with advanced casting techniques

Recent advancements have also focused on integrating AM with traditional casting techniques to combine the best features of both methods. One such approach is known as "AM-assisted casting," where AM is used to create complex cores or moulds that are then used in traditional casting processes. This hybrid method allows for the production of components with intricate internal features and complex geometries that would be challenging to achieve with conventional casting alone. For example, the use of 3D printed sand cores in investment casting has enabled the production of complex parts with detailed internal structures, which are crucial for industries like aerospace and automotive [23].

4.6   Performance and reliability

The performance and reliability of components produced via AM have been subjects of extensive research and development. Studies have shown that components made from superalloys using AM can achieve comparable, if not superior, mechanical properties compared to those produced by traditional methods. For instance, a study on nickel-based superalloys processed by SLM demonstrated improved fatigue resistance and higher tensile strength compared to conventionally cast counterparts. This enhanced performance is attributed to the fine microstructure and uniform distribution of the reinforcing phases achieved through the additive manufacturing process [24].

In a nutshell, the advances in additive manufacturing have significantly impacted the fields of casting and superalloys, offering new possibilities for design, material processing, and production efficiency. By enabling the creation of complex geometries, enhancing material properties, reducing waste, and integrating with traditional methods, AM is transforming modern engineering applications. The continued development and innovation in this field promise to drive further progress, pushing the boundaries of what is achievable in manufacturing and materials [25].

5.   Sustainability and future trends

The pursuit of sustainability in manufacturing has become a critical focus as industries seek to minimize their environmental impact while maintaining high performance and economic efficiency. In the realm of casting and superalloys, several emerging trends and practices are shaping the future of these technologies, aiming to address environmental concerns and enhance the sustainability of production processes. One of the primary sustainability challenges in casting is the management of material waste and energy consumption. Traditional casting methods often result in significant material waste, especially in processes like sand casting and die casting, where excess material must be trimmed or discarded. The adoption of advanced casting techniques, such as precision investment casting and additive manufacturing, has shown promise in reducing material waste. For instance, additive manufacturing enables near-net-shape production, which minimizes excess material and reduces the need for post-processing. By improving the accuracy and efficiency of component production, these methods contribute to more sustainable manufacturing practices [26]. Energy consumption is another critical area of concern. The production of superalloys and the casting process itself require substantial amounts of energy, contributing to greenhouse gas emissions and resource depletion. Future trends in the casting industry are focusing on the development of energy-efficient technologies and processes. Innovations such as advanced furnace designs, waste heat recovery systems, and the use of renewable energy sources are being explored to reduce the energy footprint of casting operations. For example, the integration of electric arc furnaces and induction melting technologies has demonstrated potential in lowering energy consumption compared to traditional methods. The recycling and reuse of materials are also gaining traction as part of the sustainability agenda. In the context of superalloys, the recycling of scrap metal and used components can significantly reduce the demand for virgin materials and decrease the overall environmental impact of production [27].

The development of closed-loop recycling systems, where material waste is collected and reintroduced into the production cycle, is becoming more prevalent. Such systems not only conserve resources but also reduce the environmental burden associated with mining and processing raw materials. Future trends in superalloys and casting are also driven by advancements in material science and process optimization. Researchers are exploring new alloy compositions and processing techniques that enhance the performance and longevity of components while reducing their environmental impact. For example, the development of novel superalloys with improved high-temperature strength and corrosion resistance can lead to longer-lasting components, reducing the frequency of replacements and the associated resource use. Furthermore, the integration of digital technologies and smart manufacturing practices is set to revolutionize the sustainability of casting and superalloy production [28-32]. The application of data analytics, artificial intelligence, and machine learning can optimize manufacturing processes, predict maintenance needs, and enhance overall efficiency [32-37]. These technologies enable more precise control over production parameters, reducing waste and improving resource utilization. As industries continue to evolve, the focus on sustainability will drive innovation and transformation in casting and superalloy technologies [38-41]. By addressing challenges related to material waste, energy consumption, and resource management, and by leveraging advancements in material science and digital technologies, the future of casting and superalloys promises to be more sustainable and efficient. Embracing these trends will not only contribute to environmental preservation but also enhance the economic viability and performance of modern engineering applications.

Conclusion

Casting processes and superalloys represent crucial advancements in modern engineering, each with its unique strengths and applications. From sand casting to continuous casting, the variety of casting techniques available today allows for the production of components with precise dimensions and excellent mechanical properties. Each method—whether it's sand casting's cost-effectiveness for large parts, die casting's high-volume precision, or investment casting's intricate detail—plays a pivotal role in manufacturing diverse and high-performance components. Recent innovations in these processes, including advancements in materials and automation, continue to drive improvements in efficiency and quality. Superalloys, characterized by their exceptional high-temperature strength, creep resistance, and oxidation and corrosion resistance, are indispensable in high-stress environments such as aerospace and power generation. The development of these alloys has been guided by rigorous scientific research, resulting in materials capable of withstanding extreme conditions while maintaining structural integrity. Properties like high-temperature strength and creep resistance are enhanced by precise alloying and processing techniques, contributing to their critical role in engines, turbines, and other demanding applications. The future of casting and superalloys is closely linked to advancements in sustainability and manufacturing efficiency. The integration of additive manufacturing technologies has opened new avenues for producing complex geometries and optimizing material use, significantly reducing waste and enhancing precision. Meanwhile, the focus on energy-efficient technologies and closed-loop recycling systems reflects a growing commitment to minimizing environmental impact. As the industry progresses, the adoption of digital technologies and smart manufacturing practices will further drive innovation, making processes more sustainable and efficient. In summary, the ongoing evolution in casting processes and superalloys is characterized by technological advancements that enhance performance, sustainability, and efficiency. These developments not only address the immediate needs of various industries but also pave the way for future innovations, ensuring that casting and superalloys continue to meet the rigorous demands of modern engineering applications.

References

[1] N.M. Dawood, A.M. Salim, A review on characterization, classifications, and applications of superalloys, JUBES. 29 (2021) 53-62.

[2] G.R. Thellaputta, P.S. Chandra, C.S.P. Rao, Machinability of nickel-based superalloys: a review, Mater. Today Proc. 4 (2017) 3712-3721. https://doi.org/10.1016/j.matpr.2017.02.266

[3] I.G. Akande, O.O. Oluwole, O.S.I. Fayomi, O.A. Odunlami, Overview of mechanical, microstructural, oxidation properties and high-temperature applications of superalloys, Mater. Today Proc. 43, (2021) 2222-2231. https://doi.org/10.1016/j.matpr.2020.12.523

[4] W. Betteridge, S.W.K. Shaw, Development of superalloys, Mater. Sci. Technol. 3 (1987) 682-694. https://doi.org/10.1179/mst.1987.3.9.682

[5] H. Fecht, D. Furrer, Processing of nickel-base superalloys for turbine engine disc applications, Adv. Eng. Mater. 2 (2000) 777-787. https://doi.org/10.1002/1527-2648(200012)2:12<777::AID-ADEM777>3.0.CO;2-R

[6] P.S. Gowthaman, S. Jeyakumar, A review on machining of high temperature aeronautics super-alloys using WEDM, Mater. Today Proc. 18 (2019) 4782-4791. https://doi.org/10.1016/j.matpr.2019.07.466

[7] A. Kracke, A. Allvac, Superalloys, the most successful alloy system of modern times-past, present and future, in: Proceedings of the 7th International Symposium on Superalloy. 718 (2010) pp. 13-50. https://doi.org/10.7449/2010/Superalloys_2010_13_50

[8] D.K. Ganji, G. Rajyalakshmi, Influence of alloying compositions on the properties of nickel-based superalloys: A review, in: H. Kumar, P. Jain (Eds.), Recent Advances in Mechanical Engineering, Select Proceedings of NCAME, 2019, pp. 537-555. https://doi.org/10.1007/978-981-15-1071-7_44

Materials Research Forum LLC
https://doi.org/10.21741/9781644903698-11

[9] A. Behera, A.K. Sahoo, S.S. Mahapatra, Application of Ni-based superalloy in aero turbine blade: A review, in: Proc. Inst. Mech. Eng. E: J. Process. Mech. Eng. 2023, 09544089231219104. https://doi.org/10.1177/09544089231219104

[10] Gudivada, G., & Pandey, A. K. Recent developments in nickel-based superalloys for gas turbine applications, J. Alloys Compds. (2023) 171128. https://doi.org/10.1016/j.jallcom.2023.171128

[11] F.L. VerSnyder, Keynote lecture superalloy technology today and tomorrow, in: E. Bachelet. (Eds.), High Temperature Alloys for Gas Turbines, Proceedings of a Conference held in Liège, Belgium, 4-6 October 1982 (pp. 1-49). Dordrecht: Springer Netherlands. https://doi.org/10.1007/978-94-009-7907-9_1

[12] D. Coutsouradis, A. Davin, M. Lamberigts, Cobalt-based superalloys for applications in gas turbines, Mater. Sci. Eng. 88 (1987) 11-19. https://doi.org/10.1016/0025-5416(87)90061-9

[13] J. Maniraj, F. Sahayaraj, J. Giri, T. Sathish, M.R. Shaik, Enhancing performance of prosopis juliflora fiber reinforced epoxy composites with silane treatment and syzygium cumini filler, J. Mater. Res. Technol. 31 (2024) 93-108. https://doi.org/10.1016/j.jmrt.2024.06.058

[14] R. Schafrik, R. Sprague, Superalloy technology-a perspective on critical innovations for turbine engines, Key Eng. Mater. 380 (2008), 113-134. https://doi.org/10.4028/www.scientific.net/KEM.380.113

[15] R. Darolia, Development of strong, oxidation and corrosion resistant nickel-based superalloys: critical review of challenges, progress and prospects, Int. Mater. Rev. 64 (2019), 355-380. https://doi.org/10.1080/09506608.2018.1516713

[16] Luo, A. A., Sachdev, A. K., & Apelian, D. (2022). Alloy development and process innovations for light metals casting, J. Mater. Process. Technol. 306, 117606. https://doi.org/10.1016/j.jmatprotec.2022.117606

[17] Sonar, T., Ivanov, M., Trofimov, E., Tingaev, A., & Suleymanova, I. (2024). An overview of microstructure, mechanical properties and processing of high entropy alloys and its future perspectives in aeroengine applications, Mater. Sci. Energy Technol. 7, 35-60. https://doi.org/10.1016/j.mset.2023.07.004

[18] A. Gloria, R. Montanari, M. Richetta, A. Varone, Alloys for aeronautic applications: State of the art and perspectives, Metals, 9 (2019) 662. https://doi.org/10.3390/met9060662

[19] H.A. Youssef, Machining of stainless steels and superalloys: traditional and nontraditional techniques, John Wiley & Sons, 2015. https://doi.org/10.1002/9781118919514

[20] A. Behera, A. Behera, Superalloys, in: Advanced Materials: An Introduction to Modern Materials Science, Springer, 2022, 225-261. https://doi.org/10.1007/978-3-030-80359-9_7

[21] A.F. Sahayaraj, M.T. Selvan, M. Ramesh, J. Maniraj, I. Jenish, K.J. Nagarajan, Extraction, purification, and characterization of novel plant fiber from tabernaemontana divaricate stem to use as reinforcement in polymer composites, Biomass Convers. Biorefin. (2024) 1-15. https://doi.org/10.1007/s13399-024-05352-4

Materials Research Forum LLC
https://doi.org/10.21741/9781644903698-11

[22] A.R. Paul, M. Mukherjee, D. Singh, A critical review on the properties of intermetallic compounds and their application in the modern manufacturing, Cryst. Res. Technol. 57 (2022) 2100159. https://doi.org/10.1002/crat.202100159

[23] Y.T. Tang, C. Panwisawas, J.N. Ghoussoub, Y. Gong, J.W. Clark, AA. Németh, R.C. Reed, Alloys-by-design: Application to new superalloys for additive manufacturing, Acta Mater. 202 (2021), 417-436. https://doi.org/10.1016/j.actamat.2020.09.023

[24] A.N. Jinoop, C.P. Paul, K.S. Bindra, Laser-assisted directed energy deposition of nickel superalloys: A review, Proc. Inst. Mech. Eng. Pt. L J. Mater. Des. Appl. 233 (2019), 2376-2400. https://doi.org/10.1177/1464420719852658

[25] J. Maniraj, M. Ramesh, S.G. Kumar, A.F. Sahayaraj, Introduction of biochar: sources, composition, and recent updates, in: A.K. Nadda, (Eds.), Biochar and its Composites: Fundamentals and Applications, Singapore: Springer Nature Singapore, 2023 pp. 1-17. https://doi.org/10.1007/978-981-99-5239-7_1

[24] H. Dai, A study of solidification structure evolution during investment casting of Ni-based superalloy for aero-engine turbine blades, Doctoral Dissertation, University of Leicester. 2009.

[25] G. Wu, C. Wang, M. Sun, W. Ding, Recent developments and applications on high-performance cast magnesium rare-earth alloys, J. Magnes. Alloy. 9 (2021), 1-20. https://doi.org/10.1016/j.jma.2020.06.021

[26] M. Javidani, D. Larouche, Application of cast Al-Si alloys in internal combustion engine components, Int. Mater. Rev. 59 (2014), 132-158. https://doi.org/10.1179/1743280413Y.0000000027

[27] S.F. Yang, S.L. Yang, J.L. Qu, J. Du, Y. Gu, P. Zhao, N. Wang, Inclusions in wrought superalloys: A review, J. Iron. Steel Res. Int. 28 (2021) 921-937. https://doi.org/10.1007/s42243-021-00617-y

[28] J. Du, X. Lv, J. Dong, W. Sun, Z. Bi, G. Zhao, B. Zhang, Research progress of wrought superalloys in China. Acta Metall. Sin. 55 (2019), 1115-1132.

[29] J. Maniraj, F.S. Arockiasamy, C.R. Kumar, D.A. Kumar, I. Jenish, I. Suyambulingam, I S. Siengchin, Machine learning techniques for the design and optimization of polymer composites: a review, in: E3S Web of Conferences, EDP Sciences, 428 (2023) pp. 02013. https://doi.org/10.1051/e3sconf/202342802013

[30] D. Ma, Novel casting processes for single-crystal turbine blades of superalloys, Front. Mech. Eng. 13 (2018), 3-16. https://doi.org/10.1007/s11465-018-0475-0

[31] S.K. Dewangan, A. Mangish, S. Kumar, A. Sharma, B. V. Kumar, A review on high-temperature applicability: a milestone for high entropy alloys, Eng. Sci. Technol. Int. J. 35 (2022) 101211. https://doi.org/10.1016/j.jestch.2022.101211

[32] M. Dada, P. Popoola, S. Adeosun, N. Mathe, High entropy alloys for aerospace applications. in: M.G. Bandpy, A.M. Aly (Eds.), Aerodynamics, Intech Open, 2019, 248.

[33] J. Bai, Y. Yang, C. Wen, J. Chen, G. Zhou, B. Jiang, F. Pan, Applications of magnesium alloys for aerospace: A review, J. Magnes. Alloy. 10 (2023) 3609-3619. https://doi.org/10.1016/j.jma.2023.09.015

[34] V. Dhinakaran, J. Ajith, A.F.Y. Fahmidha, T. Jagadeesha, T. Sathish, B. Stalin, Wire Arc Additive Manufacturing (WAAM) process of nickel-based superalloys - A review, Mater. Today Proc. 21 (2020), 920-925. https://doi.org/10.1016/j.matpr.2019.08.159

[35] A. Pineau, S.D. Antolovich, High temperature fatigue of nickel-base superalloys-a review with special emphasis on deformation modes and oxidation, Eng. Fail. Anal. 16 (2009) 2668-2697. https://doi.org/10.1016/j.engfailanal.2009.01.010

[36] K.P. Boopathiraja, R. Ramamoorthi, P. Hariprasad, F.S. Arockiasamy, Enhancing microstructural and mechanical properties of magnesium AZ31 matrix composites through friction stir processing incorporating silicon carbide, titanium carbide, and graphite particles, Mater. Res. Express. 11 (2024), 076507. https://doi.org/10.1088/2053-1591/ad5cdc

[37] M.V. Raja, K. Manonmani, A.F. Sahayaraj, Investigation on electrical discharge machining parameters of aluminium7075/boron carbide/titanium diboride hybrid composites by grey relational analysis, Proc. Natl. Acad. Sci. India A: Physical Sciences, (2024) 1-13. https://doi.org/10.1007/s40010-024-00879-7

[38] S. Ravichandran, R. Duraiswamy, F.S. Arockiasamy, Tool and formability multi-response optimization for ultimate strength and ductility of AA8011 during axial compression. Adv. Mech. Eng. 14 (2022), 16878132221131731. https://doi.org/10.1177/16878132221131731

[39] B.A. Kumar, M.M. Krishnan, A.F. Sahayaraj, M.R.A. Refaai, G. Yuvaraj, D. Madhesh, H.L. Allasi, Characterization of the aluminium matrix composite reinforced with silicon nitride (AA6061/Si3N4) synthesized by the stir casting route, Adv. Mater. Sci. Eng. 1 (2022) 8761865. https://doi.org/10.1155/2022/8761865

Superalloys: Fundamentals and Applications
Materials Research Foundations 178 (2025) 230-258

Materials Research Forum LLC
https://doi.org/10.21741/9781644903698-12

Chapter 12

Cutting Tool Coatings for Cryogenic Machining

Mohamed El Garah [*1] and Driss Soubane [2,3]

[1] LRC, LASMIS, Antenne de Nogent – 52, Pôle Technologique de Sud – Champagne, 52800 Nogent, France

[2] Institut National de la Recherche Scientifique - Centre Énergie, Matériaux et Télécommunications (INRS-EMT), Varennes, Québec J3X 1P7 Canada

[3] Cadi Ayyad University - Multi-Disciplinary Faculty of Safi, Department of Physics, Energy = Materials × Computation × Confinement (ε =mc²) EBCG-Lab, Safi 46 000, Morocco

mohamed.el_garah@utt.fr, driss.soubane@inrs.ca

Abstract

In recent years, there has been a strong demand for tough materials to meet the needs of modern industry. In sectors such as transport and energy, the performance of these materials has increased considerably. However, their manufacture, and particularly their machining, remains a significant challenge. Cooling fluids can be used to improve the machining process. Modern industry demands fluids that meet a range of economic, social, and environmental requirements. Unlike water-oil mixtures, cryogenic coolants are advantageous due to their favorable environmental impact and worker health benefits. However, under such conditions, conventional cutting tools are subject to severe wear, which reduces their lifespan. Therefore, improving their performance is crucial for industrial needs. Coating technology plays an important role in producing innovative and cost-effective materials. To date, while there have been reviews on cryogenic machining, little work has focused on improving cutting tools through coating techniques. This study aims to review the benefits of cryogenic machining and its various operations. The deposition techniques and different coatings used to enhance cutting tool performance are also discussed. Additionally, the review highlights process simulations, which open new prospects for the development of innovative materials.

Keywords

Machining Process, Cryogenic Coolants, Tool Lifespan, Coating Technology, Finite Element Method, Computational Fluid Dynamics, Industry, Performance

Contents

1.   Introduction

Manufacturing is a vital process in all areas of industry to produce components of required shapes and sizes. A large proportion of these products involve machining. As technology continues to progress, advancements in machining, particularly in the metallurgical sector, show potential promise and significant impact on both the environment and healthcare. One of the primary concerns in the industry in recent years has been the use of cooling lubricants, often petroleum-based. Cleaning and washing tool parts after use add extra costs to the fabrication process. Additionally, if these fluids are disposed of improperly, they can be harmful not only to the environment but also to people [1]. The use of cryogenic fluids, such as liquid nitrogen (LN2) or liquid carbon dioxide (CO_2), can address several of these issues [2]. Since these fluids evaporate quickly, the workplace remains clean, and machined parts do not require additional cleaning. Given that economy, environment, and society are the three pillars of sustainability, the economic aspect in machining covers production, costs, and process quality. The environmental aspect relates to fluid usage, energy consumption, and waste management. The societal aspect encompasses worker health and safety. Notably, machining costs can be reduced as cryogenic cooling requires less infrastructure. Cutting conditions are critical for machining performance as they determine the quality of the machined parts. One of the current industrial challenges is the temperature increase during machining, especially at the contact surface, which leads to premature tool wear. Developing innovative materials and manufacturing processes, especially for difficult-to-machine materials, remains a challenge that the industry must overcome.

Machining plays a crucial role in material cutting processes. A cutting tool is used to remove material from a workpiece [3] involving the removal of chips through plastic deformation while applying mechanical and thermal forces between the tool and the workpiece. As previously mentioned, certain materials, such as titanium and nickel alloys, are difficult to machine due to their relatively low thermal conductivity. The heat generated during the process cannot be easily dissipated through the workpiece, resulting in extremely high temperatures that lead to premature tool wear and reduced tool life.

The application of coolant during machining is a critical factor in improving performance. The primary function of this fluid is to lubricate the interfaces between the tool and the workpiece while also dissipating the heat generated during machining. Conventional fluids typically consist of water and oil mixtures [4-6]. However, as previously discussed, these mixtures can be harmful to both the environment and worker health. One solution to this issue is to use dry cooling methods [7]. For example, cryogenic fluids like liquid nitrogen (LN2) and liquid carbon dioxide (LCO$_2$) offer advantages in machining. Recent reviews have highlighted progress in cryogenic machining [2, 8, 9]. Research into cryogenically assisted machining using conventional cutting tools shows that these tools are prone to significant wear, which reduces their service life. Therefore, the development of innovative materials is essential to support technological advancements.

An important aspect to consider when developing high-performance materials is their surface treatment. It is possible to replace low-quality materials with those that meet industrial requirements and safety standards. The durability of these materials is ultimately influenced by the quality of their surface. For this reason, a wide variety of coatings have been developed to address the various causes of degradation. The development of coatings is an effective strategy for improving the functional properties of materials. Current scientific research focuses on developing high-performance coatings to enhance the properties of low-cost materials. Conventional hard coatings, such as nitrides and carbides, provide excellent hardness and improved wear resistance. Coatings with enhanced mechanical properties can be produced through two main approaches: Physical Vapor Deposition (PVD) or Chemical Vapor Deposition (CVD) [10-13]. These techniques are widely used in industry, and many coatings can be developed for cutting tools, particularly when used under cryogenic assistance.

This chapter reports on recent progress in cryogenic machining and the development of cutting tool coatings. Various processes and conditions are presented. The most used industrial techniques for the deposition of coatings are also discussed, with examples of both mono- and multi-layered coatings. Their mechanical properties, tribological behavior, and performance are examined. We have also focused on simulation models, which compute heat transfer parameters and cutting temperature. The theoretical section addresses dry cutting.

2.   Metal cutting

2.1   Cutting system

Machining plays a crucial role in modern industry, enabling the production of parts with complex shapes and precise tolerances. Machined components are of great interest to a wide range of industrial sectors, including aerospace engines, automotive suspension parts, and even surgical implants and instruments. Evolving industrial needs have led to the development of various machining techniques that offer specific advantages in terms of precision, productivity, and durability.

Cryogenic machining offers significant benefits when working with difficult-to-machine materials and in applications where surface quality and tool life are critical. This process includes several machining methods, such as turning, milling, and drilling, all of which will be discussed in detail in Section 5.

2.2 Working fluid

In general, several types of fluids are used in machining. Their primary roles include removing the heat generated during machining, lubricating to reduce friction, and removing chips from the cutting zone. The various lubricants available on the market are typically mixtures of oils (synthetic, semi-synthetic, or vegetable) with a small amount of water. Although these fluids comply with industry standards, they can still pose hazards to both humans and the environment. In dry machining, no fluids are used, which avoids pollution issues, but a major drawback is the limited lifespan of cutting tools, particularly when machining refractory alloys. Lubrication and cooling are therefore essential, which is why modern industry is increasingly turning to cryogenic machining.

3. Cryogenic technology

3.1 Cutting tools

During the machining process, the cutting tool is subjected to very high mechanical and thermal stresses. Its wear, characterized by damage and changes in geometry, occurs continuously and is a key determinant of tool life and performance. Furthermore, wear affects the quality of the machined workpiece. Various cutting tools, such as diamond and ceramic, have been utilized for cryogenic machining; however, most studies have focused on coated carbide cutting tools. There is a need for further research to improve the efficiency of this type of tool to meet current industrial demands.

Several parameters influence the cutting tool's performance, including its geometry, fluid pressure, and cutting speed. The most used tools are typically square-shaped with specific dimensions, often around 12 mm long and nearly 5 mm thick (see Figure 1). However, in milling or drilling operations, there are no specific trends regarding the tools employed.

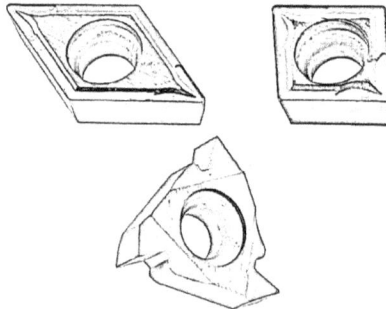

Figure 1. Some models of cutting tools used in the machining process.

In every machining process, cutting parameters play a crucial role. To achieve optimal results in terms of the quality of the machined workpiece and to ensure an efficient process, these cutting parameters must be optimized (as will be discussed below).

Superalloys: Fundamentals and Applications
Materials Research Foundations 178 (2025) 230-258

Materials Research Forum LLC
https://doi.org/10.21741/9781644903698-12

3.2 Workpiece and cooling systems (LCO₂, LN2)

Cryogenic cooling of the workpiece is a widely employed method in this type of machining. Its primary objective is to create an optimal operating environment for high-quality machining and prolonged tool life. Generally, two methods can be utilized for cooling the workpiece: cryogenic bath cooling or liquid gas spraying. Currently, machines are equipped with liquid gas nozzles, which are more practical for this type of machining process.

Two types of fluids are used in cryogenic machining: liquid nitrogen (LN2) and carbon dioxide (CO_2). Due to their differing properties, both liquid gases require specialized storage and handling. Liquid gas is typically stored in a high-pressure tank. For example, LN2 must be stored in an insulated tank at high pressure due to its temperature. At atmospheric pressure and ambient temperature, LN2 begins to boil at −196 °C. In contrast, liquid CO_2 can be stored at room temperature. When released, it can convert into CO_2 gas and form dry ice (a mixture of carbon dioxide snow). Cryogenic liquids can be delivered through internal or external nozzles. The configuration of the machine used for this type of machining significantly influences the results of the machined part and the service life of the tool. Key parameters for quality machining include the position and orientation of the nozzles, as well as the flow rate. LN2 necessitates an insulated system, and low temperatures can often pose challenges. In the case of LCO_2, being stored at room temperature eliminates the need for insulation.

4. Process parameters

4.1 Flow rate and pressure

The pressure of the cryogenic cooling during the machining operation varies depending on the fluid used. For liquid nitrogen (LN2), reported pressure values range from 1.4 to 24 bar. The flow rate has been documented to range from 0.5 kg/min to over 3 kg/min, with these values primarily reported for turning operations. Other machining processes have also been explored in the literature. For instance, in the drilling process, Ahmed and Kumar [14] reported using a pressure of 0.4 MPa for LN2 coolant during operations. In the milling process, Shokrani et al. [15] investigated the machining of Ti6Al4V alloy under cryogenic conditions. They utilized LN2 with a flow rate of 0.4 L/min at a pressure of 1.5 MPa. Furthermore, the authors compared the machining process across different environments, including flooding and dry machining, demonstrating a reduction in tool wear of up to 40% with cryogenic cooling.

4.2 Cutting parameters

The cutting speeds utilized in machining processes are highly dependent on the material of the workpiece. For instance, in turning operations, speeds ranging from 30 to 130 m/min have been tested for machining nickel-based alloys [16-18]. Other studies have reported speeds reaching up to 300 m/min [18-20].

Titanium-based alloys are notoriously difficult to machine due to their hardness, necessitating higher cutting parameters than those used for nickel-based alloys. Several studies have shown that cutting speeds between 70 and 150 m/min can be achieved, with feed rates ranging from 0.10 to 0.25 mm/rev [21]. Venkatesh et al. [22] reported a high cutting speed of 300 m/min, while Dhananchezian and his colleagues [23] noted the lowest cutting speed at 27 m/min.

5. Operations

5.1 Turning

Cryogenic turning involves cooling the cutting tool or cutting zone using a stream of cryogenic gas, typically liquid nitrogen. This cooling process reduces tool wear, enhances thermal stability, and improves the machinability of difficult materials such as titanium alloys and superalloys. In a turning operation, the workpiece, typically cylindrical, rotates around its axis while the cutting tool moves in the machining direction. This movement, known as the feed rate, is parallel to the workpiece axis in longitudinal turning (Figure 2). The cutting tool remains continuously engaged throughout the process.

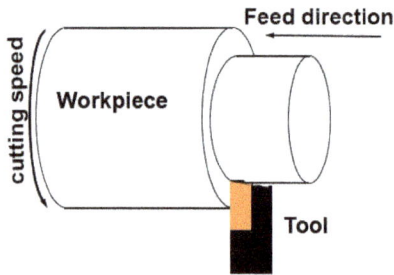

Figure 2. Illustration of turning machining process.

5.2 Milling

In milling operations, the tool rotates, bringing its cutting edges into contact with the workpiece. Unlike turning, milling involves multiple cutting edges and operates in an interrupted cutting manner. As the tool rotates, the cutting edges engage and disengage with the material to shape the workpiece (Figure 3). During face milling, the feed direction is perpendicular to the cutter's axis of rotation, allowing for the creation of flat surfaces or specific geometries.

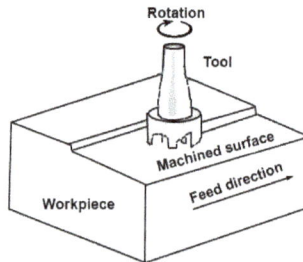

Figure 3. Illustration of milling machining process.

Superalloys: Fundamentals and Applications
Materials Research Foundations 178 (2025) 230-258

Materials Research Forum LLC
https://doi.org/10.21741/9781644903698-12

5.3 Drilling

Drilling is another essential machining process widely used across various industries. The operation involves cutting or enlarging a hole in a workpiece (Figure 4). However, there have been relatively few studies on cryogenic drilling due to the challenge of effectively delivering coolant to the cutting zone. Shah et al. [24] investigated the drilling of titanium under different cooling conditions, including liquid nitrogen (LN2), liquid carbon dioxide (LCO2), dry, and flooded environments. They evaluated several factors, such as cutting tool life, thrust force, and energy consumption. The findings revealed that LN2 cooling extended tool life, allowing 299 holes to be drilled, compared to 202 holes with LCO2. However, LCO2 cooling demonstrated lower energy consumption. Conventional drilling conditions (dry and flooded) were also tested, and cryogenic cooling outperformed these methods in terms of tool longevity and process efficiency.

Khanna et al. [9] conducted a study on drilling Inconel 718 in both cryogenic and dry environments. The results indicated improved tool life and better hole quality under cryogenic conditions compared to dry drilling, further demonstrating the benefits of cryogenic technology in demanding drilling applications.

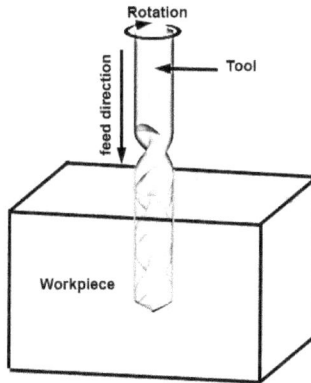

Figure 4. Illustration of milling machining process.

6.   Coatings for cryogenic machining

6.1   Deposition techniques

Chemical Vapor Deposition (CVD) and Physical Vapor Deposition (PVD) are two approaches, operating in different environments, for coating materials for applications in several fields. CVD coatings appeared in the 1960s, while PVD coatings appeared 20 years later. Today, cutting tools are coated using both approaches.

6.1.1 PVD

Vapor deposition (PVD) encompasses several methods and techniques operating in a vacuum. The objective is to physically extract the material and evaporate or spray it. Elements consisting of atoms, ions, or molecules are sputtered by plasma and deposited on the substrate to be coated. The most used PVD techniques to coat cutting tools are magnetron sputtering and arc deposition.

Magnetron sputtering is a physical vapor deposition (PVD) method widely used in coatings manufacturing. Various deposition parameters can be controlled to produce coatings with high quality (Figure 5a). The machines of this technique are adapted to industrial processes and scientific research in laboratories. To sputter the desired material, three different processes can be performed in three different ways: pulsed direct current (DC), radio frequency (RF), and High-Power Impulse Magnetron Sputtering (HiPIMS).

The cathodic arc deposition process is an approach to physical vapor deposition. It uses an electric arc to sublimate the target material. The target material is then deposited on the surface of the part to be coated, forming a thin layer (Figure 5b). Technically, an electromagnetic field can be applied to orient the point of impact of the arc to progressively and homogeneously cover the entire surface of the target. In addition, gases can be introduced to form desired compound materials such as ceramics.

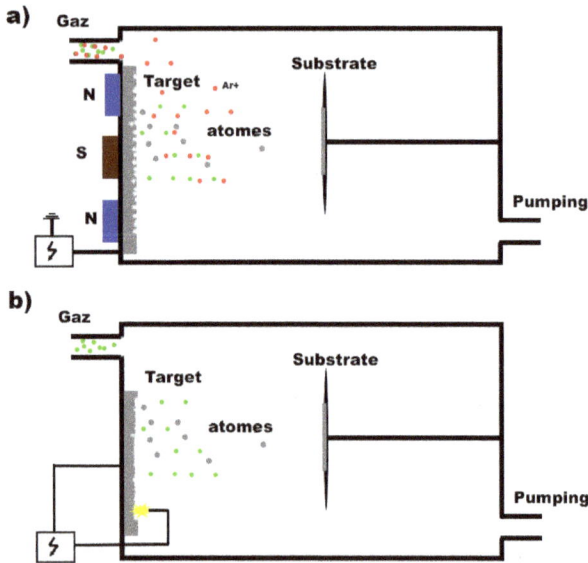

Figure 5. Illustration of PVD technique. a) Magnetron sputtering. b) Arc deposition

Superalloys: Fundamentals and Applications   Materials Research Forum LLC
Materials Research Foundations 178 (2025) 230-258   https://doi.org/10.21741/9781644903698-12

6.1.2 CVD

As mentioned above, CVD is a second approach to tool coating. Its process is based on the reaction of gaseous chemical compounds with substrates heated to high temperatures. These reactions take place at sub-atmospheric pressure, and deposition temperatures are often between 800 and 1200°C.

Extremely pure materials can be obtained using this deposition technique. It enables atomic-scale structural control through chemical reactions. Moreover, it offers the possibility of fabricating coatings with different architectures: monolayer, multilayer, or composite. It's an effective way to coat complex workpieces.

6.2 Hard coatings for cutting tools

Since the invention of cemented carbide, a very hard material compared to steel, the development of new cutting tools has continued to improve. Tungsten carbides started in 1930, followed by carbide alloys and then ceramics. Coated carbides started in 1970 [25]. Due to the increasing consumption of these tools, tool life has become a critical parameter for tool quality. To further improve the performance of these tools, coating has emerged as a potential approach to meet industrial needs. Various coatings have been developed, to reach this objective, in the form of monolayers, multilayers, and composites.

6.3 Monolayered coatings

In PVD techniques, a monolayer consists of a single, continuous, and homogeneous layer. It can be made from pure metallic elements or ceramics such as nitrides, carbides, or oxides. Various monolayers of coated cutting tools using the PVD or CVD approach have been based on aluminum or titanium materials [26-30]. TiAlN has been reported as a hard coating used under cryogenic conditions. Some reviews discuss the performance of this monolayer coating. However, these coatings present limitations, particularly in terms of weak corrosion resistance despite their tribological performance. Chromium-based coatings have also been developed and revealed some interesting properties, including good adhesion, high mechanical strength, and improved corrosion resistance.

Several Cr-based coatings have been studied, including CrN [31-33], CrAlN [34], and CrZrN [35]. The inclusion of a third element benefits the alloys, enhancing their performance. Selvan et al. [36] published a study on AlCrN coatings for machining Ti6Al4V under cryogenic conditions. The study demonstrated that a cutting tool coated with AlCrN and lubricated with a cryogenic coolant (LN2) results in a better surface finish of the machined part (Ti-6Al-4V). Additionally, a reduction in tool wear was observed. The study reports optimum parameters, suggested by the TOPSIS method, as cutting speed (Vc) = 215 m/min, feed rate (f) = 0.102 mm/rev, and depth of cut (doc) = 0.5 mm, with the following cutting forces: feed force Fx = 61.6 N, thrust force Fy = 92.9 N, cutting force Fz = 72.4 N, and surface roughness Ra = 0.78 μm.

6.4 Multilayered coatings

Multilayered coatings offer an advantage over monolayers by combining layers with different properties to meet the demands of machining, especially under cryogenic conditions. Zhang et al. [37] studied Cr/CrN/AlCrN multilayer coatings for use in the cryogenic machining of Ti6Al4V alloy. The results showed that these multilayer coatings could significantly improve tool life under cryogenic cooling conditions, offering a 33% improvement compared to non-coated tools. Additionally, the study identified a strong correlation between tool wear and machining forces. Pearson correlation analysis was used to determine the relationship between tool flank wear and other parameters, such as force and surface roughness. Using statistical analysis, models were developed to predict flank wear as a function of measured forces, utilizing the principal component analysis (PCA) methodology. Sheik Muhamad et al. [38] coated a carbide cutting tool with a multilayer coating (TiAlN/AlCrN). They used LN2 as a coolant to study the tool life and wear mechanisms of this coated tool. They found that, during milling under cryogenic conditions, tool life was extended. The results are therefore conclusive and demonstrate the importance of using this type of material with the cryogenic technique (LN2), which does not pollute the environment.

6.5   Nanocomposites

Nanocomposite coatings are emerging as critical components in enhancing the efficiency of cryogenic machining. They effectively protect cutting tools from rapid wear and improve the quality of machined surfaces. These coatings are typically composed of innovative and often complex materials. Zhang et al. [39] reported on the high-speed milling of SiCf/SiC composites in two environments: dry and cryogenic. The diamond tools used were manufactured from Polycrystalline Diamond (PCD) and Chemical Vapor Deposition (CVD) techniques. Their study investigated the effects of cryogenic coolant on tool wear and the quality of the machined surface. The use of cryogenic cooling significantly reduced wear on the diamond tools, leading to a service life increase of four times when machining cryogenically compared to dry conditions. Furthermore, for the same milling and cryogenic parameters, the tool wear on the PCD cutter was lower than that on the CVD diamond cutter. The study revealed that the cobalt metal bond in the tool could also influence the machining outcomes, with the PCD tool exhibiting a more consistent fracture pattern than the CVD diamond tool, thereby reducing wear during the machining of SiCf/SiC composites. Similarly, Sap et al. [40] examined the machining performance of Cu-based hybrid composites reinforced with 10 wt.% B-Ti-SiCp particles under various environments, including cryogenic conditions. The results indicated that LN2-assisted cryogenic cooling was the most effective method for machining, providing better tribological properties—23% and 19.5% improvements compared to dry machining and Minimum Quantity Lubrication (MQL), respectively. Overall, cryogenic cooling was found to have a significant positive impact on surface quality, making it the optimal choice for minimizing composite aggregation during milling operations.

7.   Mechanical and tribological properties of coatings

7.1   Mechanical properties

7.1.1   Hardness and Young modulus

CrN coatings, typically deposited as monolayers using PVD techniques such as magnetron sputtering or arc deposition, have been developed to meet the demands of cryogenic machining processes. The reported hardness and Young's modulus of CrN coatings are approximately 20 GPa and 200-400 GPa, respectively [41-43]. However, lower hardness values have also been

Superalloys: Fundamentals and Applications   Materials Research Forum LLC
Materials Research Foundations 178 (2025) 230-258   https://doi.org/10.21741/9781644903698-12

observed in some studies [44-46]. These discrepancies may arise from variations in deposition parameters across different investigations.

In general, nanocomposite coatings exhibit high hardness levels. For instance, TiN-SiN coatings deposited by magnetron sputtering demonstrate a hardness of about 38 GPa, compared to 27 GPa for pure TiN, attributable to the inclusion of 5 at. % Si [47]. Jiang et al. reported a hardness of 35 GPa for TiN-SiN$_x$ coatings. When the same alloy is deposited via CVD, promising mechanical properties are noted, with hardness reaching 50 GPa and Young's modulus at 500 GPa [48]. The addition of elements like Si can significantly enhance the mechanical properties of coatings intended for cryogenic machining.

Multilayered coatings have also been investigated for their mechanical properties. Zhang et al. compared monolayered and multilayered coatings, specifically CrAlN and Cr/CrN/CrAlN. Their findings revealed that the hardness of the multilayered coatings ranges from approximately 16 to 24 GPa, while Young's modulus varies between 429 and 670 GPa. The improved mechanical properties in multilayers are attributed to their architectural design, which allows for the superposition of monolayers. Consequently, many researchers have concentrated on increasing the number of interfaces and adjusting parameters such as layer periods. These studies indicate that multilayers possess superior mechanical properties and hold great promise for the development of innovative cutting tools.

7.2   Wear and mechanisms

The wear of a cutting tool represents a gradual deterioration that affects its mechanical properties and ultimately reduces its service life. Various types of wear can be observed on cutting tools, and in cryogenic machining—a specialized environment—the wear mechanisms can differ from those in conventional processes due to the lower temperatures involved.

7.2.1   Abrasive wear

Cryogenic machining mitigates abrasive wear by limiting the plastic deformation of the tool, which reduces the contact with hard or abrasive materials. The use of cryogenic fluids enhances tool rigidity and minimizes the effects of frictional forces. Abrasive wear involves material loss due to the presence of hard particles in contact with the tool. These particles may originate from the microstructure of the material being machined, debris generated during the machining operation, or chemical reactions between the coolant and the chips produced.

7.2.2   Adhesive wear

Cryogenic machining helps prevent cold welding between the chips and the tool by lowering the cutting temperature. This reduction is particularly beneficial when machining sticky or difficult-to-machine materials, such as titanium. Adhesive wear results from material loss on the cutting tool during the machining process, where the interaction between the cutting tool and the workpiece leads to deformation and shearing, ultimately causing wear due to bonding.

7.2.3   Diffusive wear

Diffusive wear occurs when atomic diffusion leads to material loss from the tool surface. This type of wear is often influenced by temperature and time; higher temperatures can accelerate diffusion processes. In cryogenic machining, lower temperatures may slow down diffusion rates,

Superalloys: Fundamentals and Applications   Materials Research Forum LLC
Materials Research Foundations 178 (2025) 230-258   https://doi.org/10.21741/9781644903698-12

thereby potentially extending tool life by reducing this mechanism of wear. Understanding the interplay between temperature, material properties, and wear mechanisms is crucial for optimizing the performance of cutting tools in cryogenic environments.

8. Tool lifespan prediction models

In cryogenic machining, cutting tools are exposed to extremely low temperatures, making tool life prediction crucial for both efficiency and cost-effectiveness. This task becomes more complex when using coated tool materials, as the coatings can dissolve during contact with the workpiece and under cryogenic conditions. To address these challenges, researchers have developed two primary types of models: empirical and analytical. These models provide a foundation for understanding and predicting tool performance in cryogenic machining. While both types of models are useful in practice, they each have limitations in terms of precision and accuracy. However, advancements in integrating these models have led to significant improvements in tool design and machining practices, resulting in substantial reductions in manufacturing costs. Figure 6 provides a graphical summary of the sequence of our discussion.

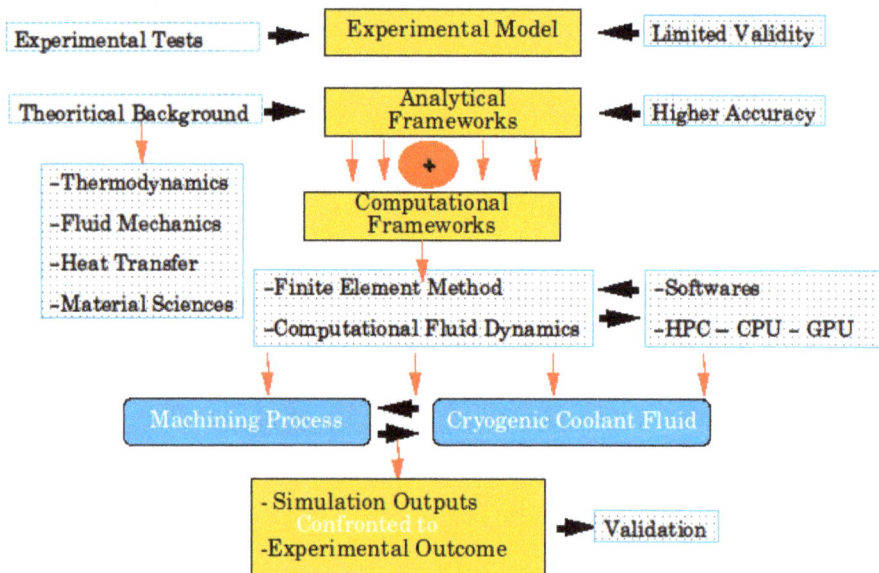

Figure 6. Graphical representation of the theoretical-empirical-computational approach in cryogenic machining.

8.1 Empirical models versus analytical frameworks and computational models

The development of empirical models in machining processes relies on the progression of physical laws and trial-and-error experiments. The accumulation of experimental data facilitates the creation of these models. For instance, the Taylor equation is commonly used to predict tool life [49]:

$$T = \frac{c}{V^n} \qquad\qquad \text{Equation} - 1$$

where T represents tool lifetime, V is the cutting speed, and n is the tool sensitivity to speed changes.

To accurately estimate tool lifespan, researchers have expanded this basic equation to incorporate factors such as tool coating, cryogenic cooling, and the hardness of the workpiece. However, empirical models often depend on costly experiments conducted under specific conditions, limiting their generalizability and accuracy.

Analytical frameworks address the limitations of empirical models. They are grounded in principles from fluid mechanics, thermodynamics, material science, and heat transfer. Cryogenic machining involves physical phenomena at very low temperatures, such as low-temperature thermal conductivity. Analytical models offer valuable insights into these phenomena and the effects of cryogenic coolant fluids on the machining process, providing a more comprehensive understanding compared to empirical models.

8.2 Simulation methods

Machining processes are crucial for advancing various industrial sectors, including healthcare, welfare, aviation, and space. Modern tools must be lighter, stronger, biocompatible, and resistant to corrosion. Cryogenic cooling techniques, which utilize non-toxic fluids such as nitrogen to lower the temperature of cutting tools, help extend tool lifespan by reducing friction between the tool and the workpiece. Traditional lubricants like oil and water often contaminate the workpiece and pose environmental risks [50-52].

Before conducting successful and reproducible experiments, it is essential to develop a theoretical understanding of the phenomena involved. Most physical systems can be described by differential equations, but solving these equations to understand system behavior can be challenging.

Historically, the Finite Element Method (FEM) [53-57] has proven to be an ideal numerical approach for approximating solutions to partial differential equations. Developed in the 1930s and widely adopted about forty years later due to advancements in computer technology, FEM involves discretizing a continuous domain into small elements. Smaller elements provide more accurate solutions but require longer computation times. When combined with Computational Fluid Dynamics (CFD), FEM becomes a powerful tool for predicting cryogenic fluid machining processes [58-60].

This section discusses three key aspects to provide a comprehensive understanding of machining processes involving coolant fluids:

a. Mechanical Fluid Dynamics, Heat Transfer, and Thermodynamics

A solid grasp of these principles is essential for accurately predicting the behavior of coolant fluids during machining and understanding interactions among machining components.

b. Simulations

Finite Element Methods and Computational Fluid Dynamics are critical for conducting detailed simulations of machining processes. These methods will be discussed in detail later in this section.

c. Analytical Methods

This includes deriving mathematical equations from theoretical models and utilizing empirical and semi-empirical formulas. These simplified approaches provide an overview of the machining system and its interaction with coolant fluids.

8.2.1 Computational fluid dynamics

Computational Fluid Dynamics has developed into a powerful method for observing and mapping fluid motion and its time evolution through computational techniques. Starting from the conservation equations of momentum, mass, and energy, CFD is crucial for predicting heat transfer and fluid flow, where macroscopic behavior is derived from microscopic variables such as temperature, velocity, pressure, and density. Its applications extend across various industries, including aerospace, automotive, and healthcare.

CFD typically follows three stages:

- **Domain Identification:** Defining the physical region of interest.

- **Discretization:** Dividing this domain into a mesh or grid.

- **Resource Allocation:** Assigning computational resources to different regions of the mesh.

Historically, fluid flow modeling has relied on the Navier-Stokes equations [61], which serve as the foundation of CFD since the early 1990s. With advancements propelled by Moore's Law, CFD transitioned from simple calculations in the 1970s to enhanced numerical methods that broadened its capabilities. By the 2000s, the rise of high-performance computing (HPC) enabled researchers to conduct complex and detailed simulations with unprecedented efficiency.

8.2.2 Finite element methods

The Finite Element Method fundamentally relies on the discretization of continuum problems. Two primary approaches exist: the Eulerian method, where elements remain fixed in space while the material or fluid flows through them, and the Lagrangian method, which tracks particles as they move with the material. The Lagrangian approach maintains the shape of elements over time, making it particularly valuable in fluid mechanics.

Choosing the appropriate finite element method for machining analysis involves a trade-off. The updated Lagrangian method is computationally intensive but offers greater accuracy in tracking material evolution over time. In contrast, Eulerian methods, while simpler, struggle to account for the dynamic surface movements typical of chip formation.

8.2.3 Simulating cryogenic coolant fluid machining process requisite

Simulating machining processes is essential due to the high costs associated with experimental setups, which can often exceed $100 USD per hour [62]. When incorporating cryogenic methods, the expenses related to coolant fluids further compound the costs. These considerations highlight the necessity of simulation prior to experimental validation. FEM serves this purpose by facilitating the analysis of critical factors such as tool geometry, cutting coefficients, and temperature.

The origins of FEM can be traced back to the foundational work of Zienkiewicz in 1971 [57]. Subsequent advancements in CPU speed in the late 1980s enabled the exploration of non-steady chip formation, notably in studies by Strenkowski and Carroll [63]. By the 1990s, 3D analysis began to gain traction, with 2D FEM modeling of cutting mechanisms becoming viable [64, 65].

The computational cost of FEM simulations is well-documented in the literature. Notably, the maximum reported simulation time is 1 second, achieved using a simplified rigid body model with a tight mesh and an artificially imposed heat exchange coefficient [66-68].

When modeling the complex machining process, several parameters must be considered, including:

- The behavior of the workpiece material.
- The shape of the object.
- The surface contact between the tool and workpiece.
- Heat exchange at the tool-chip interface.
- Heat exchange between the workpiece and coolant.

Material deformation is primarily governed by the Johnson-Cook empirical law [69-71]:

$$\sigma = (A + B\epsilon^n)\left[1 + Cln\frac{\dot{\epsilon}}{\epsilon_0}\right]\left[1 - \frac{(T-T_0)}{(T_m-T_0)^m}\right] \qquad \text{Equation} - 2$$

Where:

A: Yield stress of the material.

B: Hardening modulus.

n: Strain hardening exponent.

C: Strain rate sensitivity coefficient.

m: Thermal softening exponent.

$\dot{\epsilon}$: Strain rate.

ϵ_0: Reference strain rate.

T: Material temperature.

T_0: Reference temperature (usually room temperature).

T_m: Melting temperature of the material.

The formula governing the heat transfer Q during cryogenic area machining, as influenced by coolant area interaction, is expressed as follows:

Superalloys: Fundamentals and Applications
Materials Research Foundations 178 (2025) 230-258

Materials Research Forum LLC
https://doi.org/10.21741/9781644903698-12

$$Q = h(T_a - T_c)$$
Equation - 3

In this equation, h represents the area heat transfer coefficient, while T_a and T_c denote the temperatures of the area and the coolant, respectively.

Elfanti et al. [25] utilized FORGENxT-based FEM and Solid Edge CAD software to design the initial geometry of the workpiece. Cooling with cryogenic liquids, particularly nitrogen, enhanced the machinability of titanium. Evidence includes increased tool lifespan and improved surface quality due to the low temperatures' effect on Ti6Al4V. This widely used titanium alloy, composed of 90% titanium, 6% aluminum, and 4% vanadium, has been the focus of extensive research. The cryogenic fluid also acted as a lubricant, improving surface quality and cleanliness, which in turn increased the fatigue life of the workpiece. However, it is important to note that directing extremely cold fluid into a rotating tool is challenging, explaining the scarcity of research on cryogenic cooling-based milling systems.

The simulation results conducted by Elfanti et al. [72] provide significant insights into the effects of cryogenic cooling on the cutting process, especially when compared to experimental data. The simulation showed a modest 5% reduction in maximum temperature during cryogenic cutting compared to dry cutting, although this reduction was not particularly substantial. However, temperature contour maps revealed that high-temperature zones—represented by areas nearing red—were notably reduced in cryogenic cutting. This suggests that cryogenic cooling is effective at reducing overall temperatures across the cutting zone, even if it does not significantly reduce peak temperatures. This reduction is expected to increase tool life by lowering the heat flow toward the tool, which in turn decreases the likelihood of wear mechanisms such as chemical diffusion.

One limitation is the weaker penetration of liquid nitrogen in the contact zone between the tool and workpiece or chip. This effect was accounted for in the simulation, where one assumption was that no heat exchange occurs on directly contacting surfaces, and the cryogenic cooling effect was only active on the free surfaces. The friction power was lower at reduced friction coefficients, with liquid nitrogen also cooling the system. The simulation highlighted that the actual cutting temperature would be lower than simulated, as the mechanical action of nitrogen flow over the chip would have further reduced the temperature.

The comparison of the simulation conducted by Elfanti et al. [72] with experimental observations showed a reasonable match regarding chip morphology. Chips from cryogenic cutting appeared more damaged and exhibited a saw-like pattern on the edges, with a smaller curl radius compared to actual chips. These characteristics were likely caused by the damage model in the simulation, where element removal exceeded the damage threshold. This resulted in the observed chip deformation. However, the morphology of lamellae and folds was not well represented due to the larger mesh size used in the simulation—a tighter mesh would likely better display realistic chip morphology.

In a study conducted by Umbrello [70], three material model sets from the literature were compared using 2D FEM simulations and validated through orthogonal cutting tests. The first two sets of material constants were derived from Split Hopkinson Pressure Bar (SHPB) tests—one obtained at high temperatures ranging from 700 to 1100°C and the other at room temperature. The third set was developed using analytical modeling and metal-cutting experiments. The FEM simulation results indicated that the first model, based on high-temperature SHPB tests, provided more accurate predictions of cutting forces compared to the other models.

Furthermore, it was emphasized that material models are more realistic and reliable when the testing conditions—such as temperature, strain, and strain rate—closely match actual cutting conditions [68]. It was also noted that variations in material constants significantly impacted the simulated cutting forces without altering chip geometry [67].

A notable limitation of the Johnson-Cook material law, when applied to simulate the flow behavior of Ti6Al4V, is its inability to fully reproduce the mechanisms responsible for saw-tooth chip formation and adiabatic shearing. This shortcoming arises from the absence of a criterion that accounts for the effects of tensile stress on chip segmentation, which is necessary for replicating these phenomena.

Several advanced studies [73-77] have researched the optimum conditions for cryogenic machining of the aerospace titanium alloy Ti-6Al-4V, focusing on the influence of nozzle positioning. These studies employed a combination of FEM (Finite Element Method) and CFD (Computational Fluid Dynamics) simulations, as well as experimental methodologies, to examine how variations in the nozzle's distance from the tool-chip interface and its inclination angle affect machining performance. The results clearly indicated that nozzle position plays a critical role in cutting temperature and forces. Specifically, a smaller separation distance and reduced inclination angle were found to enhance the cooling efficiency of the process. Experimentally, cryogenic machining demonstrated reductions in cutting temperatures and forces by approximately 45% and 46%, respectively.

Higher velocity and turbulence around the tool-chip interface, as well as static pressure, contributed to increased cooling efficiency by decreasing the thickness of the thermal boundary layer. A reduction in cutting forces of up to 48%—compared to dry machining—was also observed. This improvement was attributed to reduced thermal softening, which allowed for a sharper cutting tool edge with less wear. Cryogenic cooling also minimized adhesion and diffusion-related wear, resulting in improved surface finish and tool life. Surface roughness decreased by 44%, highlighting cryogenic machining's effectiveness in maintaining tool sharpness and improving workpiece quality. This data underscores the potential for further optimization of cryogenic machining, offering prospects for further analysis and greater industrial applicability of nozzle parameters.

Additionally, Kanellos et al. [77] employed FEM coupled with CFD to simulate the dry turning of Ti-6Al-4V. They used the Abaqus/Explicit module, designed to model highly nonlinear, transient dynamic events such as metal cutting, to handle severe distortions and complex contact conditions [78-80]. The explicit time integration approach in Abaqus/Explicit efficiently manages complex interactions, despite requiring small time steps for accuracy. This hybrid approach allowed the thermal and mechanical behavior of the machining process to be simulated with acceptable accuracy.

In the first stage, Abaqus/Explicit was used to calculate the maximum temperature during the cutting process, which averaged 1048.57 K. This value was then incorporated into an ANSYS CFX model for CFD analysis, to compute the convective heat transfer coefficient. Through an iterative process, adjusting both temperature and the convective heat transfer coefficient, the simulation converged on a coefficient of 196.68 W/m²K. Two case studies followed: one analyzed the effect of cutting speed on temperature, revealing that increased cutting speed leads to higher temperatures due to increased material deformation. The second study examined the impact of feed rate, finding a similar rise in temperature with higher feed rates.

The study demonstrated that the hybrid FEM-CFD model is a powerful tool for simulating dry turning processes and provides valuable information about the thermal behavior of machining. However, the addition of cooling fluids to the process was not extensively investigated.

Tahmasebi et al. [60] explored cryogenic cooling using liquid nitrogen for enhanced milling performance, combining CFD simulations with experimental validation. They found that operating liquid nitrogen at pressures between 2 and 4 bar increased cooling efficiency and reduced coolant consumption. Proper insulation was critical to minimizing coolant dissipation. Their research also suggested that low-pressure cryogenic jets are more effective than high-pressure jets in delivering efficient cooling. The experimental validation confirmed that effective insulation and proper coolant management are key to reliable cryogenic milling performance [60, 81-84].

9. Applications

9.1 Aerospace components

Aerospace materials are known for their strength and corrosion resistance, as well as their ability to maintain properties in harsh environments. This allows them to meet the stringent safety standards of the aerospace industry. However, these materials are extremely difficult to machine. Their low thermal conductivity creates a significant thermal load on the tool, which can reduce its lifespan.

Gonzalez et al. reported on a study of machining rotors with integrated Ti6Al4V blades using CO_2 for cryogenic cooling. They developed a new CO_2 supply device to inject CO_2 directly into the cutting zone, aiming to prevent any formation of dry ice [85]. Promising results were achieved, particularly in terms of surface integrity and other benefits.

MAG Europe GmbH has developed a new liquid nitrogen cooling system, reporting various advantages in the cryogenic machining of aircraft turbine components. The cutting life of the tools was doubled, cutting speeds increased, and a significant rise in production was observed [86].

9.2 Automotive components

The sale and production of electric cars have increased worldwide in recent years. With this growth comes a demand for new equipment. For example, battery cases and other critical components made from lightweight materials, such as aluminum, can be machined using cryogenic methods. The surface finish of these components must be high-precision and damage-free. Various other equipment can also be fabricated using the cryogenic machining process, such as turbocharger turbines, drive shafts, brake discs, and more. The benefits of cryogenic cooling in automotive machining are numerous. It extends tool life, improves the surface finish of machined parts (such as engine components), and increases cutting speed with better material removal rates. Cryogenic cooling is increasingly being used to machine critical, high-precision automotive components.

Superalloys: Fundamentals and Applications   Materials Research Forum LLC
Materials Research Foundations 178 (2025) 230-258   https://doi.org/10.21741/9781644903698-12

9.3 Power generation components

Cryogenic machining plays an important role in the production of high-precision components for power generation systems. For example, stators and rotors, often made from rolled steel or other high-strength materials, can be machined using cryogenic cooling. This method ensures high precision and reduces thermal deformation of the machined parts. In addition, heat exchangers, particularly those used in nuclear power plants, are often made from tube plates composed of high-performance alloys. The drilling of these plates must be carried out with high precision. Cryogenic machining of this equipment offers several advantages, particularly in terms of precision during machining operations.

10. Future trends and research directions

10.1 Innovations in tool materials

During machining operations, heat is generated, which can damage the processed surface, increase cutting tool wear, and reduce machinability. Lubricating or coolant fluids help address these issues by significantly extending the lifespan of cutting tools, though they are often expensive. However, the use of coolant fluids at high-temperature machining processes may show little noticeable impact. Additionally, the cost of cooling fluids can be exorbitant, sometimes accounting for up to 30% of total machining expenses [1], and cryogenic fluids can have negative environmental impacts.

Minimum Quantity Lubrication (MQL), unlike dry machining processes, is based on spraying a small amount of lubricant through a nozzle onto the cutting zone [87]. This approach has shown promising potential [88], except when using excessive cutting speeds [89]. Electro Hydrodynamic Atomization (EHD) MQL offers a promising alternative to tackle these challenges. Even after nearly two decades, MQL has not yet been widely adopted on an industrial scale. Electrically charged MQL droplets, however, have emerged as a novel and advantageous perspective [90-96]. Simultaneously, achieving productivity and sustainability is the driving goal for upcoming technologies.

10.2 New coatings and composite tools

Innovative tool materials are crucial for the future of this important technological and engineering sector. A significant avenue for exploration lies in advancing composite materials and novel coatings, such as diamond, a carbon allotrope known for enhancing functionality, lifecycle, and performance. The unique incorporation of various carbon allotrope nanostructures, including 2D graphene and 1D carbon nanotubes, presents an interesting approach to improving tool materials.

To effectively handle extremely high temperatures, minimize friction, and enhance wear resistance, novel metal matrix composites and ceramic composites, particularly for machining Inconel 718— a nickel superalloy [94] are promising options. Exploring lightweight yet robust novel polymers and carbon fibers also offers new avenues for development. Furthermore, employing sophisticated techniques to deposit wear-resistant, multifunctional coatings will help protect against corrosion and extreme thermal conditions. To extend tool lifespan, a strategic trend involves utilizing sensors for real-time monitoring of tool wear [96-98].

10.3 Process automation and control

10.3.1 Smart machining systems

Machine learning and smart machining processes can lead to the automation, control, and optimization of machining systems. The emergence of big-data analytics, supported by advancements in information technology and large memory storage, alongside artificial intelligence, is paving the way for a new era in the machining industry. These technologies not only help in detecting machining defects and enable fast intervention to address issues but also assist in managing decision-making processes swiftly and effectively. The advent of 5G networks is another breakthrough, providing effective and fast real-time monitoring of machining process parameters. Moreover, the introduction of quantum computing, with Qbit algorithms, holds the potential to boost dynamic maintenance and drastically enhance production efficiency.

Collaborative robotics will also be a reliable avenue for ensuring high-accuracy machining and supporting large-scale industrial production. Additionally, the Internet of Things (IoT) offers an innovative path to connect a vast number of machining systems, fostering seamless communication and integration. To better serve cyber-physical systems, real-time control and optimization of machining systems can be achieved, although concerns around security and reliability must be addressed.

10.4 Sustainable concerns

10.4.1 Energy

The transition to green and sustainable energy sources is crucial for reducing energy consumption, extending tool lifetimes, and recycling byproducts from cryogenic coolant fluid machining processes. Addressing these urgent environmental issues has become a top priority.

To enhance economic efficiency, machines should be designed to integrate optimal cutting coefficients within the machining process, thereby lowering production costs. Two key aspects in this regard include the pursuit of renewable energy to power machines and the development of next-generation energy storage solutions.

10.4.2 Environment and societal considerations

Addressing environmental concerns involves a dual strategy: first, minimizing waste generated from machining processes, and second, recycling materials derived from fabrication. When it comes to cryogenic fluids, it is crucial to control emissions throughout the entire machining process.

Adhering to green machining practices has become an urgent obligation. There is a pressing need to develop engineering tools that have minimal environmental impact while also prioritizing human safety in workplaces and workshops.

Future perspectives must consider these factors to enhance the quality of life for the younger generation. Advancements in machining processes and next-generation technologies promise an abundance of job creation, fostering a skilled workforce. To meet these requirements, it is essential to implement crash courses and practical training programs that enable adaptation to novel machining technologies.

Ultimately, these technological advancements will not only improve the quality of life but also find applications in healthcare equipment and aerospace technology. It is imperative that we uphold ethical standards alongside the positive economic implications of these developments.

Conclusion

In this chapter, we have examined recent advancements in cryogenic machining and its distinctive advantages over traditional machining processes. Notably, the cryogenic method has demonstrated a significant reduction in surface roughness by approximately 56%, while also halving cutting forces. This technique effectively minimizes thermal softening, reduces tool wear, and enhances tool sharpening.

We also explored how FEM-CFD models contribute to achieving accurate dry turning simulations by computing heat motion parameters alongside cutting temperatures. These simulations have proven effective in forecasting experimental outcomes, although adjustments to models are sometimes necessary to align with these results. Moreover, the integration of FEM with CFD has enhanced the accuracy of simulations, opening pathways for detailed insights into both mechanical and thermal evolution during the machining process.

However, more comprehensive investigations are needed to better understand the cooling effects of cryogenic fluids. The incorporation of cryogenic machining has positively impacted various sectors, including aerospace, power generation, and automotive industries. Its benefits extend beyond extending tool longevity—such as in aeolian blades and gas turbine components—to enhancing overall performance.

Additionally, the reduction in operational costs has been achieved by lowering maintenance expenses, while cryogenic machining also promotes improved accuracy and sustainability. Investigating composite materials, including carbon allotropes, diamond multifunctional coatings, carbon nanotubes, and graphene, presents promising opportunities for prolonging tool lifespan and preserving their properties.

Furthermore, the integration of machine learning, big data analytics, and 5G technology is anticipated to shape the future of advanced cryogenic machining. These innovations will facilitate real-time monitoring, guidance, and operation throughout the machining process. Considering environmental factors, recycling all byproducts of the process and the coolant fluids, coupled with minimizing energy consumption, will undoubtedly pave the way for a brilliant, eco-friendly future.

Acknowledgments

This work is supported by ANR, POLARISS Project (ANR-22-CE08-0021). Authors thank the University of Technology of Troyes (UTT) and Cadi Ayyad University.

References

[1] F. Pusavec, P. Krajnik, J. Kopac, Transitioning to sustainable production-Part I: application on machining technologies, J. Clean. Prod. 18 (2010) 174-184. https://doi.org/10.1016/j.jclepro.2009.08.010

[2] I.S. Jawahir, H. Attia, D. Biermann, J. Duflou, F. Klocke, D. Meyer, S.T. Newman, F. Pusavec, M. Putz, J. Rech, V. Schulze, D. Umbrello, Cryogenic manufacturing processes, CIRP Ann. 65 (2016) 713-736. https://doi.org/10.1016/j.cirp.2016.06.007

[3] E.M. Trent, P.K. Wright, Metal cutting, Fourth ed., Butterworth-Heinemann, Boston, 2000. https://doi.org/10.1016/B978-075067069-2/50007-3

[4] W. Grzesik, Advanced machining processes of metallic materials: theory, modelling and applications, Second ed., Elsevier, 2008.

[5] S. Pervaiz, S. Anwar, I. Qureshi, N. Ahmed, Recent advances in the machining of titanium alloys using minimum quantity lubrication (MQL) based techniques, Int. J. Precis. Eng. Manuf. - Green Technol. 6 (2019) 133-145. https://doi.org/10.1007/s40684-019-00033-4

[6] J. Singh, S.S. Gill, M. Dogra, R. Singh, A review on cutting fluids used in machining processes, Eng. Res. Express. 3 (2021) 012002. https://doi.org/10.1088/2631-8695/abeca0

[7] P. Bhat, C. Agrawal, N. Khanna, Development of a sustainability assessment algorithm and its validation using case studies on cryogenic machining, J. Manuf. Mater. Process. 4 (2020) 42. https://doi.org/10.3390/jmmp4020042

[8] A. Kale, N. Khanna, A review on cryogenic machining of super alloys used in aerospace industry, Procedia Manuf. 7 (2017) 191-197. https://doi.org/10.1016/j.promfg.2016.12.047

[9] N. Khanna, C. Agrawal, M.K. Gupta, Q. Song, Tool wear and hole quality evaluation in cryogenic drilling of Inconel 718 superalloy, Tribol. Int. 143 (2020) 106084. https://doi.org/10.1016/j.triboint.2019.106084

[10] K. Bobzin, High-performance coatings for cutting tools, CIRP J. Manuf. Sci. Technol. 18 (2017) 1-9. https://doi.org/10.1016/j.cirpj.2016.11.004

[11] M. El Garah, D. Soubane, F. Sanchette, Review on mechanical and functional properties of refractory high-entropy alloy films by magnetron sputtering, Emerg. Mater. 7 (2024) 77-101. https://doi.org/10.1007/s42247-023-00607-8

[12] H. Ichou, N. Arrousse, E. Berdimurodov, N. Aliev, Exploring the advancements in physical vapor deposition coating: a review, J. Bio- Tribo- Corros. 10 (2024) 3. https://doi.org/10.1007/s40735-023-00806-0

[13] M. Vorobyova, F. Biffoli, W. Giurlani, S.M. Martinuzzi, M. Linser, A. Caneschi, M. Innocenti, PVD for decorative applications: A review, J. Mater.16 (2023) 4919. https://doi.org/10.3390/ma16144919

[14] L.S. Ahmed, M.P. Kumar, Cryogenic drilling of Ti-6Al-4V alloy under liquid nitrogen cooling, Mater. Manuf. Process. 31 (2016) 951-959. https://doi.org/10.1080/10426914.2015.1048475

[15] A. Shokrani, V. Dhokia, S.T. Newman, Investigation of the effects of cryogenic machining on surface integrity in CNC end milling of Ti-6Al-4V titanium alloy, J. Manuf. Process. 21 (2016) 172-179. https://doi.org/10.1016/j.jmapro.2015.12.002

[16] J. Kenda, F. Pusavec, J. Kopac, Analysis of residual stresses in sustainable cryogenic machining of nickel based alloy-Inconel 718, J. Manuf. Sci. Eng. 133 (2011) 041009. https://doi.org/10.1115/1.4004610

[17] F. Pušavec, J. Kopač, Sustainability assessment: cryogenic machining of Inconel 718, Strojniški vestnik- J. Mech. Eng. 57 (2011) 637-647. https://doi.org/10.5545/sv-jme.2010.249

[18] Z. Wang, K. Rajurkar, Cryogenic machining of hard-to-cut materials, Wear. 239 (2000) 168-175. https://doi.org/10.1016/S0043-1648(99)00361-0

[19] C. Machai, D. Biermann, Machining of β-titanium-alloy Ti-10V-2Fe-3Al under cryogenic conditions: cooling with carbon dioxide snow, J. Mater. Process. Technol. 211 (2011) 1175-1183. https://doi.org/10.1016/j.jmatprotec.2011.01.022

[20] A. Stoll, K. Busch, C. Hochmuth, B. Pause, Modern Cooling Strategies for Machining of high temperature materials. Innovations of Sustainable Production for Green Mobility: Energy-Efficient Technologies in Production. (2014) 299-316.

[21] S.Y. Hong, I. Markus, W.C. Jeong, New cooling approach and tool life improvement in cryogenic machining of titanium alloy Ti-6Al-4V, Int. J. Mach. Tools Manuf. 41 (2001) 2245-2260. https://doi.org/10.1016/S0890-6955(01)00041-4

[22] V. Venkatesh, S. Izman, T. Yap, P. Brevern, N. El-Tayeb, Precision cryogenic drilling, turning and grinding of Ti-64 alloys, Int. J. Precis. Technol. 1 (2010) 287-301. https://doi.org/10.1504/IJPTECH.2010.031658

[23] M. Dhananchezian, M.P. Kumar, Cryogenic turning of the Ti-6Al-4V alloy with modified cutting tool inserts, Cryogenics. 51 (2011) 34-40. https://doi.org/10.1016/j.cryogenics.2010.10.011

[24] P. Shah, N. Khanna, Comprehensive machining analysis to establish cryogenic LN2 and LCO2 as sustainable cooling and lubrication techniques, Tribol. Int. 148 (2020) 106314. https://doi.org/10.1016/j.triboint.2020.106314

[25] M.I. Sadik, An introduction to cutting tools: materials and applications, Elanders, Sweden, 2014.

[26] R. Haubner, M. Lessiak, R. Pitonak, A. Köpf, R. Weissenbacher, Evolution of conventional hard coatings for its use on cutting tools, Int. J. Refract. Met. Hard Mater. 62 (2017) 210-218. https://doi.org/10.1016/j.ijrmhm.2016.05.009

[27] T. Leyendecker, O. Lemmer, S. Esser, J. Ebberink, The development of the PVD coating TiAlN as a commercial coating for cutting tools, Surf. Coat. Technol. 48 (1991) 175-178. https://doi.org/10.1016/0257-8972(91)90142-J

[28] Z.J. Liu, Z.K. Liu, C. McNerny, P. Mehrotra, A. Inspektor, Investigations of the bonding layer in commercial CVD coated cemented carbide inserts, Surf. Coat. Technol. 198 (2005) 161-164. https://doi.org/10.1016/j.surfcoat.2004.10.112

[29] A. Matthews, Titanium nitride PVD coating technology, Surf. Eng. 1 (1985) 93-104. https://doi.org/10.1179/sur.1985.1.2.93

[30] W.D. Münz, Titanium aluminum nitride films: a new alternative to TiN coatings, J. Vac. Sci. Technol. A. 4 (1986) 2717-2725. https://doi.org/10.1116/1.573713

[31] K. Aouadi, B. Tlili, C. Nouveau, A. Besnard, M. Chafra, R. Souli, Influence of substrate bias voltage on corrosion and wear behavior of physical vapor deposition CrN coatings, J. Mater. Eng. Perform. 28 (2019) 2881-2891. https://doi.org/10.1007/s11665-019-04033-y

[32] B. Tlili, C. Nouveau, G. Guillemot, Elaboration, characterization of CrN-based coatings, AIP Conf. Proc. American Institute of Physics. (2011) 1311-1316. https://doi.org/10.1063/1.3552365

[33] M. Yousfi, J. Outeiro, C. Nouveau, B. Marcon, B. Zouhair, Tribological behavior of PVD hard coated cutting tools under cryogenic cooling conditions, Procedia CIRP. 58 (2017) 561-565. https://doi.org/10.1016/j.procir.2017.03.269

[34] C. Nouveau, B. Tlili, H. Aknouche, Y. Benlatreche, B. Patel, Comparison of CrAlN layers obtained with one (CrAl) or two targets (Cr and Al) by magnetron sputtering, Thin Solid Films. 520 (2012) 2932-2937. https://doi.org/10.1016/j.tsf.2011.11.049

[35] L. Aissani, M. Fellah, C. Nouveau, M. Abdul Samad, A. Montagne, A. Iost, Structural and mechanical properties of Cr-Zr-N coatings with different Zr content, Surf. Eng. 36 (2020) 69-77. https://doi.org/10.1080/02670844.2017.1338378

[36] L. Selvam, P.K. Murugesan, D. Mani, Y. Natarajan, Investigation of AlCrN-coated inserts on cryogenic turning of Ti-6Al-4V alloy, Metals. 9 (2019) 1338. https://doi.org/10.3390/met9121338

[37] Y. Zhang, J.C. Martins do Outeiro, C. Nouveau, B. Marcon, L. Denguir, Performance of new cutting tool multilayer coatings for machining Ti-6al-4v titanium alloy under cryogenic cooling conditions, SSRN. (2024) 4915309. https://doi.org/10.2139/ssrn.4915309

[38] S.S. Muhamad, J.A. Ghani, C.H. Che Haron, H. Yazid, Wear mechanism of multilayer coated carbide cutting tool in the milling process of AISI 4340 under cryogenic environment, J. Mater. 15 (2022) 524. https://doi.org/10.3390/ma15020524

[39] B. Zhang, Y. Du, H. Liu, L. Xin, Y. Yang, L. Li, Experimental study on high-speed milling of SiCf/SiC composites with PCD and CVD diamond tools, J. Mater. 14 (2021) 3470. https://doi.org/10.3390/ma14133470

[40] S. Şap, Ü.A. Usca, M. Uzun, M. Kuntoğlu, E. Salur, Performance evaluation of AlTiN coated carbide tools during machining of ceramic reinforced Cu-based hybrid composites under cryogenic, pure-minimum quantity lubrication and dry regimes, J. Compos. Mater. 56 (2022) 3401-3421. https://doi.org/10.1177/00219983221115846

[41] X.-M. He, N. Baker, B. Kehler, K. Walter, M. Nastasi, Y. Nakamura, Structure, hardness, and tribological properties of reactive magnetron sputtered chromium nitride films, J. Vac. Sci. Technol. A. 18 (2000) 30-36. https://doi.org/10.1116/1.582154

[42] H. Ichimura, I. Ando, Mechanical properties of arc-evaporated CrN coatings: Part I-nanoindentation hardness and elastic modulus, Surf. Coat. Technol. 145 (2001) 88-93. https://doi.org/10.1016/S0257-8972(01)01290-7

[43] C. Rebholz, H. Ziegele, A. Leyland, A. Matthews, Structure, mechanical and tribological properties of nitrogen-containing chromium coatings prepared by reactive magnetron sputtering, Surf. Coat. Technol. 115 (1999) 222-229. https://doi.org/10.1016/S0257-8972(99)00240-6

[44] E. Broszeit, C. Friedrich, G. Berg, Deposition, properties and applications of PVD CrxN coatings, Surf. Coat. Technol. 115 (1999) 9-16. https://doi.org/10.1016/S0257-8972(99)00021-3

[45] N. Panich, Y. Sun, Effect of substrate rotation on structure, hardness and adhesion of magnetron sputtered TiB2 coating on high speed steel, Thin Solid Films. 500 (2006) 190-196. https://doi.org/10.1016/j.tsf.2005.11.055

[46] A. Wilson, A. Matthews, J. Housden, R. Turner, B. Garside, A comparison of the wear and fatigue properties of plasma-assisted physical vapour deposition TiN, CrN and duplex coatings on Ti-6Al-4V, Surf. Coat. Technol. 62 (1993) 600-607. https://doi.org/10.1016/0257-8972(93)90306-9

[47] M. Diserens, J. Patscheider, F. Levy, Mechanical properties and oxidation resistance of nanocomposite TiN-SiNx physical-vapor-deposited thin films, Surf. Coat. Technol. 120 (1999) 158-165. https://doi.org/10.1016/S0257-8972(99)00481-8

[48] S. Veprek, Design of novel nanocrystalline composite materials by means of plasma CVD, Pure Appl. Chem. 68 (1996) 1023-1027. https://doi.org/10.1351/pac199668051023

[49] F.W. Taylor, On the art of cutting metals, Trans. ASME. 28 (1906) 281-350. https://doi.org/10.1115/1.4060389

[50] E. Benedicto, D. Carou, E. Rubio, Technical, economic and environmental review of the lubrication/cooling systems used in machining processes, Procedia Eng. 184 (2017) 99-116. https://doi.org/10.1016/j.proeng.2017.04.075

[51] G.W.A. Kui, S. Islam, M.M. Reddy, N. Khandoker, V.L.C. Chen, Recent progress and evolution of coolant usages in conventional machining methods: a comprehensive review, Int. J. Adv. Manuf. Technol. 119 (2022) 3-40. https://doi.org/10.1007/s00170-021-08182-0

[52] Q.J. Wang, Y.-W. Chung, Encyclopedia of tribology, Springer, US, 2013. https://doi.org/10.1007/978-0-387-92897-5

[53] K.-J. Bathe, Finite element procedures, Second ed., Klaus-Jurgen Bathe, 2006.

[54] E. Burman, S. Claus, P. Hansbo, M.G. Larson, A. Massing, CutFEM: discretizing geometry and partial differential equations, Int. J. Numer. Methods Eng.104 (2015) 472-501. https://doi.org/10.1002/nme.4823

[55] W.K. Liu, S. Li, H.S. Park, Eighty years of the finite element method: birth, evolution, and future, Arch. Comput. Methods Eng. 29 (2022) 4431-4453. https://doi.org/10.1007/s11831-022-09740-9

[56] J.N. Reddy, An introduction to the finite element method, Fourth ed., McGraw-Hill, New York, (1993)

[57] O.C. Zienkiewicz, The finite element methods in engineering science, London, McGraw-Hill, New York, 1971.

[58] R. Bejjani, E. Bachir, C. Salame, Advanced manufacturing of titanium alloy ti-6al-4v by combining cryogenic machining and ultrasonic-assisted turning, J. Mater. Eng. Perform. 33 (2024) 6507-6527. https://doi.org/10.1007/s11665-023-08430-2

[59] O. Pereira, A. Rodríguez, J. Barreiro, A.I. Fernández-Abia, L.N.L. de Lacalle, Nozzle design for combined use of MQL and cryogenic gas in machining, Int. J. Precis. Eng. Manuf.-Green Technol. 4 (2017) 87-95. https://doi.org/10.1007/s40684-017-0012-3

[60] E. Tahmasebi, P. Albertelli, T. Lucchini, M. Monno, V. Mussi, CFD and experimental analysis of the coolant flow in cryogenic milling, Int. J. Mach. Tools Manuf. 140 (2019) 20-33. https://doi.org/10.1016/j.ijmachtools.2019.02.003

[61] D. McLean, Understanding aerodynamics: arguing from the real physics, John Wiley & Sons, 2012. https://doi.org/10.1002/9781118454190

[62] C. Constantin, S.M. Croitoru, G. Constantin, E. Strãjescu, FEM tools for cutting process modelling and simulation, UPB Sci. Bull. D. Mech. Eng. 74 (2012) 149-162.

[63] J.S. Strenkowski, J.T. Carroll III, A finite element model of orthogonal metal cutting, J. Manuf. Sci. Eng. 107 (1985) 349-354. https://doi.org/10.1115/1.3186008

[64] P.J. Arrazola, T. Özel, D. Umbrello, M. Davies, I.S. Jawahir, Recent advances in modelling of metal machining processes, CIRP Annu. 62 (2013) 695-718. https://doi.org/10.1016/j.cirp.2013.05.006

[65] K. Ueda, K. Manabe, Rigid-plastic FEM analysis of three-dimensional deformation field in chip formation process, CIRP Annals 42 (1993) 35-38. https://doi.org/10.1016/S0007-8506(07)62386-5

[66] E. Chiappini, S. Tirelli, P. Albertelli, M. Strano, M. Monno, On the mechanics of chip formation in Ti-6Al-4V turning with spindle speed variation, Int. J. Mach. Tools Manuf. 77 (2014) 16-26. https://doi.org/10.1016/j.ijmachtools.2013.10.006

[67] L. Filice, D. Umbrello, S. Beccari, F. Micari, On the FE codes capability for tool temperature calculation in machining processes, J. Mater. Process. Technol. 174 (2006) 286-292. https://doi.org/10.1016/j.jmatprotec.2006.01.012

[68] D. Umbrello, A. Bordin, S. Imbrogno, S. Bruschi, 3D finite element modelling of surface modification in dry and cryogenic machining of EBM Ti6Al4V alloy, CIRP J. Manuf. Sci. Technol. 18 (2017) 92-100. https://doi.org/10.1016/j.cirpj.2016.10.004

[69] M. Calamaz, D. Coupard, F. Girot, A new material model for 2D numerical simulation of serrated chip formation when machining titanium alloy Ti-6Al-4V, Int. J. Mach. Tools Manuf. 48 (2008) 275-288. https://doi.org/10.1016/j.ijmachtools.2007.10.014

[70] D. Umbrello, Finite element simulation of conventional and high speed machining of Ti6Al4V alloy, J. Mater. Process. Technol. 96 (2008) 79-87. https://doi.org/10.1016/j.jmatprotec.2007.05.007

[71] H. Wu, S. Zhang, 3D FEM simulation of milling process for titanium alloy Ti6Al4V, Int. J. Adv. Manuf. Technol. 71 (2014) 1319-1326. https://doi.org/10.1007/s00170-013-5546-0

[72] A. Elefanti, P. Albertelli, M. Strano, M. Monno, Estimation of cutting and friction coefficients in dry and cryogenic milling through experiments and simulations, AIP Conf. Proc. 2113 (2019). https://doi.org/10.1063/1.5112614

[73] X. Liang, Z. Liu, Tool wear behaviors and corresponding machined surface topography during high speed machining of Ti-6Al-4V with fine grain tools, Tribol. Int.121 (2018) 321-332. https://doi.org/10.1016/j.triboint.2018.01.057

[74] S. Manel, S. Kumar, Heat generation and temperature in orthogonal machining, Int. J. Sci. Eng. Res. 8 (2017) 31-33.

[75] P. Olander, J. Heinrichs, On wear of WC-Co cutting inserts in turning of Ti6Al4V-a study of wear surfaces, Tribol. Mater. Surf. Int. 15 (2021) 181-192. https://doi.org/10.1080/17515831.2020.1830251

[76] Y. Zhao, Y. Zhao, Y. Shao, T. Hu, Q. Zhang, X. Ge, Research of a smart cutting tool based on MEMS strain gauge, J. Phys. Conf. Series. (2018) 012016. https://doi.org/10.1088/1742-6596/986/1/012016

[77] D. Zindani, K. Kumar, A brief review on cryogenics in machining process, SN Appl. Sci. 2 (2020) 1107. https://doi.org/10.1007/s42452-020-2899-5

[78] C. Salame, R. Bejjani, P. Marimuthu, A better understanding of cryogenic machining using CFD and FEM simulation, Procedia CIRP. 81 (2019) 1071-1076. https://doi.org/10.1016/j.procir.2019.03.255

[79] C. Tahri, P. Lequien, J. Outeiro, G. Poulachon, CFD simulation and optimize of LN2 flow inside channels used for cryogenic machining: application to milling of titanium alloy Ti-6Al-4V, Procedia CIRP. 58 (2017) 584-589. https://doi.org/10.1016/j.procir.2017.03.230

[80] X. Yang, T. Xi, Y. Qin, H. Zhang, Y. Wang, Computational fluid dynamics-discrete phase method simulations in process engineering: a review of recent progress, Appl. Sci. 14 (2024) 3856. https://doi.org/10.3390/app14093856

[81] P. Albertelli, V. Mussi, M. Strano, M. Monno, Experimental investigation of the effects of cryogenic cooling on tool life in Ti6Al4V milling, Int. J. Adv. Manuf. Technol. 117 (2021) 2149-2161. https://doi.org/10.1007/s00170-021-07161-9

[82] Y. De Bondt, I. Liberloo, C. Roye, E.J. Windhab, L. Lamothe, R. King, C.M. Courtin, The effect of wet milling and cryogenic milling on the structure and physicochemical properties of wheat bran, Foods. 9 (2020) 1755. https://doi.org/10.3390/foods9121755

[83] K. Gutzeit, M. Berndt, J. Schulz, D. Müller, B. Kirsch, E. von Harbou, J.C. Aurich, Optimization of the cooling strategy during cryogenic milling of Ti-6Al-4 V when applying a sub-zero metal working fluid, Prod. Eng. 17 (2023) 501-510. https://doi.org/10.1007/s11740-022-01178-z

[84] M. Nalbant, Y. Yildiz, Effect of cryogenic cooling in milling process of AISI 304 stainless steel, Trans. Nonferrous Met. Soc. China. 21(1) (2011) 72-79. https://doi.org/10.1016/S1003-6326(11)60680-8

[85] H. Gonzalez, O. Pereira, L.N. López de Lacalle, A. Calleja, I. Ayesta, J. Muñoa, Flank-milling of integral blade rotors made in Ti6Al4V using Cryo CO2 and minimum quantity lubrication, J. Manuf. Sci. Eng.143 (2021) 091011. https://doi.org/10.1115/1.4050548

[86] https://www.industrial-production.de/firma/mag-europe-gmbh.htm.

[87] K. Weinert, I. Inasaki, J.W. Sutherland, T. Wakabayashi, Dry machining and minimum quantity lubrication, CIRP Annu. 53 (2004) 511-537. https://doi.org/10.1016/S0007-8506(07)60027-4

[88] P. Ball, L.H. Huatuco, R.J. Howlett, R. Setchi, Sustainable design and manufacturing, 2019: Proc. 6th Int. Conf. Sustain. Des. Manuf. (KES-SDM 19). Springer, (2019). https://doi.org/10.1007/978-981-13-9271-9

[89] Y. Liao, H. Lin, Mechanism of minimum quantity lubrication in high-speed milling of hardened steel, Int. J. Mach. Tools Manuf. 47 (2007) 1660-1666. https://doi.org/10.1016/j.ijmachtools.2007.01.007

[90] R.K. Gunda, S.K.R. Narala, Evaluation of friction and wear characteristics of electrostatic solid lubricant at different sliding conditions, Surf. Coat. Technol. 332 (2017) 341-350. https://doi.org/10.1016/j.surfcoat.2017.08.073

[91] S. Huang, T. Lv, M. Wang, X. Xu, Effects of machining and oil mist parameters on electrostatic minimum quantity lubrication-EMQL turning process, Int. J. Precis. Eng. Manuf. - Green Technol. 5 (2018) 317-326. https://doi.org/10.1007/s40684-018-0034-5

[92] S. Huang, T. Lv, M. Wang, X. Xu, Enhanced machining performance and lubrication mechanism of electrostatic minimum quantity lubrication-EMQL milling process, Int. J. Adv. Manuf. Technol. 94 (2018) 655-666. https://doi.org/10.1007/s00170-017-0935-4

[93] S. Huang, T. Lv, X. Xu, Y. Ma, M. Wang, Experimental evaluation on the effect of electrostatic minimum quantity lubrication (EMQL) in end milling of stainless steels, Mach. Sci. Technol. 22 (2018) 271-286. https://doi.org/10.1080/10910344.2017.1337135

[94] R. M'Saoubi, T. Larsson, J. Outeiro, Y. Guo, S. Suslov, C. Saldana, S. Chandrasekar, Surface integrity analysis of machined Inconel 718 over multiple length scales, CIRP Ann. 61 (2012) 99-102. https://doi.org/10.1016/j.cirp.2012.03.058

[95] U.R. Paturi, N. Suresh Kumar Reddy, Investigation on wear behavior of electrostatic micro-solid lubricant coatings under dry sliding conditions, ASME Int. Mech. Eng. Congress Expo. 3 (2012) 2105-2110. https://doi.org/10.1115/IMECE2012-87201

[96] C. Salame, R. Rapold, B. Tasdelen, A. Malakizadi, Sensor-based identification of tool wear in turning, Procedia CIRP. 121 (2024) 228-233. https://doi.org/10.1016/j.procir.2023.09.252

[97] Y. Cheng, X. Gai, R. Guan, Y. Jin, M. Lu, Y. Ding, Tool wear intelligent monitoring techniques in cutting: a review, J. Mech. Sci. Technol. 37 (2023) 289-303. https://doi.org/10.1007/s12206-022-1229-9

[98] N. Cook, Tool wear sensors, Wear. 62 (1980) 49-57. https://doi.org/10.1016/0043-1648(80)90036-8

Superalloys: Fundamentals and Applications
Materials Research Foundations 178 (2025) 259-278

Materials Research Forum LLC
https://doi.org/10.21741/9781644903698-13

Chapter 13

Materials for High Temperature: A General Concept

Adarsh Kumar Arya[1], Ashish Kapoor[1]*, Devanshi Srivastava[1], Murali Pujari[2]

[1]Department of Chemical Engineering, Harcourt Butler Technical University, Kanpur, Uttar Pradesh, India

[2]Department of Chemical Engineering, School of Engineering, University of Petroleum & Energy Studies, Energy Acres Building, Bidholi, Dehradun, India

ashishkapoorchem@gmail.com

Abstract

High-temperature materials play a crucial role in advancing sustainable engineering, which seeks to reduce environmental impact and enhance resource efficiency. The effective use of such materials, including ceramics, polymers, and metals, enables advancements in several commercial applications. This work explores the selection criteria for these materials, highlights their distinctive properties, and discusses their diverse applications. Furthermore, it reviews essential characterization techniques used to analyze their material properties.

Keywords

High-Temperature Materials, Characterization, Applications, Characterization, Sustainable Engineering

Contents

Superalloys: Fundamentals and Applications
Materials Research Foundations 178 (2025) 259-278

Materials Research Forum LLC
https://doi.org/10.21741/9781644903698-13

1. Introduction

Since the inception of fire and the subsequent utilization of thermal energy, humans have consistently sought out and employed various materials that can withstand and endure exceptionally high temperatures without succumbing to degradation. In the earliest stages of material usage, it was primarily the indigenous populations who relied on naturally occurring stones that were readily sourced from their immediate local environments, demonstrating a profound understanding of the properties of these materials [1]. A vast array of metallic alloys, have emerged as suitable for use in a wide range of high-temperature applications due to their unique properties. This diversity in heat-resistant materials is a result of significant advancements in recent decades. The utilization of such elevated-temperature materials spans a wide range of critical applications, including the demanding environments of aircraft jet engines, the intricate systems of nuclear power stations, the intense conditions found within combustion engine furnaces, and even the construction of various lighting devices which require materials that can withstand significant heat. A high-temperature aerogel's categorization and potential uses are shown in Figure 1.

Figure 1. Aerogels that operate at high temperatures: categorisation and potential uses
(Reproduced from [2] under CC BY license)

It is essential to underscore that the specific type of material chosen, along with the methods of heat processing and the application of protective coatings, is paramount to ensuring the efficiency and safety of high-temperature applications in numerous industrial contexts. A previous investigation in this field has established a significant correlation between the heat treatment process performed in an oxygen-rich atmosphere and the subsequent fatigue characteristics exhibited by zirconium alloy specimens, providing valuable insights into how material performance can be tailored through specific processing conditions. This research investigation meticulously observed that implementing oxygen gas treatment significantly enhanced fatigue life during both stretching and tensile testing procedures. The remarkable enhancement in fatigue life was primarily attributed to forming a robust layer adjacent to the roughened surface, contributing to improved durability and performance. In a subsequent study referenced by [3], experimental examinations were conducted to analyse strain localisation phenomena in heat-resistant nanocoated steel, particularly during critical stages such as crack nucleation, clumping, and the fragmentation of the nanocoating itself. Among materials scientists, there is a consensus that a temperature of two-thirds or higher than a solid's melting point is considered very high. This is a crucial criterion for evaluating the performance and integrity of materials in high-temperature applications.

Nevertheless, Meetham and Voorde [4] contribute to the discourse by providing further classifications based on application, thereby specifying which materials are particularly suitable for resisting temperatures that surpass the notable threshold of 500°C. The increase in temperature invariably leads to a reduction in the strength of various substances; thus, these materials must possess high strength and an adequate margin of safety at the requisite temperature settings, ensuring that they remain functional and cost-effective under operational conditions.

Furthermore, substances subjected to elevated temperatures must exhibit resilience against potential degradation factors, including but not limited to corrosion and combustion, exacerbated by the heightened thermal environment. As the temperature rises, numerous consequential impacts ensue, including the degradation of the material framework, which can compromise the overall structural integrity and performance of the material in question. Consequently, understanding these phenomena is paramount for advancing materials science and engineering, as it allows for developing more robust materials that can withstand the challenges posed by high-temperature applications [5]. Ultimately, the findings from these studies underscore the critical importance of further research in this field to enhance our comprehension of material behaviour under extreme conditions. The integration of sophisticated high-temperature materials into the practices of environmentally sustainable engineering is a transformative approach that significantly amplifies the efficiency with which resources are harnessed and utilized while concurrently diminishing the emissions that are detrimental to the environment and greatly enhancing the overall sustainability of a a vast variety of tools, machinery, and consumer goods used in a myriad of sectors vital to contemporary life. These advanced high-temperature materials are crucial in the domain of sustainable energy generation technologies, which encompass a wide-ranging assortment of systems including, but not limited to, gas turbines, steam turbines, nuclear reactors, aviation engines, and propulsion systems that are essential to the functioning and advancement of contemporary energy infrastructures [3]. The distinctive properties exhibited by these materials, which empower them to endure extreme thermal conditions and severe operational environments, are instrumental in fostering the development of energy generation systems that demonstrate a considerable enhancement in efficiency and reliability compared to their traditional counterparts, which often lack such advanced capabilities. The improvements realized in operational efficiency as a direct consequence of employing these high-temperature materials result in a tangible reduction in the consumption of precious natural resources along with a significant decrease in the number of harmful emissions generated per unit of energy produced, thus promoting a more sustainable paradigm for energy production that aligns with contemporary environmental objectives. In this extensive and detailed study, a comprehensive identification and thorough investigation of various high-temperature materials were meticulously conducted, particularly on the diverse application domains wherein such materials find utility.

In contrast, particular emphasis was placed on their critical applications within the aerospace sector, a crucial innovation area. Furthermore, an exhaustive analysis of the manufacturing processes utilized in fabricating these essential materials was executed, alongside a conscientious evaluation of the methodologies employed to assess their performance and overall suitability for various applications in demanding environments. Additionally, the research encompassed a wide-ranging examination of prospective challenges and considerations that may emerge within this dynamic field, thereby providing an all-encompassing perspective on the current landscape and the future advancements anticipated in the applications of high-temperature materials. This holistic approach highlights the significance of these advanced materials[6]. It underscores the critical need for ongoing research and development to address the complex challenges of evolving technological demands and environmental considerations. Ultimately, the findings of this research contribute valuable insights that could inform future innovations and strategies aimed at enhancing the sustainability and efficiency of energy generation systems across multiple sectors. Thus, the interplay between high-temperature materials and sustainable engineering practices emerges as a focal point of interest for researchers, engineers, and policymakers alike, necessitating a concerted effort to harness their potential to pursue a more sustainable and environmentally friendly future.

2. Selection Criteria

High-temperature material selection depends on various factors. Some of these are discussed below:

2.1 Operating Temperature

The selection of an appropriate alloy is often predominantly dictated by temperature considerations, which frequently emerge as the primary—if not the exclusive—factor influencing the decision-making process regarding alloy choice in various engineering applications. However, it is imperative to acknowledge that a comprehensive and satisfactory selection of an alloy cannot be effectively achieved through an exclusive focus on temperature alone, as this approach would neglect other critical variables that also significantly impact alloy performance and suitability for specific applications. In this context, one of the essential criteria for selecting alloys involves thoroughly evaluating the maximum temperatures at which a particular alloy can sustain its advantageous long-term technical properties without succumbing to degradation or loss of functionality.

The role of molybdenum during this thermal exposure period cannot be understated, as it significantly influences the formation and stability of the sigma phase within the alloy matrix. The presence of the sigma phase is particularly concerning because it effectively reduces the material's flexibility and stiffness when assessed at room temperature, impacting its overall applicability in various structural and mechanical contexts. The degree of embrittlement a metal experience is mainly contingent upon the quantity and specific morphology of the sigma phase present within the alloy structure. It is important to note that metals generally exhibit brittle characteristics predominantly at room temperature; conversely, they often retain a sufficient level of flexibility when subjected to operational temperatures that fall within the range of 600 to 1,000°F (315 to 540°C), which is a critical consideration for their performance in high-temperature applications.

Nickel-based alloys of higher grades, such as N08811, N08330, N06600, or N06601, have been demonstrated to possess notable resistance to the embrittlement effects that are induced by the presence of the sigma phase, which makes them particularly valuable for applications requiring high performance under thermal stress. Furthermore, it is essential to highlight that cast heat-resistant alloys, characterized by elevated carbon content, tend to experience a notable decline in elasticity during service, primarily due to the precipitation of carbide phases that occur due to the high carbon levels present.

Superalloys: Fundamentals and Applications Materials Research Forum LLC
Materials Research Foundations 178 (2025) 259-278 https://doi.org/10.21741/9781644903698-13

2.2 Strength

In materials science and engineering, it is common for a diverse array of manufacturers to furnish detailed information regarding the creep-rupture properties of their materials when subjected to high-temperature conditions. It is noteworthy that a wide variety of alloys are systematically addressed and governed by the stringent standards outlined in the Boiler and Pressure Vessel Code, promulgated by the esteemed American Society of Mechanical Engineers (ASME), which serves as a critical benchmark for ensuring the safety and reliability of pressure-retaining components in engineering applications.

2.3 Oxidation

Chromium stands out as the singular element that is universally present in all varieties of heat-resistant alloys, and it is the protective chromium oxide scale that fundamentally supports resistance to high-temperature environmental conditions, thereby playing a pivotal role in enhancing the durability and longevity of these materials. Following chromium in terms of importance, nickel holds a significant position, which is subsequently succeeded by elements such as silicon, aluminium, and various rare earth elements that collectively contribute to the overall performance of these alloys. Heat cycling and creep notably impact the oxidation rates during these materials' operational service. This factor can significantly exacerbate the issue of scale spalling, thereby compromising the structural integrity of the alloys. Furthermore, impurities, particularly alkali metal salts, can adversely affect the grain size of the protective chromium scale, which in turn can influence the rates of chromium diffusion. At the same time, the specific environmental conditions surrounding the alloys further intensify the oxidation processes at play. In addition to these factors, it is observed that elevated levels of water vapour within the environment typically tend to enhance the oxidation rates, leading to more rapid degradation of the alloy's protective features. Consequently, understanding the interplay between these elements and environmental conditions is crucial for advancing high-temperature alloy technology, as it allows for developing more resilient materials to withstand extreme operational scenarios. Thus, ongoing research and investigation into these complex interactions are essential for improving the performance and reliability of heat-resistant alloys in demanding applications.

2.4 Sulfidation

The presence of nickel in low to moderate concentrations, combined with elevated levels of chromium, has been demonstrated to significantly mitigate sulfidation corrosion when subjected to high-temperature environments, enhancing materials' overall durability and longevity. It is noteworthy, however, that except for the specific alloy known as HR-160, it is generally advantageous to maintain a nickel concentration that does not exceed 20%, as this has been found to optimize the performance of various alloys in corrosive conditions. Therefore, carefully considering the compositional balance between nickel and chromium is essential to achieve the desired resistance to sulfidation corrosion while ensuring that the material's structural integrity is preserved under elevated thermal stresses.

2.5 Fabricability

The issue of fabricability does not emerge as a critical factor when discussing the routine melting processes associated with wrought metals, which are commonly utilized in various applications across multiple industries. Furthermore, specific grades that have been enhanced through the incorporation of oxide dispersion, exemplified by the notable material MA956®, exhibit an exceptional combination of mechanical strength alongside remarkable resistance to oxidative degradation even when subjected to elevated temperature conditions; however, it is essential to note that these advanced materials pose significant challenges in terms of their manufacturability when approached through conventional production methods that are typically employed for more standard alloys. In light of these considerations, it becomes evident that while the advantages of such specialized materials are considerable, the complexities involved in their manufacturing processes warrant careful examination and potentially the development of innovative techniques to facilitate their effective production and application.

2.6   Design

The design codes established by the American Society of Mechanical Engineers (ASME) play a crucial role in determining the allowable stress levels for various materials used in engineering applications. In the case of most thermal processing systems and equipment, the design stress is typically defined as being equivalent to half of the fracture strength that can be sustained over 10,000 hours, or it may be calculated as half of the stress required to produce an absolute minimum creep rate of 1% when subjected to the exact duration of 10,000 hours. When the temperature exceeds approximately 1,000°F, equivalent to 540°C, the phenomena of creep or material breach become the primary considerations for establishing design stresses, as these factors critically influence the integrity and longevity of the materials involved. At temperatures within this elevated range, materials that were once capable of exhibiting elastic properties begin to lose their ability to return to their original shape, leading to a scenario where they undergo a gradual and permanent deformation over time, which is a significant concern for engineers and designers alike. Thus, understanding the implications of these temperature thresholds on material behaviour is essential for ensuring thermal processing equipment's safe and efficient operation. Consequently, the interplay between temperature, stress, and material performance becomes a focal point of analysis in the design process, warranting comprehensive evaluation and adherence to established guidelines to mitigate the risks associated with material failure.

2.7   Thermal Expansion

A significant contributor to the distortion and cracking observed in equipment designed to operate under high-temperature conditions can be attributed to the insufficient attention given to the critical issue of thermal expansion and the complexities associated with differential thermal expansion. Notably, seemingly modest temperature gradients, such as those measuring merely 200°F (equivalent to 110°C), can induce strains in metallic materials that exceed their yield point, compromising their structural integrity and functionality. Consequently, this highlights the need for engineers and designers to meticulously consider and address these thermal factors during high-temperature equipment's design and operation phases to mitigate potential failure risks.

2.8   Molten Metals

In the realm of industrial applications, the utilization of low-melting point metals, which include but are not limited to copper and silver brazing alloys, as well as zinc and aluminium, tends to precipitate a variety of significant challenges and complications that can adversely affect the integrity and performance of the materials involved. It is generally observed that these low-melting metals exhibit a pronounced tendency to rust and degrade higher nickel alloys with greater efficiency and severity compared to their interactions with low-nickel alloys or ferritic grades, thereby necessitating a nuanced understanding of their behaviour in various metallurgical contexts. Consequently, this phenomenon underscores the critical importance of material selection and compatibility in industrial processes, mainly when dealing with alloys with differing nickel content and thermal properties, which can have profound implications for the longevity and reliability of the final products.

2.9   Galling

Austenitic nickel-based alloys exhibit a pronounced tendency to experience galling phenomena when they undergo sliding interactions with one another, thereby raising concerns regarding their tribological performance in various applications. At elevated thermal conditions, it has been observed that cobalt oxide possesses a certain degree of lubricating properties, which can potentially mitigate some of the adverse effects associated with friction and wear. Moreover, cobalt itself, along with alloys that contain a significant proportion of cobalt, such as the cast variant known as Super-therm, demonstrates a remarkable resistance to galling, particularly at high-temperature conditions that could be classified as red heat. In the context of heat treatment furnace applications that operate at temperatures reaching as high as 1650°F, the alloy Nitronic® 6010, designated by its UNS code S21800, has shown a commendable performance in resisting galling, making it a suitable material choice for such demanding environments. It is crucial to choose the right alloy for high-temperature applications based on their combined material attributes and performance characteristics to guarantee operational efficiency and endurance. Therefore, a comprehensive understanding of the interactions between these materials under varying conditions is essential for optimizing their usage in industrial settings.

3.   Applications of High-Temperature Materials

A comprehensive and in-depth comprehension of the myriad properties associated with materials—encompassing mechanical, physical, and chemical attributes—coupled with an astute selection of production techniques tailored to yield materials that possess the requisite characteristics, has the potential to facilitate the development of groundbreaking materials that can be utilized in entirely novel and innovative applications that were previously inconceivable[7]. However, it is imperative to acknowledge that the characterization process for traditional materials is typically conducted using established and widely accepted testing methodologies. In contrast, exploring novel materials along with their prospective applications necessitates implementing unconventional or entirely innovative testing approaches and methodologies specifically designed for the nuanced characterization and thorough understanding of these materials [1–4]. Thus, one must carefully consider the implications of the evolution of material science, as it not only impacts the advancement of materials technology but also influences the broader fields of engineering, manufacturing, and applied sciences, ultimately paving the way for future innovations that could significantly enhance various industrial applications and societal needs[5].

High-temperature materials find utility in various tasks and sectors that pose very particular and stringent requirements for material properties, which are often challenging to attain and maintain under operational conditions. Illustrative examples of such industries include, but are not limited to, thermal processing, which encompasses metallurgical engineering, the chemical sector, marine applications, road construction, aviation, and the aerospace industry. In thermal processing, it is

Superalloys: Fundamentals and Applications
Materials Research Foundations 178 (2025) 259-278

Materials Research Forum LLC
https://doi.org/10.21741/9781644903698-13

noteworthy that high-temperature techniques are extensively implemented across many industries, including but not restricted to chemical, metallurgical or allied engineering, along with the manufacturing sectors. The tools utilized in such diverse arena comprises an array of specialized apparatus such as various types of furnaces, reactors, kilns, and heat exchangers, all of which share a fundamental requirement for materials that can withstand elevated temperatures to ensure durability and efficacy in their construction [8].

3.1 Chemical Industry

The realm of chemical engineering encompasses a wide array of processes and treatments conducted under elevated temperatures and significant pressures, a phenomenon particularly prevalent within the chemical industry. When it comes to the construction of materials utilized for chemical reactors, these materials must inherently possess an impressive degree of resistance not only to extreme temperatures but also to various forms of corrosion, as well as to oxidation and reduction processes, high-pressure environments, and a multitude of other specific environmental conditions that may arise during operation. A substantial number of materials that are employed in these demanding applications are categorized as thermal barrier coatings and include specially engineered alloys that exhibit desirable properties, such as stainless steel, spatial aluminium alloys, and heat-resistant alloys, along with a diverse range of ceramic and composite materials designed for high-performance applications. In the context of energy generation, the fuels utilized to produce the heat essential for these processes often result in the emission of flue gases at elevated temperatures, which presents a significant challenge that is typically addressed through the implementation of heat exchangers aimed at enhancing energy efficiency and facilitating cost savings, in addition to the use of filters and various waste gas treatment technologies and equipment designed to mitigate environmental impacts. Given the critical nature of these applications, high-temperature materials must be employed, particularly in industries such as oil and gas, where operational temperatures in specific catalytic processes can escalate to the range of 950 to 1050 degrees Celsius, thereby necessitating the use of materials that can withstand such extreme conditions[9].

3.2 Aircraft and Space Vehicles

The advancement and fabrication of innovative materials that possess specifically engineered properties are intrinsically linked to the contemporary industrial landscape and the various demands that arise from it. Among the plethora of modern industries, the aerospace and space exploration sectors stand out as noteworthy examples, imposing stringent and unique requirements regarding the selection and application of materials utilized in every conceivable aircraft or space vehicle component. In response to these rigorous demands, a multitude of contemporary materials has been meticulously engineered and developed, including but not limited to high-performance superalloys, refractory metals that can withstand extreme conditions, tungsten with its remarkable density and thermal properties, as well as specialized forms of graphite and advanced carbide and nitride ceramic and composite materials. The distinctive characteristics of these advanced materials encompass exceptional thermal resistance, the ability to endure thermal shock, high thermal stress tolerances, as well as robust resistant properties in the context of corrosion and erosion, making them indispensable in applications that require reliability and performance under challenging conditions [10].

Superalloys: Fundamentals and Applications   Materials Research Forum LLC
Materials Research Foundations 178 (2025) 259-278   https://doi.org/10.21741/9781644903698-13

3.3   High-Temperature Steam Turbines

Such turbines typically function effectively at elevated operating temperature conditions that hover near to the substantial threshold of 600°C. Yet, it is essential to note that these temperatures may exceed this baseline under certain operational conditions, presenting additional challenges and requirements for material performance. To ensure the integrity and longevity of these turbines, the materials employed must possess exceptional thermal stability under such extreme conditions and exhibit a high degree of resistance to both corrosive environments and cavitation phenomena that can occur during operation. By selecting and utilizing optimized materials that are specifically tailored to accommodate the unique demands of the operational conditions, various critical components, including but not limited to rotor, blade, nozzle, and pipe, can be manufactured to achieve enhanced levels of reliability and an extended economically viable service duration, which is crucial for the overall efficiency and cost-effectiveness of the turbine system. Consequently, meticulous attention to material selection and engineering design plays a pivotal role in enhancing steam turbines' operational performance and durability, ensuring their effectiveness in various demanding applications within the energy sector.

3.4   Nuclear Reactors

In contrast to the various materials that have been elucidated in the preceding chapters and the operating conditions prevalent across a wide range of industries, the materials employed in nuclear reactors must exhibit a remarkable degree of resistance to particular cooling fluids, notably helium and sodium, which are utilized in the heat exchange processes within these systems. Furthermore, in addition to possessing the ability to withstand elevated temperatures, the materials designated for use in nuclear reactors must also demonstrate a formidable resilience against both corrosive and erosive attacks, thereby ensuring their longevity and performance under the demanding conditions characteristic of nuclear environments, as referenced in the literature [11].

3.5   Marine Industry

Throughout the extensive history of the marine industry, the conventional materials that have predominantly been utilized include organic substances such as wood and metals like steel and substantial composites like concrete, each chosen for their unique properties and suitability for various applications. The harsh influence of the saline environment, coupled with the particular requirements dictated by marine vessels, necessitates that the materials employed exhibit specific characteristics including, but not limited to, exceptional resistance to corrosion, a favourable strength-to-weight ratio that optimizes performance, the capability to be moulded into complex geometries as demanded by engineering specifications, and economic viability to ensure cost-effectiveness in production. In contrast to these traditional materials, a range of advanced materials has emerged, encompassing innovative substances such as sulphur concrete [12,13], various forms of cement and composites based on it [14,15], glass-reinforced plastic substances that offer enhanced durability, ferrocement which combines properties of concrete and steel, and composites made from aramid fibres that provide strength and resilience. Adopting these novel materials necessitates the formulation of new and sophisticated testing methodologies aimed at evaluating the entire life cycle of the materials used, ensuring they meet the rigorous demands of marine applications [16]. Notably, despite advancements in material science, specific categories of vessels continue to rely on diesel engines that are invariably subjected to extreme stress conditions from mechanical and thermal perspective and aggressive corroding environments at elevated

Superalloys: Fundamentals and Applications   Materials Research Forum LLC
Materials Research Foundations 178 (2025) 259-278   https://doi.org/10.21741/9781644903698-13

temperatures throughout their operational lifespan [17]. These operational temperatures can reach approximately 800 degrees Celsius in specific components, particularly in critical areas such as pre-combustion chamber, exhaust valve, along with other components subjected to extreme loads and stressful operating conditions.

3.6   Rail and Road

Moreover, thoroughly considering environmental conditions is imperative, as these conditions encompass extensive temperature variations, including operational requirements that necessitate effective performance at extremely low and exceedingly high temperatures, thereby underscoring the essential need for robust corrosion resistance. Consequently, selecting materials for such applications must prioritize wear resistance and incorporate factors that address the challenges posed by varying environmental conditions and the types of fuels in use, further complicating the decision-making process. Ultimately, this multifaceted approach to material selection is critical in ensuring the longevity and reliability of transportation infrastructure, particularly in contexts where both road and rail systems interface with varying operational demands and environmental challenges.

4.   Characterization

In scientific inquiry, image examination is profoundly and extensively applied to the comprehensive study of physical construction, allowing researchers to gain invaluable insights into the intricate details of material structures. When it comes to the meticulous analysis of the surface characteristics of various materials, the informal yet direct accessibility to the surface facilitates the application of numerous scientific instruments and methodologies. It enhances the precision and accuracy of the data collected. Furthermore, the examination of the internal constructions that lie beneath the surface is carried out utilizing specialized techniques that are specifically designed to enable the understanding and quantification of crucial interior properties, which include, but are not limited to, stress and strain as well as the chemical configuration of the materials in question. Notably, many scientific methodologies have been developed and employed to thoroughly investigate high-temperature materials, often subjected to extreme conditions that can significantly alter their properties. In this context, some of the most significant techniques utilized in the field are reviewed and discussed, providing a comprehensive overview of the methodologies that have proven effective in high-temperature material analysis. This review highlights the advancements made in the field and emphasizes the importance of these techniques in enhancing our understanding of material behaviour under varying conditions. Integrating these sophisticated methods into studying high-temperature materials is essential for developing innovative applications and technologies in various scientific and industrial domains.

4.1   Materials Analysis

Materials engineered for applications at high temperature require thorough assessment throughout their entire lifecycle, encompassing stages from research and development to decommissioning. Such approaches can promote advancements in materials, ensure pre-service quality verification, facilitate in-service analysis, or investigate failures. Characterisation is generally performed (a) during the production phase, (b) before the commencement of service, (c) throughout the operational lifespan, and (d) post-failure to determine root causes. Material characterisation includes chemical, mechanical, microstructural, macroscopic, destructive, as well as nondestructive evaluations and testing. Evaluation methods typically involve assessing elemental composition, conducting microstructural analysis, and identifying material phases, and validating adherence to defined criteria. An appraisal may also examine fracture mechanisms, creep behaviour, fatigue resistance, wear evaluation, distortion assessment, corrosion analysis, and comprehensive material review and optimisation. Moreover, evaluations might involve microstructural investigations, grain size measurement, appraisal of heat treatment outcomes, and the documentation of porosity and phase characteristics alongside hardness assessments. Exhaustive microscopic studies can be executed to illuminate the microstructural properties, the processes of carburisation and decarburisation, coating metrics, intergranular corrosion issues, surface discrepancies, and the measurement of nodules. To determine the mechanical properties of the samples under scrutiny, specimens of varying geometries are produced and subjected to mechanical testing. Assessments include thermal exposure, cooling rates, humidity impacts, strain assessment, plastic deformation, failure modes, and other factors.

4.1.1 Surface Chemical Analysis Techniques

The analysis of surface chemistry is of critical significance for the characterisation of materials intended for high-temperature applications for several reasons:

a) The interface of a material's surface functions as the point of interaction within mechanical systems, and the surface chemistry may demonstrate considerable variation from the chemistry of the bulk substance. Surface phenomena may lead to distinctly altered surface morphology along with other characteristics.

b) The impurities on surface often lead to undesirable malfunctioning in crucial operations. A careful examination can yield crucial data on the material's effectiveness, and signs of detrimental wear and tear.

c) Establishing grain boundaries is pivotal in augmenting material strength in metallic substances and various conductive materials. Surface analytical methodologies, including field emission Auger electron spectroscopy (FEAES) provide significant insights regarding material surfaces.

4.2 Imaging and Visual Analysis

The evaluation of visual characteristics of examination specimens and constructs constitutes the most essential and comprehensive methodology for gathering critical information pertaining to various aspects of morphology, pigmentation, texture, as well as other significant physical attributes that are inherent to these specimens. A multitude of sophisticated instruments is utilized in this process, which includes but is not limited to magnifying lenses and diverse forms of microscopes, all intended to enable the observation of materials that transcend the typical limitations of human visual perception. In a broader context, these advanced instruments can be systematically classified into three primary categories of microscopy: namely, electron microscopy, scanning probe microscopy, and optical microscopy. Electron microscopes harness the sophisticated interactions that take place between electron beams and assorted materials, consequently facilitating the retrieval of extensive data regarding how the specimen engages with, absorbs, refracts, and scatters various types of radiation. The electron beam applied in this context may be judiciously aimed at the specimen via techniques such as broad-area irradiation or through the execution of a raster scan approach that utilizes a finely concentrated beam for greater accuracy. Alternatively, scanning probe microscopes fundamentally rely on the subtle interactions between a fine probe tip and the surface of the specimen subjected to analysis, integrating complex phenomena such as tunnelling current, magnetic attraction, lateral force, and adhesion to procure high-resolution imaging and data.

4.3   Material Analysis Techniques

4.3.1   X-ray Diffraction Examination

X-ray Diffraction (XRD) analysis serves as a critical technique utilized for the comprehensive investigation and characterization of the crystallographic attributes of various materials, enabling researchers to extract valuable insights concerning the elemental composition as well as the physical properties of diverse substances. This intricate process involves the precise measurement of the scattering phenomena that occur when an X-ray beam is directed towards a material, with the resulting interactions varying according to factors such as the incident angle, scattering angle, polarization, and the specific wavelength of the X-ray employed, which can be influenced by whether the material is in film or powder form. Recognizing the profound capability of XRD to elucidate the intricate characteristics of substances, a sophisticated and compact XRD system was thoughtfully integrated into the array of scientific instruments aboard the Curiosity Rover, which is part of the ambitious Mars Science Laboratory mission that successfully achieved its landing on the Martian surface in August of the year 2012. The integration of such advanced analytical techniques aboard the rover underscores the significance of XRD in planetary exploration, as it facilitates the in-situ analysis of Martian materials, thus contributing to our broader understanding of the planet's geology and potential for past habitability.

4.3.2   X-Ray Fluorescence Spectrometry

The phenomenon known as fluorescence emission of X-rays occurs when different substances are bombarded with X-rays that have high energy levels. . This scientific process is extensively utilized to characterize and analyze different materials, as referenced in scholarly work. This bombardment results in the ejection of electrons that possess binding energy relatively lower in comparison to energy levels of the incoming X-rays, thereby facilitating the subsequent detection of emitted radiation. The radiation released during this process, which exhibits distinct properties corresponding to the atomic structure of the elements present, is effectively captured and measured using comparative methodologies or by deploying multiple solid-state detectors designed for such purposes. X-ray fluorescence (XRF) devices have gained widespread acceptance. Due to their efficiency and reliability in producing accurate results, they are extensively employed for the fundamental examination and analysis of various materials, including metals, glass, and ceramics. Furthermore, it is noteworthy that both XRF and X-ray diffraction (XRD) devices were integral components of the Mars Curiosity rover's comprehensive instrument suite, specifically within its CheMin analyzer, which successfully landed on the Martian surface on August 5, 2012, to conduct in-situ analyses. This dual application of XRF and XRD technologies on extraterrestrial terrain underscores their versatility. It reflects the significant advancements in analytical methodologies that enable scientists to explore and understand the composition of planetary materials beyond Earth. Thus, integrating such sophisticated analytical instruments into space exploration missions exemplifies the profound impact of X-ray-based techniques in expanding our knowledge of the universe.

Superalloys: Fundamentals and Applications   Materials Research Forum LLC
Materials Research Foundations 178 (2025) 259-278   https://doi.org/10.21741/9781644903698-13

4.3.3   Low Energy Electron Diffraction

LEED, an acronym for Low Energy Electron Diffraction, embodies a sophisticated methodology that is meticulously utilised to ascertain the spatial configuration of crystalline substances, which involves the intricate process of directing a collimated stream of low-energy electrons, generally within an energy spectrum ranging from 20 to 200 electron volts (eV), onto the crystal surface, followed by a comprehensive analysis of the subsequently diffracted electrons, which are represented as distinct spots on a specialised fluorescent display. Applying this LEED technique is critically important as it allows researchers to conduct a detailed examination of the resulting diffraction patterns, thereby facilitating a deeper understanding of the homogeneity and intricacies of the surface structure inherent to the crystalline material under investigation. In addition to this, a thorough assessment of the intensity of diffracted electron beam, which varies with respect to energy levels of incident beam, is scrupulously performed for extracting highly accurate and nuanced information about the specific atomic arrangements and positions on the surface of the material being analysed. The depiction of these low-energy electron diffraction patterns is illustrated in Figure 2, which serves as a visual representation of the complex interactions that occur during the diffraction process. This intricate interplay between the incident electron beam and the crystal surface elucidates the underlying physical principles of electron diffraction and significantly contributes to the broader field of materials science. Ultimately, the insights gained from this technique are invaluable for advancing our comprehension of crystalline materials and their applications across various scientific disciplines.

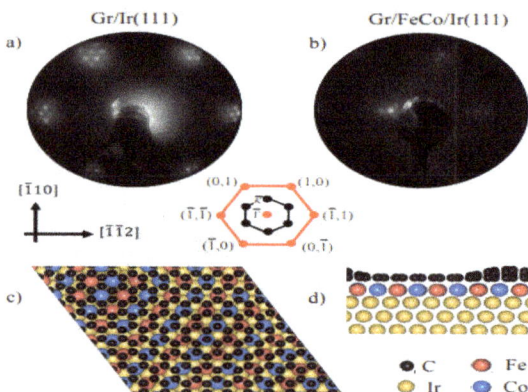

Figure 2. LEED patterns (electron beam energy 90 eV) of (a) Gr/Ir(111) and (b) Gr/FeCo/Ir(11)1 (c) Gr/FeCo/Ir(111) (d) Gr/FeCo/Ir(111)(Reproduced from [18] CC BY license)

4.3.4   Neutron Diffraction

Neutron diffraction is an essential methodology in determining various substances' atomic structure and arrangement, as referenced in scholarly work. In this sophisticated technique, a beam of either cold or thermal neutrons is directed at the material undergoing scrutiny. Subsequently, the diffraction pattern that emerges from this interaction is meticulously analyzed to extract critical data regarding the internal structure of the material in question. Similar to the established X-ray Diffraction (XRD) principle, scattering of neutron beam adheres to Bragg's Law, thereby providing complementary and balancing information crucial for understanding the structural characteristics of the substance. Notably, for specific light atoms, the intensity of the diffraction pattern can remain remarkably robust even in the presence of elements that possess significantly higher atomic numbers, underscoring the unique advantages of neutron diffraction in specific analytical contexts. However, one of the primary limitations associated with neutron diffraction lies in the relatively restricted availability of the neutron radiation source, which can pose8 challenges for extensive application.

Additionally, the specimens subjected to neutron diffraction techniques typically require using crystals that are considerably larger than those commonly employed in XRD analysis, thus imposing further constraints on the experimental design. It is also noteworthy that neutrons possess an intrinsic property known as spin, enabling them to interact effectively with the magnetic moments of electrons present within the material, thereby allowing researchers to investigate the microscopic magnetic arrangement of the materials being analyzed in a detailed manner. This unique interaction enhances the understanding of the material's structural properties and provides insights into its magnetic characteristics, which can be critical in various research fields. Consequently, the application of neutron diffraction represents a powerful tool in material science, as it facilitates a more profound comprehension of atomic structures and magnetic arrangements that would be challenging to elucidate through other techniques. Overall, the integration of neutron diffraction into the toolkit of analytical methods underscores its significance in advancing the frontiers of knowledge about material properties at the atomic level.

4.3.5 Electron Microprobe

EMP, or electron microprobe, represents a sophisticated system for determining the elemental makeup of solid materials, particularly those in minimal volumes that typically measure no more than 10 to 30 cubic micrometres. These advanced instruments are frequently referred to in the literature as electron microprobe analyzers (EMPA) [18], highlighting their significance in material science and analysis. In a manner akin to scanning electron microscopes (SEMs), these devices operate by directing a focused electron beam onto the sample in question, and they subsequently analyze the discrete wavelengths of the resultant x-rays that are emitted, which allows for a foundational understanding of the elemental composition of the substance being studied. The continual advancements and enhancements in the functionality and precision of EMP systems have led to remarkable improvements in measuring trace elements, even those present in minute concentrations that were previously challenging to detect and quantify. As a result, these developments have refined our analytical capabilities and expanded the potential applications of EMP analysis across various scientific disciplines, including geology, materials science, and environmental studies. Consequently, detecting and analyzing these trace elements with high precision opens up new avenues for research and innovation, contributing significantly to our understanding of complex materials and their properties.

Superalloys: Fundamentals and Applications   Materials Research Forum LLC
Materials Research Foundations 178 (2025) 259-278   https://doi.org/10.21741/9781644903698-13

5. Challenges in High-Temperature Materials

High-temperature materials are critical in numerous industries, such as aerospace, automobile, and energy. Such materials must withstand extreme conditions while maintaining their structural integrity and performance. This report discusses the challenges of high-temperature materials, drawing from recent advancements and ongoing research.

5.1 Environmental Degradation

High-temperature materials, especially those used in engines, face severe environmental conditions. These materials operate in pressurized gaseous mixtures at temperatures up to 1600°C, accelerating oxidation, sulfidation, and corrosion. The presence of elements like sodium, sulfur, vanadium, and halides in the environment exacerbates these degradation processes. Understanding the atomic-scale degradation behaviour and the role of deformation-induced defects is crucial for developing materials that can better withstand these harsh conditions

5.2 Material Deformation and Defects

The dynamic nature of material deformation during operation, including the formation of dislocations, stacking faults (SFs), anti-phase boundaries (APBs), and micro-twins, significantly impacts the corrosion behaviour of high-temperature materials. However, there is a lack of specific experiments evaluating these effects, highlighting a gap in current research that needs to be addressed to improve material performance

5.3 Structural Integrity and Longevity

Ensuring the durability and longevity of materials in high-temperature applications is paramount. This involves a deep understanding of material properties and their environmental interactions. Materials must withstand high temperatures and maintain their structural integrity over time, which requires continuous research and innovation.

5.4 Advanced Materials

Recent advancements have led to the developing new materials and technologies to address high-temperature challenges. For instance, NASA has developed an Oxide Dispersion Strengthened Medium Entropy Alloy (ODS-MEA) employing additive manufacturing to enhance mechanical properties at extreme temperatures. New nickel-based superalloys and polymer systems have also been created to improve structural performance and high-temperature capability.

5.5 Lightweight and High-Performance Composites

In the aerospace industry, lightweight materials such as Ceramic Matrix Composites (CMCs) and Carbon-Carbon composites are increasingly used due to their ability to handle extreme thermal conditions while reducing overall aircraft weight and fuel consumption. Refractory metals like tungsten, molybdenum, and niobium are also employed for their outstanding heat resistance in critical components like rocket nozzle throats and re-entry vehicle leading edges.

5.6 Thermal Barrier Coatings

Thermal barrier coatings (TBCs) are essential for protecting components from high temperatures, particularly in gas turbines and aircraft engine parts. These coatings help extend the life of the

Superalloys: Fundamentals and Applications
Materials Research Foundations 178 (2025) 259-278

Materials Research Forum LLC
https://doi.org/10.21741/9781644903698-13

underlying materials by providing a thermal insulation layer that reduces the thermal load on the components.

5.7 Service Temperature and Material Selection

Understanding the service temperature, which is the range in which a material retains its desired mechanical properties, is crucial for material selection and overall design.Factors influencing this range include material composition, crystalline structure, impurity levels, and manufacturing processes.Accurate determination of service temperature requires adherence to standards and testing methodologies.

5.8 Temperature Cycling and Thermal Shock

Refractory materials used in industries like metals, glass, and ceramics must endure repeated heating and cooling cycles, known as temperature cyclin. This cycling causes expansion and contraction, leading to stress and durability issues. Understanding thermal shock resistance and implementing design innovations that account for thermal expansion can help extend the life of these materials.

5.9 High-Temperature Electronics

Silicon-based devices face considerable reliability concerns for temperatures exceeding 125°C, necessitating the development of new material solutions for electronics applications at elevated temperatures. Future technologies in space research, geothermal power, nuclear power, and autonomous systems would depend on control systems, sensing, and communication devices that operate at temperatures of 500°C and more. This requires a united effort to develop non-silicon-based logic and memory technologies, unconventional metals for interconnects, and ceramic packaging technologies.

6.   Future Prospects

High-temperature materials are essential for advanced applications, particularly in the aerospace, defence, and energy sectors. The future research scope in this field is vast and multifaceted, focusing on enhancing material properties, developing new manufacturing techniques, and understanding material behaviour under extreme conditions. This report synthesizes critical points from recent research to outline the future directions in high-temperature materials [18].

6.1   Advanced Alloys and Nanomaterials

Research into advanced alloys and nanomaterials is paving the way for groundbreaking innovations. These materials are being developed to withstand extreme temperatures and stresses, which are critical for aerospace and defence applications. The development of new alloys, such as Oxide Dispersion Strengthened Medium Entropy Alloy (ODS-MEA), demonstrates significant improvements in mechanical properties and resistance to stress cracking at high temperatures.

6.2   Additive Manufacturing Techniques

Additive manufacturing (AM), mainly 3D printing, is a beacon of innovation in high-temperature materials research. AM allows for the fabrication of complex geometries and the development of materials with enhanced properties, such as improved creep rupture life and increased strength.

Superalloys: Fundamentals and Applications   Materials Research Forum LLC
Materials Research Foundations 178 (2025) 259-278   https://doi.org/10.21741/9781644903698-13

This technology is crucial for producing high-performance components for aerospace and other high-temperature applications.

6.3 High-Temperature Ceramics and Composites

Developing high-temperature ceramics and composites, such as ultrahigh-temperature ceramics (UHTCs) and carbon-carbon composites, is a crucial area of research. These materials are essential for applications that require extreme thermal stability and resistance to oxidation and thermal shock. Improving the processing and oxidation resistance of these materials will extend their service life and enhance their performance in high-temperature environments.

6.4 Thermal and Chemical Extremes

Research programs like the Office of Naval Research's Materials for Thermal and Chemical Extremes aim to develop materials operating in chemically active environments at high temperatures and stresses. This includes the development of novel processing techniques and computational tools to optimize material properties and design. Understanding and controlling material responses in transient thermal gradients and high heat flux environments are also critical research areas.

6.5 High-Temperature Heat Treatment Technology

High-temperature thermal treatment technology becomes crucial to enhance the properties of materials such as metals. Research in this area focuses on understanding the mechanisms of heat treatment and developing new equipment and processes to enhance material quality and performance. This technology has broad prospects and high market conversion rates.

6.6 High-Temperature Electronics

Developing high-temperature electronics is essential for future technologies in space exploration, geothermal energy, and autonomous systems. Research is focused on creating technologies based on innovative materials to enable devices to operate at temperatures ranging beyond 500°C.

6.7 Refractory Metals and Alloys

Advancements in refractory metal processing and alloying have significantly improved their high-temperature properties. These materials are crucial for efficient and reliable propulsion systems and other high-temperature applications. Ongoing research aims to enhance these materials' creep resistance, fatigue life, and thermal stability.

6.8 Thermal Barrier Coatings

Developing novel thermal barrier coatings with improved adhesion and durability is a critical area of research. Thermal barrier coatings (TBCs) protect components from high temperatures, particularly gas turbines and aircraft engines. Enhancing the performance of TBCs will extend the service life of these components and improve overall system efficiency.

6.9 High-Temperature Aluminum Alloys

Research into high-temperature aluminium alloys, such as the oxide dispersion strengthened aluminium alloy, is expanding the service temperature range and improving tensile strength and

Superalloys: Fundamentals and Applications   Materials Research Forum LLC
Materials Research Foundations 178 (2025) 259-278   https://doi.org/10.21741/9781644903698-13

stability. These advancements are significant for aerospace engines and critical components, with ongoing collaborations between research institutions and industry leaders to drive development.

Conclusion

The challenges in high-temperature materials are multifaceted, involving environmental degradation, material deformation, structural integrity, and the need for advanced material solutions. Continuous research and innovation are essential to develop materials that can withstand extreme conditions while maintaining their performance and longevity. Understanding these materials' specific requirements and behaviours under high temperatures will drive future advancements in various high-temperature applications.

References

[1] F. Findik, High-temperature materials processing, High Temp. Mater. Mech. (2014) 179–208. https://doi.org/10.1201/b16545-10

[2] C. Wang, L. Bai, H. Xu, S. Qin, Y. Li, G. Zhang, A review of high-temperature aerogels: composition, mechanisms, and properties, Gels. 10 (2024) 286. https://doi.org/10.3390/gels10050286

[3] P.O. Maruschak, S.V. Panin, S.R. Ignatovich, I.M. Zakiev, I.V. Konovalenko, I.V. Lytvynenko, V.P. Sergeev, Influence of deformation process in material at multiple cracking and fragmentation of nanocoating, Theor. Appl. Fract. Mech. 57 (2012) 43–48. https://doi.org/10.1016/j.tafmec.2011.12.007

[4] G.W. Meetham, M.H. Van de Voorde, Materials for high temperature engineering applications, Eng. Mater. (2000). https://doi.org/10.1007/978-3-642-56938-8

[5] K. Spear, E. Wuchina, E.D. Wachsman, High temperature materials, Electrochem. Soc. Interface. 15 (2006) 48–51. https://doi.org/10.1149/2.f14061if

[6] S.V. Raj, Comparison of the isothermal oxidation behavior of As-cast Cu–17% Cr and Cu–17% Cr–5% Al psart I: Oxidation kinetics, Oxid. Met. 70 (2008) 85–102. https://doi.org/10.1007/s11085-008-9110-5

[7] V. Trush, P. Maruschak, M. Student, S. Lavrys, A. Luk'yanenko, Effect of heat treatment in oxygen-containing medium on fatigue life of zirconium alloy, strojnícky casopis - J. Mech. Eng. 72 (2022) 211–218. https://doi.org/10.2478/scjme-2022-0030

[8] S. Marenovic, M. Dimitrijevic, T.V. Husovic, B. Matovic, Thermal shock damage characterization of refractory composites, Ceram. Int. 34 (2008) 1925–1929. https://doi.org/10.1016/j.ceramint.2007.07.021

[9] M.M. Vuksanović, M. Gajić-Kvaščev, M. Dojčinović, T.V. Husović, R.J. Heinemann, New surface characterization tools for alumina based refractory material exposed to cavitation - Image analysis and pattern recognition approach, Mater. Charact. 144 (2018) 113–119. https://doi.org/10.1016/j.matchar.2018.07.003

[10] M. Dimitrijevic, M. Posarac, J. Majstorovic, T. Volkov-Husovic, B. Matovic, Behavior of silicon carbide/cordierite composite material after cyclic thermal shock, Ceram. Int. 35 (2009) 1077–1081. https://doi.org/10.1016/j.ceramint.2008.04.029

Superalloys: Fundamentals and Applications   Materials Research Forum LLC
Materials Research Foundations 178 (2025) 259-278   https://doi.org/10.21741/9781644903698-13

[11]   B. Matović, D. Bučevac, V. Urbanović, N. Stanković, N. Daneu, T. Volkov-Husović, B. Babic, Monolithic nanocrystalline SiC ceramics, J. Eur. Ceram. Soc. 36 (2016) 3005–3010. https://doi.org/10.1016/j.jeurceramsoc.2015.10.031

[12]   M.M. Vlahović, P.B. Jovanić, S.P. Martinović, T.Đ. Boljanac, T.D. Volkov-Husović, Quantitative evaluation of sulfur–polymer matrix composite quality, Compos. Part B Eng. 44 (2013) 458–466. https://doi.org/10.1016/j.compositesb.2012.04.005

[13]   M.M. Vlahović, M.M. Savić, S.P. Martinović, T.Đ. Boljanac, T.D. Volkov-Husović, Use of image analysis for durability testing of sulfur concrete and Portland cement concrete, Mater. Des. 34 (2012) 346–354. https://doi.org/10.1016/j.matdes.2011.08.026

[14]   I.S. Ignjatović, S.B. Marinković, Z.M. Mišković, A.R. Savić, Flexural behavior of reinforced recycled aggregate concrete beams under short-term loading, Mater. Struct. 46 (2012) 1045–1059. https://doi.org/10.1617/s11527-012-9952-9

[15]   D. Jevtic, D. Zakic, A. Savic, Investigation of cement based composites made with recycled rubber aggregate, Hem. Ind. 66 (2012) 609–617. https://doi.org/10.2298/hemind111203010j

[16]   R.A. Shenoi, J.M. Dulieu-Barton, S. Quinn, J.I.R. Blake, S.W. Boyd, Composite materials for marine applications: Key challenges for the future, Compos. Mater. (2011) 69–89. https://doi.org/10.1007/978-0-85729-166-0_3

[17]   S. Martinovic, M. Vlahovic, T. Boljanac, M. Dojcinovic, T. Volkov-Husovic, Cavitation resistance of refractory concrete: Influence of sintering temperature, J. Eur. Ceram. Soc. 33 (2013) 7–14. https://doi.org/10.1016/j.jeurceramsoc.2012.08.004

[18]   D. Pacilè, C. Cardoso, G. Avvisati, I. Vobornik, C. Mariani, D.A. Leon, P. Bonfà, D. Varsano, A. Ferretti, M.G. Betti, Narrowing of d bands of FeCo layers intercalated under graphene, Appl. Phys. Lett. 118 (2021). https://doi.org/10.1063/5.0047266

Superalloys: Fundamentals and Applications
Materials Research Foundations 178 (2025) 279-290

Materials Research Forum LLC
https://doi.org/10.21741/9781644903698-14

Chapter 14

Superalloys for Blade Applications

N.V. Ramesh Maganti

Department of Mechanical Engineering, Nalla Malla Reddy Engineering College, Hyderabad-500088, India

ramesh.mnv@gmail.com

Abstract

Superalloys are engineered materials that provide exceptional performance in extreme environments by exhibiting high temperature strength, creep resistance, oxidation and corrosion resistance, and thermal fatigue resistance which make them indispensable in aerospace, power generation, rocket engines and various industrial applications. Thus, the superalloys enable the development of advanced technologies and improve the reliability and efficiency of critical systems. The current chapter focuses on types, properties of superalloys, applications in manufacturing of blades which are used in aerospace, power generation, rocket engine and chemical processing industries, manufacturing methods, failure analysis of blades and future trends. The important challenges such as new superalloy development, composition tailoring, waste blade recycling, prediction models, microstructure degradation, deformation behaviour, durability, residual lifespan etc have to be explored.

Keywords

Blade, Corrosion, Failure, Manufacturing, Nickel, Superalloys, Temperature, Turbine

Contents

Superalloys: Fundamentals and Applications   Materials Research Forum LLC
Materials Research Foundations 178 (2025) 279-290   https://doi.org/10.21741/9781644903698-14

1. Introduction

Blades are the vital components in different machines and systems and based on material composition, function and operational environment, they are broadly categorized as turbine, compressor, fan, rocket engine, hydroelectric, wind turbine and industrial blades [1, 2]. The turbine blades are used in turbines to extract energy from a fluid such as gas or steam and converting it into mechanical energy and are crucial in aerospace and power generation sectors. They are operated under enhanced temperature, elevated pressure and high rotational speed [3]. Further, the gas turbine blades are classified as high pressure or low pressure depending on the operational pressure and temperature of the gas exiting the combustion chamber of the engine [4, 5]. While the steam turbine blades are divided as high pressure (Fig 1), intermediate (Fig 2) and low pressure (Fig 3) steam turbines depending on the operational pressure and temperature of the steam [6, 7]. The compressor blades are utilized for compressing the air or gas in engines and compressors to increase the pressure of the working fluid before entering into the combustion chamber or further processing stages. They are further divided into axial and centrifugal types based on the flow/movement of the air or gas [8]. The fan blades are used to move either air or gas for creating air flow or increasing the pressure and are typified as axial and centrifugal [1]. The rocket engine blades are exploited in different components of rocket engines such as turbo pumps which operate under high centrifugal forces, temperature and aerodynamic pressure for pumping fuel and oxidizer into the combustion chamber of the engine [9]. The hydroelectric and wind turbine blades convert the kinetic and energy of the flowing water or wind into mechanical and rotational energy, respectively (Fig. 4) [1]. The industrial blades utilized for cutting, mixing and chemical processing of materials in various machinery [6, 10].

The blades used in wide range of applications should have the high strength to withstand mechanical stress and durability at variable environmental conditions. Whereas blades used in turbines and engines that operate at elevated temperatures should have high temperature resistance. While the blades utilized in gas turbines, rocket engines and aggressive chemical processing should exhibit corrosion and oxidation resistance [11]. The blades used in high performance cutting tools should have high durability and wear resistance. For ensuring efficiency and performance, the blades used in aerospace engines should be light weight and thin walled [11-13]. Thus, the choice of material is crucial in achieving the efficient and reliable performance in their respective applications where high melting point, thermal stability and microstructural integrity, low density, elevated stiffness, superior fabricability, reasonable price, reproducible performance, oxidation and corrosion resistance and creep resistance at high operating temperature and high pressure environment under high centrifugal force and mechanical stress is needed [2, 3, 7]. In this scenario, the use of superalloys comes into the picture in fabrication of blades for diverse applications. The current chapter focuses on types, properties of superalloys, applications in manufacturing of blades which are used in aerospace, power generation, rocket engine and chemical processing industries, manufacturing methods, failure analysis of blades and future trends.

Superalloys: Fundamentals and Applications   Materials Research Forum LLC
Materials Research Foundations 178 (2025) 279-290   https://doi.org/10.21741/9781644903698-14

Fig. 1 *Photograph showing the blades of high pressure steam turbine of a thermal power plant.*

Fig. 2 *Photograph showing the blades of intermediate pressure steam turbine of a thermal power plant.*

Superalloys: Fundamentals and Applications   Materials Research Forum LLC
Materials Research Foundations 178 (2025) 279-290   https://doi.org/10.21741/9781644903698-14

2.   Superalloys and characteristic properties

Superalloys are a class of high performance alloys designed to maintain their strength, stability, resistance to oxidation and corrosion at elevated temperatures [14]. They are commonly used in applications where materials are exposed to extreme conditions such as high temperature, high stress and aggressive oxidative and reducing environments [15]. Superalloys are commonly used in gas turbines, jet engines, rocket engines and other high temperature machines [15, 16]. One of the critical characteristics of the superalloys are the mechanical properties such as high tensile strength to withstand tensile forces without breaking; superior fatigue strength to resist failure under cyclic loading; and excellent impact toughness to absorb energy without fracturing [17]. The other important feature is high temperature strength/stability to resist deformation at elevated temperatures via an amalgamation of solid solution strengthening, precipitation hardening and other microstructural enhancements [2, 18]. Another characteristic is thermal fatigue resistance to thermal cycling by maintaining structural integrity and durability at varying temperatures [19]. The other important feature is the creep resistance to constant stress over time/prolonged stress achieved through the microstructure such as the formation of fine, stable precipitates that hinder dislocation movement [2, 14, 20, 21]. Another feature is the resistance to oxidation and corrosion which is critical for longevity and performance, which is achieved via formation of surface protective oxides of alloying elements [14, 22].

Fig. 3 Photograph showing the blades of low pressure steam turbine of a thermal power plant.

Superalloys: Fundamentals and Applications
Materials Research Foundations 178 (2025) 279-290

Materials Research Forum LLC
https://doi.org/10.21741/9781644903698-14

Fig. 4 Photograph showing the blades of wind turbine.

3. Materials used as superalloys

The superalloys used for manufacturing various blades in aerospace, power generation and industrial sectors can be classified as nickel based, cobalt based, iron based and tungsten based [14, 23]. The Ni based superalloys are crucial for turbine blade functions because of their exceptional mechanical properties (superior strength, creep resistance and oxidation resistance at enhanced temperatures) accounted to precipitation hardening and solid solution strengthening mechanisms [2, 24]. They are mainly composed of nickel, chromium and various other elements. The important sub categories are gamma (γ) and gamma prime (γ') and single crystal superalloys. The γ and γ' superalloys are mainly composed of Ni (50-60%) and Al (3-6%), in which gamma phase is a face centered cubic structure of Ni and the gamma prime phase is a precipitate of Ni^3(Al, Ti) within the gamma matrix [16, 25-27]. The single crystal superalloys (DD5, DD6, CMSX-4) are cast in a single crystal structure and engineered to minimize/no grain boundaries, thereby enhancing creep resistance [8, 11, 14, 23, 28]. The important examples of Ni based superalloys are Inconel 718, Inconel 625, Incoloy 825, Nimonic 80A, Hynes 282, Superni 263 etc [19, 29]. Inconel 718 [Ni (50-55%), Cr (17-21%), Nb (4.75-5.5%), Mo (2.8-3.3%), Ti (0.65-1.15%), Co (1%) Al (0.2-0.8%), Fe (balance)] is known for good weldability and high temperature stability up to 700 °C with protective silicon nitride coating used in blades of high pressure and high temperature steam, gas and jet turbine engines; compressor blades and aerospace components [30]. The other

Superalloys: Fundamentals and Applications   Materials Research Forum LLC
Materials Research Foundations 178 (2025) 279-290   https://doi.org/10.21741/9781644903698-14

one is Inconel 625 [Ni (58%), Cr (20-23%), Mo (8-10%), Fe (5%), Nb + Tl (3.15-4.15%), Co (1%), Si (0.5%), Mn (0.5%)] with superior thermal fatigue strength and excellent oxidation and corrosion resistance and is used in aerospace engines, turbine blades, exhaust systems and chemical processing equipment[22]. The other example, Rene 41 [Ni (55-57%), Mo (15-17%), Cr (14.5-16.5%), Fe (4-7%), W (3-4.5%), Co (2.5-10%), Mn (1%), Va (0.35%), C (0.06-0.12%), Si (0.2%), B (0.003-0.01%)] is known for high temperature oxidation resistance and exploited in advanced jet engines and high performance gas turbines; and CMSX-4, single crystal superalloy (Ni, Cr, Co, Rh) is known for superior creep resistance arising from anisotropic properties, an essential property for industrial gas turbine automotive turbocharger turbine and jet engine blades that experience high stress and temperature (700-1100 °C) during operation [28, 31, 32].

The cobalt based superalloys are chiefly composed of Co (60%), Cr (26-29%), W (6-8%), Mo (5.5%), Ni (2-3%), Fe (1.5%), Si (0.75%) and C (0.35%) and is known for excellent hardness, wear resistance and thermal stability and the examples are Stellite 21, Stellite 6B, Stellite 25, L605 etc which are used in turbine blades, jet engine parts and valve seats [11, 33]. The iron based superalloys are less commonly used in turbine blades compared to Ni based superalloys and are utilized in low to moderate temperature applications where cost effectiveness is a criterion. The example is Fe-Cr-Al alloy, Kovar, composed of Fe, Cr, Al and other additional alloying elements and provides good oxidation resistance and reasonable high temperature strength [18, 34]. The refractory tungsten based superalloys are primarily composed of W, Mo, Tl with additions of Ni and Fe and offer excellent high temperature and compressive strength and thermal stability and utilized in applications requiring high density and high temperature strength [35, 36].

4. Superalloy manufacturing techniques

The important factors influencing the material choice of blades used in steam, gas, marine, industrial turbines; jet engines and rocket engines are the operating conditions such as temperature, pressure, stress and environmental corrosion [19, 23]. While designing and manufacturing the blades, the aerodynamic and mechanical design principles have to be considered [13]. Usually, the superalloys are made through conventional techniques such as casting, forging, powder metallurgy, wrought stock methods and these methods are limited by complex steps, material wastage etc [27, 37, 38]. The advanced techniques such as vacuum induction melting, additive manufacturing etc are preferred due to excellent precision, effective material usage, cost effectiveness, design flexibility, customization amenability, rapid prototyping etc [23, 24, 27, 28, 39]. Using different protective coatings based on diffusion, overlaying and thermal barrier, the surfaces of superalloys are made resistant against temperature, oxidation, corrosion and microstructure depletion [2, 11, 16, 21, 23, 25, 30, 40, 41]. After manufacturing the superalloys, they have to be tested and evaluated systematically at high temperature for various mechanical (elastic, plastic, residual stress, creep, vibration) properties such as tensile and yield strength; compressive strength, hardness, elastic plastic behaviour, work hardening behaviour, creep and fatigue resistance, residual stress, hydrogen embrittlement etc [24, 29, 42]. Also, microstructural characteristics, oxidation and pitting corrosion resistance [18, 23, 37, 43], thermal cycling and failure analysis under different modes such as creep, fatigue and corrosion have to be done utilizing standard methods to comply with national and international standards and guidelines [9, 14, 19, 42]. Also, the machining methods such as rigid machining, belt grinding, robotic belt grinding etc induce deformations and surface defects in superalloys [44]. Another important aspect towards utilization of superalloy for manufacturing industrial gas turbine engines is the amenability to welding to meet the design demands. The commonly used welding processes are gas tungsten arc, electron beam, laser powder deposition and friction welding. The welding process leads to defects such as deformation, post wedding cracking etc which are dependent on alloy composition, grain size, pre and post welding treatment, type of welding process etc [10, 15, 21, 30]. For example, a turbine blade made with CM247LC superalloy by directional solidification casting process was evaluated for its weldability by gas tungsten arc welding and hot cracking susceptibility. The generated cracks were recognized as liquation cracks evident from the microstructural analyses and thermodynamic calculations [21].

The major producers and institutions of superalloys in India are Hindustan Aeronautics Limited (HAL), National Aerospace Laboratories (NAL), Defence Metallurgical Research Laboratory (DMRL), Steel Authority of India Limited (SAIL), Mishra Dhatu Nigam Limited (MIDHANI), National Metallurgical Laboratory (NML), Bharat Heavy Electricals Limited (BHEL), Nuclear Fuel Complex (NFC), Gas Turbine Research Establishment (GTRE), Bharat Forge, Jindal Stainless etc for defence, aerospace, power and industrial sectors.

5. Failure analysis of superalloy blades

The failure analysis and case studies on superalloy made turbine blades which are used in various commercial jet engines, gas turbines etc are discussed in this subsection [9, 45]. The life time of a blade depends upon various factors such as microstructure, design, operation environment of elevated temperature, fuel, air contamination; elevated mechanical stress arising from centrifugal force, vibratory and flexural stress; and enhanced thermal stress from thermal gradients [6-8]. The high pressure combustion turbine 1st stage blade of aero engine manufactured from c 6 y Ni based superalloy showed premature tip cracking resulted from a combination of environmental and thermal stress which was studied through liquid penetrant method. The service induced tip cracking can be rectified by applying oxidation and abrasion resistance coating in blade tip during maintenance and refurbishment for engine functioning and safety of the aircraft [7]. Using metallographic fractal analysis, the microstructure damage identified from carbide precipitation, cuboidal γ' presence and structural changes was evaluated in Ni superalloy matrix of gas turbine blade [46]. The static stress failure of gas turbine blade composed of Ni superalloy was accounted to fatigue strength reduction arising from hot corrosion, erosion and fatigue. It was probed using metallographic methods, including microstructure, fractography, X-ray crystal studies; and lab and bench strength tests and strength calculations [47]. The failure analysis of 1st stage turbine blade of gas turbine manufactured with Ni superalloy IN738 was carried out using visual examination, scanning electron microscopy (SEM) fractography, chemical and microstructure analyses, X-ray diffraction (XRD) and Finite Element Software ANSYS based stress evaluation. The combined mechanisms of erosion, oxidation, overheating and hot corrosion were accounted for blade failure [48]. The fatigue failure of aluminized Ni based superalloy turbine blades was evaluated based on result obtained from cross sectional microstructure, elemental composition and advanced intelligent algorithms and micro defect transfer mechanism was accounted for the surface cracking failure [41]. The cracking behaviour of Ni based superalloy, K403 turbine blade evaluated under joint high and low cycle fatigue (CCF) states indicated various microstructural factors, grain boundary interaction and cracking mode transformation from transgranular to intergranular and the data was obtained from optical microscopy (OM), SEM and energy dispersive X-ray spectroscopy (EDS) studies [49]. The creep and accelerated failure of K403 superalloy turbine blade was accounted to dendrite separation, γ' phase rafting behaviour and development of void migration mechanism [3]. The failure evaluation of 1st stage blade of gas turbine fabricated with CM88Y Ni based superalloy was studied by visual inspection, stereomicroscopy and erosion, thermos barrier coating damage and microstructure degradation were observed [2]. In a case study carried out on failure of 2nd stage Ni based superalloy (Udimet 500) gas turbine blade which experienced uncontrolled combustion and overheating in power plant indicated dynamic fracture, microstructure deterioration of airfoil, which were investigated employing OM, field emission SEM and EDS [5]. In another study carried out on large scale industrial 3rd stage gas turbine blades made from MCrAlY coated Ni based superalloy operated for 47,000 h, the microstructure characterization, surface oxidation, interfacial diffusion; and elemental diffusion on microstructure was studied with SEM, EDS and electron probe microanalyzer (EPMA), respectively and non uniform microstructural degradation was observed [6]. In Ni based superalloy made turbine blade operated for 10,000 h at 800-1000 °C inlet temperature, the surface pitting and cracking was noted through SEM, XRD and transmission electron microscopy (TEM) techniques. The turbine blade failure was accounted to high temperature oxidation and corrosion and thermal fatigue resulting in the formation of topologically dense phase of σ [4].

Conclusions

Superalloys are advanced materials designed to withstand extreme conditions, making them crucial for high performance applications like turbine blades in jet engines, gas turbines, and other high temperature environments. They are engineered to maintain strength and resist degradation under high stress and temperature. An overview on the types of superalloys commonly used for blade applications, their manufacturing techniques and blade failure analysis are detailed. In response to growing national/domestic and international demand in aerospace and power generation sectors, more research is needed in development of new superalloys, composition tailoring/customization/custom alloy development, improvements in manufacturing techniques, compliance with national and international standard certification agencies, stringent quality control, potential in renewable energy technologies, waste blade recycling [28] etc. Also, simulation, prediction and machine learning models on turbine blade working environment, service conditions, service evaluation, microstructure degradation, deformation behaviour, durability, residual lifespan etc are needed [8, 20, 50].

References

[1] I. Alrowwad, X. Wang, N. Zhou, Numerical modelling and simulation analysis of wind blades: A critical review, Clean Energy. 8 (2024) 261-279.https://doi.org/10.1093/ce/zkad078

[2] M. Najmi, S.M.H. Mirbagheri, CM88Y superalloy blade metallurgical degradation in a gas turbine, Eng. Fail. Anal. 146 (2023) 107110.https://doi.org/https://doi.org/10.1016/j.engfailanal.2023.107110

[3] L. Han, P. Li, S. Yu, C. Chen, C. Fei, C. Lu, Creep/fatigue accelerated failure of Ni based superalloy turbine blade: Microscopic characteristics and void migration mechanism, Int. J. Fatigue. 154 (2022) 106558.https://doi.org/https://doi.org/10.1016/j.ijfatigue.2021.106558

[4] S. Han, B. Choe, D. Kim, J. Kim, K. Choi, Y. Kim, The effect of TCP-σ precipitates on surface pitting and cracking in a Ni-based superalloy turbine blade, Eng. Fail. Anal.158 (2024) 107989. https://doi.org/https://doi.org/10.1016/j.engfailanal.2024.107989

[5] H.Mohammadi, S. Khani Moghanaki, A. Fallah Sheykhlari, Effects of uncontrolled combustion in a power plant gas turbine on fracture surface features and microstructural evolutions of a nickel-based superalloy blades, Eng. Fail. Anal. 160 (2024) 108175.https://doi.org/https://doi.org/10.1016/j.engfailanal.2024.108175

[6] Y. Chen, Z. Yao, J. Wang, J. Dong, M. Ren, J. Peng, H. Yang, L. Leng, Microstructural evolution and diffusion mechanism of MCrAlY coated nickel-based superalloy turbine blades after serviced for 47,000 h, Surf. Coat. Technol. 493 (2024) 131288.https://doi.org/https://doi.org/10.1016/j.surfcoat.2024.131288

[7] Y.J. Xie, M.C. Wang, G. Zhang, M. Chang, Analysis of superalloy turbine blade tip cracking during service, Eng. Fail. Anal. 13 (2006) 1429-1436.https://doi.org/10.1016/j.engfailanal.2005.07.022

[8] K. Liu, Y. Tan, F. Wang, J. Zhao, Y. Chen, Z. Zhang, H. Wei, The unique influence of1000°C high-speed rotation simulated turbine blade working conditions on microstructural degradation of single-crystal superalloy NiAlReRu, J. Mater. Res. Technol. 32 (2024) 2386-2394.https://doi.org/https://doi.org/10.1016/j.jmrt.2024.08.081

[9] V. Infante, J.M. Silva, M. De Freitas, L. Reis, Failures analysis of compressor blades of aeroengines due to service, Eng. Fail. Anal. 16 (2009) 1118-1125.https://doi.org/https://doi.org/10.1016/j.engfailanal.2008.07.005

[10] M.B. Henderson, D. Arrell, R. Larsson, M. Heobel, G. Marchant, Nickel based superalloy welding practices for industrial gas turbine applications, Sci. Technol. Weld. Join. 9 (2004) 13-21. https://doi.org/10.1179/136217104225017099

[11] G. Gudivada, A.K. Pandey, Recent developments in nickel-based superalloys for gas turbine applications: Review, J. Alloys Compd. 963 (2023) 171128-171128.https://doi.org/10.1016/j.jallcom.2023.171128

[12] G. Prashar, K. Thakur, S. Singh, P. Singh, V.K. Srivastava, Superalloys for high temperature applications: An overview, AIP Conf. Proc. 2986 (2024).https://doi.org/10.1063/5.0192482

[13] H. Xu, T. Huang, C. Ai, C. Zhang, J. Zhang, L. Liu, Effect of thickness on microstructure of thin-walled nickel-based single-crystal superalloy castings, J. Mater. Res. Technol. 29 (2024) 2912-2917.https://doi.org/https://doi.org/10.1016/j.jmrt.2024.02.043

[14] M. Motamedi, M. Nikzad, M. Nasri, Molecular dynamics simulation of superalloys: A review, Arch. Comput. Methods Eng. 31 (2024) 2417-2429.https://doi.org/10.1007/s11831-023-10051-w

[15] N. Qian, M. Jamil, W. Ding, Y. Fu, J. Xu, Thermal management in grinding of superalloys – A critical review, JIMSE. 5 (2024) 3-33. https://doi.org/10.1108/JIMSE06-2024-0019

[16] Q. Tan, K. Liu, J. Li, S. Geng, L. Sun, V. Skuratov, A review on cracking mechanism and suppression strategy of nickel-based superalloys during laser cladding, J. Alloys Compds. 1001 (2024) 175164.https://doi.org/https://doi.org/10.1016/j.jallcom.2024.175164

[17] Dilkush, M. Raffi, G.M. Reddy, K.S. Rao, Comparative studies on microstructure, mechanical and pitting corrosion of post weld heat treated IN718 superalloy GTA and EB

welds, IOP conference series: Mater. Sci. Eng. 330 (2018)
012051.https://doi.org/10.1088/1757-899X/330/1/012051

[18] Dilkush, M. Raffi, G.M. Reddy, K.S. Rao, Effect of PWHT on microstructure, mechanical and corrosion behaviour of gas tungsten arc welds of IN718 superalloys, IOP conference series: Mater. Sci. Eng. 330 (2018) 012030.https://doi.org/10.1088/1757-899X/330/1/012030

[19] G. Mustafa, B. Li, S. Zhang, Cutting condition effects on microstructure and mechanical characteristics of Ni-based superalloys—a review, The Int. J. Adv. Manuf. Technol. 130 (2024) 3179-3209. https://doi.org/10.1007/s00170-023-12910-z

[20] X. Yang, M. Wang, D. Shi, Z. Li, Y. Fan, A multi-scale framework for life reduction assessment of turbine blade caused by microstructural degradation, Chin. J. Aeronaut. 37(2024) 186-200. https://doi.org/https://doi.org/10.1016/j.cja.2023.07.021

[21] E.J. Chun, Y.S. Jeong, K.M. Kim, H. Lee, S.M. Seo, Suppression of liquation cracking susceptibility via pre-weld heat treatment for manufacturing of CM247LC superalloy turbine blade welds, J. Adv. Join. Process. 4 (2021) 100069.https://doi.org/https://doi.org/10.1016/j.jajp.2021.100069

[22] R. Darolia, Development of strong, oxidation and corrosion resistant nickel-based superalloys: Critical review of challenges, progress and prospects, Int. Mater. Rev. 64(2018) 355-380. https://doi.org/10.1080/09506608.2018.1516713

[23] I.G. Akande, O.O. Oluwole, O.S.I. Fayomi, O.A. Odunlami, Overview of mechanical, microstructural, oxidation properties and high-temperature applications of superalloys, Mater. Today Proc. 43 (2021) 2222-2231.https://doi.org/https://doi.org/10.1016/j.matpr.2020.12.523

[24] H. Hamdi, H.R. Abedi, Thermal stability of Ni-based superalloys fabricated through additive manufacturing: A review, Journal of Mater. Res. Technol. 30 (2024) 44244476. https://doi.org/https://doi.org/10.1016/j.jmrt.2024.04.161

[25] X.Z. Pan, X.M. Chen, M.T. Ning, Analysis of hot tensile fracture and flow behaviors of Inconel 625 superalloy, Materials 17 (2024) 473

[26] J.I. Sword, A. Galloway, A. Toumpis, Analysis of environmental impact and mechanical properties of Inconel 625 produced using wire arc additive manufacturing, Sustainability. 16 (2024) 4178.

[27] W. Song, J. Yang, J. Liang, N. Lu, Y. Zhou, X. Sun, J. Li, A new approach to design advanced superalloys for additive manufacturing, Addit. Manuf. 84 (2024) 104098.https://doi.org/https://doi.org/10.1016/j.addma.2024.104098

[28] Y. Li, Y. Tan, J. Zhao, G. Dong, P. Li, J. Qiang, Technical research on recycling waste blades of single crystal superalloy through ultrasonic alkali cleaning combined with electron beam smelting, Sep. Purif. Technol. 331 (2024) 125585.https://doi.org/https://doi.org/10.1016/j.seppur.2023.125585

[29] S.S. Kumar, M.S. Rao, I. Balasundar, A.K. Singh, T. Raghu, G.M. Reddy, Compressive behaviour of a nickel superalloy Superni 263 honeycomb sandwich panel, J. Sandw. Struct. Mater. 22 (2020) 1426-1449. https://doi.org/10.1177/1099636218786438

[30] V. Ajay, M. Naveen Kumar, N. Kishore Babu, T. Mahesh Kumar, K. Vamshi Krishna,G.M. Reddy, Rotary friction welding of Inconel 718– AISI 304 stainless steel dissimilar joint, Mater. Sci. Technol. 39 (2023) 1950-1960.https://doi.org/10.1080/02670836.2023.2187146

[31] F. Wang, Y. Liu, Q. Yang, D. Ma, D. Li, Microscale stray grains formation in single crystal turbine blades of Ni-based superalloys, J. Mater. Sci. Technol. 191 (2024) 134145. https://doi.org/https://doi.org/10.1016/j.jmst.2024.01.012

[32] R.A. MacKay, Alloy design challenge development of low density superalloys for turbine blade applications, NASA, Glenn Research Center, Cleveland, Ohio, 2009.

[33] B.H. Prasad, G.M. Reddy, A.K. Das, K.G. Prashanth, Fiber laser welded cobalt super alloy L605: Optimization of weldability characteristics, Materials. 15(2022)7708.

[34] S.C. Krishna, N.K. Gangwar, A.K. Jha, B. Pant, P.V.Venkitakrishnan, On the direct aging of iron based superalloy hot rolled plates, Mater. Sci. Eng. A 648 (2015) 274279. https://doi.org/https://doi.org/10.1016/j.msea.2015.09.073

[35] N. Parkes, R. Dodds, A. Watson, D. Dye, C. Hardie, S.A. Humphry-Baker, A.J.Knowles, Tungsten-based bcc-superalloys: Thermal stability and ageing behaviour, Int.J. Refract. Met. Hard Mater. 113 (2023) 106209.https://doi.org/https://doi.org/10.1016/j.ijrmhm.2023.106209

[36] A.J. Knowles, D. Dye, R.J. Dodds, A. Watson, C.D. Hardie, S.A. Humphry-Baker,Tungsten-based bcc-superalloys, Appl. Mater. Today. 23 (2021) 101014.https://doi.org/https://doi.org/10.1016/j.apmt.2021.101014

[37] C. Zhang, P. Wang, Z. Wen, Z. Xu, P. He, Z. Yue, Study on creep properties of nickel based superalloy blades based on microstructure characteristics, J. Alloys Compds. 890(2022) 161710. https://doi.org/https://doi.org/10.1016/j.jallcom.2021.161710

[38] P.H. Huang, L. Shih, H.M. Lin, C.I. Chu, C.S. Chou, Novel approach to investment casting of heat-resistant steel turbine blades for aircraft engines, The Int. J. Adv. Manuf. Technol. 104 (2019) 2911-2923. https://doi.org/10.1007/s00170-019-04178-z

[39] V. Dhinakaran, J. Ajith, A.F.Y. Fahmidha, T. Jagadeesha, T. Sathish, B.Stalin, Wire arc additive manufacturing (WAAM) process of nickel based superalloys– A review, Mater. Today Proc. 21 (2020) 920-925.https://doi.org/https://doi.org/10.1016/j.matpr.2019.08.159

[40] Z. Alam, C. Parlikar, M. Kumawat, S.G. Lakshmi, D. Das, High temperature resistant coatings for strategic aero space applications, Def. Sci. J. 73 (2023) 171-181.

[41] J. Wen, C. Fei, S.Y. Ahn, L. Han, B. Huang, Y. Liu, H.S. Kim, Accelerated damage mechanisms of aluminized superalloy turbine blades regarding combined high-and-low cycle fatigue, Surf. Coat. Technol. 451 (2022) 129048.https://doi.org/https://doi.org/10.1016/j.surfcoat.2022.129048

[42] K. Gopinath, A.K. Gogia, S.V. Kamat, R. Balamuralikrishnan, U. Ramamurty, Tensile properties of Ni-based superalloy 720Li: Temperature and strain rate effects, Metall. Mater. Trans. A. 39 (2008) 2340-2350. https://doi.org/10.1007/s11661-008-9585-3

[43] K. Sahithya, I. Balasundar, P. Pant, T. Raghu, H.K. Nandi, V. Singh, P. Ghosal, M.Ramakrishna, Deformation behaviour of an as-cast nickel-base superalloy during primary

hot working above and below the gamma prime solvus, Mater. Sci. Eng. A. 54 (2019) 521-534. https://doi.org/https://doi.org/10.1016/j.msea.2019.03.083

[44] Z. Wang, L. Zou, L. Duan, X. Liu, C. Lv, M. Gong, Y. Huang, Study on passive compliance control in robotic belt grinding of nickel-based superalloy blade, J. Manuf. Process. 68 (2021) 168-179.https://doi.org/https://doi.org/10.1016/j.jmapro.2021.07.020

[45] P. Caron, O. Lavigne, Recent studies at Onera on superalloys for single crystal turbine blades, Onera Aeros. Lab. J. 3 (2011) 1-14

[46] M. Tarafder, M. Sujata, V.R. Ranganath, S. Tarafder, S.K. Bhumik, Microstructural damage evaluation in Ni-based superalloy gas turbine blades by fractal analysis, Proced. Eng. 55 (2013) 289-294.https://doi.org/https://doi.org/10.1016/j.proeng.2013.03.256

[47] R. Dewangan, J. Patel, J. Dubey, P.K. Sen, S. Bohidar, Gas turbines blades — A critical review of failure on first and second stages, Int. J. Mech. Eng. Robot. Res. 4 (2015)216-223.

[48] S. Rani, A.K. Agrawal, V. Rastogi, Failure investigations of a first stage Ni based superalloy gas turbine blade, Mater. Today Proc. 5 (2018) 477-486.https://doi.org/https://doi.org/10.1016/j.matpr.2017.11.108

[49] L. Han, Y. Wang, Y. Zhang, C. Lu, C. Fei, Y. Zhao, Competitive cracking behavior and microscopic mechanism of Ni-based superalloy blade respecting accelerated CCF failure, Int. J. Fatigue. 150 (2021) 106306.https://doi.org/https://doi.org/10.1016/j.ijfatigue.2021.106306

[50] S. Lu, Y. Shao, W. Zheng, L. Li, Q. Feng, A prediction method for local creep strain of directionally solidified superalloys and turbine blades, Int. Symp. Superalloys. SpringerNature Switzerland, Cham, 2024, pp. 512-521

Superalloys: Fundamentals and Applications
Materials Research Foundations 178 (2025) 291-315

Materials Research Forum LLC
https://doi.org/10.21741/9781644903698-15

Chapter 15

Superalloys for Space Application

Soheila Asadzadeh-Khaneghah[1], Aziz Babapoor[1*], Seyyed Mojtaba Mousavi[2**]

[1]Department of Chemical Engineering, University of Mohaghegh Ardabil, Ardabil, Iran

[2]Department of Chemical Engineering, National Taiwan University of Science and Technology, Taipei 10607, Taiwan

*babapoor@uma.ac.ir

Abstract

During the last decades, there has been remarkable advancement in the use of high temperature applications in the aerospace industry. The requirement for materials that can withstand harsh conditions during their usage has attracted considerable consideration in multiple applications. For instance, the temperature of gas in parts of the aircraft engine may increase to 1090 °C. Hence, different procedures have been represented to impressively reduction the metallic components temperature. For this purpose, superalloys introduced as group of high-temperature advanced alloys, which are in high demand due to their high durability and strength. Superalloys have helped us in the world conquering from the depths of the ocean and the earth to space. These superalloys are often utilized in equipment with high-temperature such as a jet engine, aircraft engine and gas turbines engines. Superalloys were introduced as heat-resisting alloys, which are resistant to surface destruction at temperatures higher than 650 degrees Celsius, and also have of great mechanical strength. Thus, in this chapter, we presented detailed descriptions about the types of applications of superalloys in the space industry. The results of various prominent reports show that the demand for superalloys is currently increasing due to them much ability and strength at high temperatures and creep withstand in aircraft and aerospace applications.

Keywords

Superalloy, Aerospace Industry, High Temperature Alloys, Jet Engines, Aircraft, Aircraft Gas Turbine Engines

Contents

1.   Introduction

When safe operation at high temperatures is a required, most of the metallic materials and alloys such as ferritic steels, ceramics, titanium alloys, silicon carbide and silicon nitride which are widely utilized in industry, has lost their excellence and do not meet design standards [1, 2]. Because, high temperature also stimulates and accelerates corrosion that negatively affects component integrity in the passage of time [3].

Recently, there has been substantial enhance in usage of high temperature applications in numerous industries, containing petrochemical, aerospace, power plant, nuclear reactor sectors and energy [4–6]. Hence, without any considerable analogues, superalloys become the prime option when the safe performance and strength must be met at high temperatures. Superalloys are surely "super" because they present unique properties in elevated temperature that enable engineers to be cost-effectiveness and reliability. Moreover, high operative temperatures, high mechanical strength, great fatigue resistance, creep and usually oxidation resistance and high corrosion even at high temperatures are the main improved characteristics of the superalloys. These characteristics make superalloys appropriate for utilizing in submarines, ships, hot metals working dies, industrial gas turbines, components of aircraft engines, high-temperature chemical processing industries, coal conversion equipment, marine turbine engines, jet engine, heat exchangers and reactors [7, 8].

Although it is not convenient to draw a definite boundary for the temperature above which the use of superalloys becomes unavoidable, these materials are often used at temperatures above 650 degrees Celsius [9]. Others introduce superalloys as a heat-resisting alloy capable of resistance to surface degradation and excellent mechanical strength even at temperatures more than 650 °C (1200 °F) for a long time [10, 11]. It is worth noting that, when a constructed alloy can be introduced as a superalloy if its absolute melting temperature be about 0.7-folds lower than its operating temperature [12]. It is good to know that superalloys are also known as powder metallurgy, wrought, and cast superalloy.

Given that Co, Ni and Fe have excellent resistance to corrosion and thermal creep deformation; therefore, these elements are used as superalloy base alloying elements [13, 14]. In general, superalloys are classified to three kinds based on the parent metal in the alloy: Co-based, Fe-based, and Ni-based superalloys [7].

1. Superalloys based on Nickel:

✓ High strength due to maintaining its face-centered cubic structure.

✓ Greater resistance against to creep and corrosion.

2. Superalloys based on Cobalt:

- ✓ Superior melting point compared to iron or nickel based alloys
- ✓ Higher corrosion and thermal resistance in comparison to iron or nickel based alloys
- ✓ Higher heat fatigue resistance and weldability compared to nickel based alloys

3. Superalloys based on Iron:

- ✓ High firmness at a temperature of 25 degrees Celsius (room condition).
- ✓ High resistance to corrosion, wear, creep, oxidation [15].

Most of the superalloy tonnage utilized is: gas turbines in aircraft, space vehicles, medical applications, heat-treating tool, petrochemical and chemical industries, metals processing mills, and liquefaction systems, etc. [8].

Based on published literatures, thanks to their features, nowadays, superalloys demonstrate the highest percentage of materials used in construction of gas turbine engines especially in the aerospace industry [7, 16]. A review of the most important space applications of superalloys in aircraft gas turbine engine as a space vehicle is indicated in Figure 1 [8].

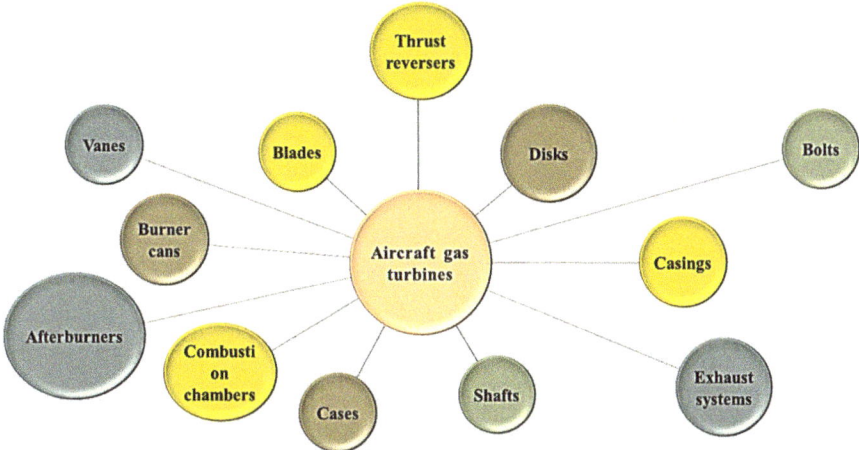

Figure 1. *Applications of superalloys in in different components of the aircraft gas turbine engine.*

Engine turbine is one of researchers' favorite parts in aerospace industry. Aircrafts engines turbine receives fuel and combustion air from combustion chamber directly. This gas is under high pressure and it contains corrosive substances and it also has high temperature. In current engines of civilian aircrafts, the temperature can reach up to 1500°C that can be a hostile environment for most substances. Suitable substances should have good strength in high temperatures. The superalloys have been used for such functions since the 1940s. In addition to the above, there is great pressure on the aerospace industry to reduce emissions as ecological targets across the globe become more stringent. Lowering fuel consumption also benefits airlines by reducing fuel costs and/or increasing the number of passengers that can be transported [17, 18].

The purpose of this chapter is to provide data on the latest scientific achievements about space applications of superalloys. Figure 2 shows the trends of using superalloys in space applications with passing of time, from 2010 to 2024. As noticed, the "superalloy" keyword demonstrates the most relevant with "nickel-based superalloys" and "high temperature applications" keywords, which demonstrates usage of superalloys in the field of high temperature applications. These results were obtained from Scopus database and drawn with VOSviewer software.

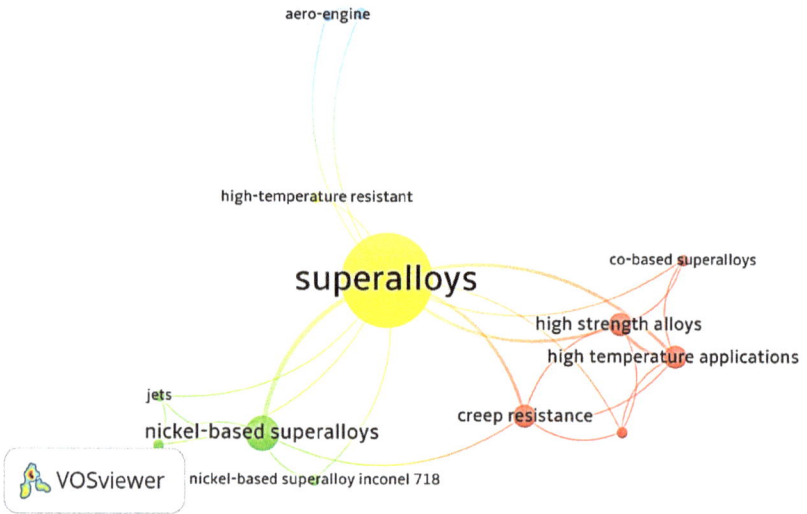

Figure 2. *Keywords co-occurrence analysis of space applications of superalloys from 2010 to 2024.*

2.   Overview of superalloys

First, it is better to have a general definition of superalloys; superalloys are alloys which are full of at least one of the elements iron, cobalt and nickel, which Figure 3 shows the superalloys available in each category [20–22]. Also superalloys maintain surface and structural stability at high temperatures, in severe environment and under high stress [19].

Superalloys: Fundamentals and Applications
Materials Research Foundations 178 (2025) 291-315

Materials Research Forum LLC
https://doi.org/10.21741/9781644903698-15

Types of Superalloys	Examples
Ni based Superalloys	• Nimonic (75, 80A, 90, 115, 115, 263, 942, PE.11, PE.16, PK.33) • Inconel (587, 597, 600, 617, 625, 706, 718, X750) • Udimet (400, 500, 520, 630, 700, 710, 720) • Hastelloy (C-22, G-30, S, X) • Rene (41, 95) • Pyromet 860 • Haynes 230 • Cabot 214 • Waspaloy • Astroloy • Unitemp • Monel
Fe based superalloys	• Incoloy (800, 801, 802, 807, 825, 903, 907, 909) • Haynes 556 • Alloy 901 • Discaloy • A-286 • H-155 • V-57
Co based superalloys	• Haynes 188 • MAR-M98 • Stellite 6B • MP35N • MP35N • MP159 • Elgiloy • L-605

Figure 3. Types of Ni, Co, Fe-based superalloys.

High temperatures are a bigger problem than most people realize. If you want something to go faster, it also gets hotter. For this reason, metallurgists and design engineers are all the time challenged by the industry's need to develop and design high-strength materials with resistance against wear and good corrosion for elevated-temperature applications. In recent decases, the application and improvement of stainless steel laid the groundwork for meeting high-temperature engineering needs. Yes, there are materials that can survive warmer temperatures (such as ceramics and tungsten), but those can only work under certain condition. Therefore, researchers soon realized that they have a limited power [15]. Hence, the metallurgical industry created something worth mentioning as "superalloys" of stainless types in response to growing demands. Superalloys are the best in the oxidation resistance, strength, and more at high temperatures [23]. According to concept of Carnot efficiency and the second law of thermodynamic, temperature difference is related to energy efficiency. In other words, higher temperature means less energy loss lost, which superalloys have this characteristic and implement this concept completely.

In addition to the above, creep is usually a critical property when designing high temperature alloys. Creep is how a material deforms at long times and high temperatures. You might have a higher-melting point and stronger alloy than a superalloy, but it won't last long. Because, at high temperatures, the materials is practically destroyed because, over hundreds of hours of use, the materials creep and creep each time until it is no longer usable. But, superalloys materials don't destroy because they experience excessive force or very high temperature [24]. In the following, we will examine the properties of superalloys in more detail.

2.1 Properties of superalloys

Superalloys have been usually utilized for types of industrial applications. These materials have been known as high-performance alloy owing to their unique properties compared to most alloys, which among these properties, the following can be mentioned:

- ✓ Excellent mechanical strength and wear resistance at high temperature
- ✓ High strength and thermal resistant at elevated temperature
- ✓ Excellent corrosion resistance at very elevated temper
- ✓ Unique oxidation resistance at very high temperature
- ✓ Higher resistance to thermal creep deformation
- ✓ Excellent cryogenic temperature properties
- ✓ High toughness and ductility
- ✓ Good surface stability.

Superalloys are expensive systems, but they are very unique for elevated temper applications. Air transportation and modern power plants would not be possible without superalloys. Hence, nowadays, the most widely used is in aircraft turbine components that must withstand exposure to high temperatures and highly oxidizing environments for reasonable periods. For this reason, in the next parts, space applications of superalloys will discuss in detail.

3. Space applications of superalloys

It is certainly true that the aerospace field has always had a particular focus on the development of systems and materials that are unmatched in terms of reliability and safety. But, these standards need significant systemic research efforts to deeply study issues and to find optimal solutions. Therefore, systems and materials optimized and/or developed in the field of aerospace are the most excellent options for applications also in other fields. Given that, superalloys are the pinnacle of modern metallurgy and thanks to their unique properties, they are nowadays used in many cases. But superalloys have appeared more successful in the aerospace industry than in other industries. As observe in Figure 4, two-thirds of superalloy productions are used by the aerospace industry for the produce of jet/rocket engines, aircraft gas turbine engine and mainly in the turbine blades. Hence, in this chapter, some applications of superalloys involved in aerospace will be discussed.

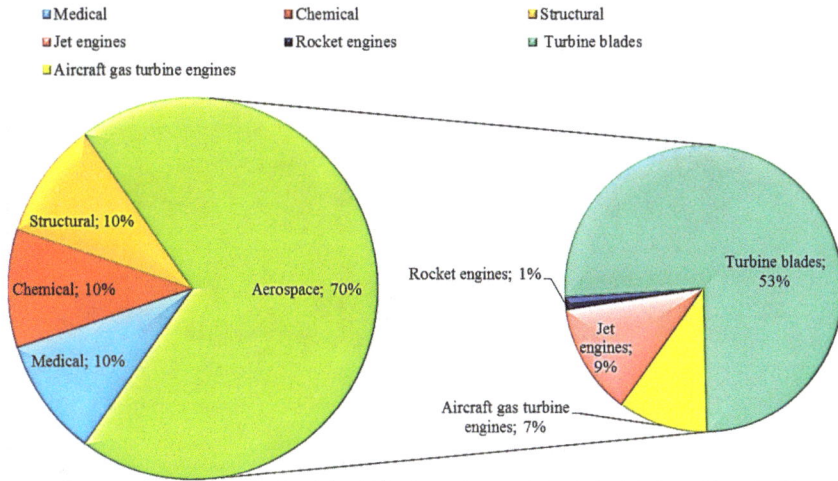

Figure 4. Percentage of superalloys used in different industries.

3.1   Application of superalloys in space vehicles

Many recent material researches and development has focused on inventions that would allow a space vehicle to re-enter a high-temperature atmosphere from orbit. The superalloys have been converted to the primary materials for critical components in other aerospace applications and re-entry vehicles [25–29].

For instance, in the research done with Lee and co-workers [30], three HIP superalloys (Rene 95, Inconel 718 and Astroloy) which are usually utilized for hot pressurized chambers, turbine disks and gas generators were investigated at different tempers for improving a constitutive equation.

3.1.1   Jet engine

Since the combustion cycle is more complete; hence, it makes the jet engines and new gas turbine more efficient at operating at higher tempers. However, working at high tempers has its own complications. Hence, the major purpose is to careful selection materials that can withstand at tempers higher than their melting point, destruction of material during their lifetime and extreme corrosion [2]. Resistance capability of superalloys in extreme operating environments, where high-temperature creep, oxidation resistance and corrosion are important, has converted them selected materials for gas turbine engines which operate at elevated tempers [31]. Many other agents like service efficiency and economic feasibility also affect the superalloys popularity in jet engines and gas turbine. The important point is that the usage of superalloys in gas turbine enhances the working temper region from 490 ℃ to 704 ℃ [32]. It is worth mentioning that high-temper working not only enhances power output but also reduces emissions impressively [33]. Figure 5 demonstrates main components of level exposure of conventional gas turbine engines to heat.

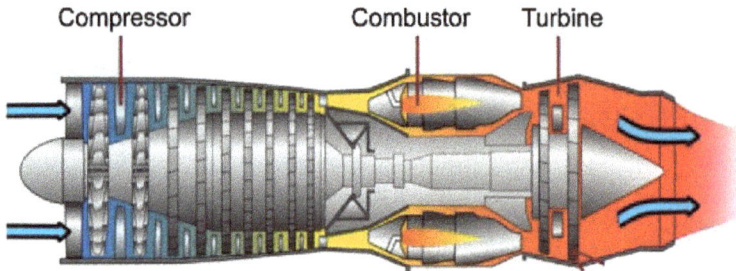

Figure 5. Schematic image of a gas turbine engine (Copyright [34]).

In the Second World War, the improvement of superalloys experienced rapid growth in the 1940s, which this leads to development of the jet engine. This striking development was due to industry's requirement for reliable, heavy and robust gas turbines and early applications of jet engines in military aircraft [2].

To make a jet engine, materials resistant to high temperature with a balance of good toughness and strength are required. Furthermore, utilized materials in hot sections of jet engines must maintain structural stability and high performances for long periods under terms of service including corrosive agents and high mechanical stresses [35].

In the mid-1940s, this led to the creation of a novel category of metal alloys called superalloys, which were maintained even at high temperatures due to their considerable structural features. Indeed, it can be stated like this, the history of the jet engine has been intrinsically linked to development of superalloys for which they were designed; so that, a modern jet aero plane could not fly without them [1].

In this regard, Miura and Kondo [36], in order to estimate stress distribution, temperature distribution and stress direction in service, the morphological change of γ' deposits in IN100 superalloy before and after aging in service in the first low pressure turbine blade of an engine was investigated. Finally, the front edge of 1st low pressure turbine blade tip was subjected to highest temperatures in service, but even so stresses were very low in all parts of blade.

In other study, Bennett et al. [37] investigated the environmental resistance of two superalloys CMSX-10 and CMSX-4 utilized in blade materials in air jet engines at temperature range between 800-1000 °C. Oxidation of CMSX-10 demonstrated dominant NiO formation on outer surface, while CMSX-4 demonstrated various oxidation regimes among interdendritic regions and dendrite core.

In another research, Sadbrock and colleagues [38] in 2015 conducted a systematic evaluation of surface preparation effect on oxidation at 815 degree Celsius after 440 hours of CMSX-4+Y and LDS-1101+Hf superalloys, which are usually in jet engine turbine blades. It has been observed that surface finishing has an obvious effect on formation of oxide at 815°C. Machined ground surfaces with low-stress after 440 h, generated polished surfaces with a mirror finishing, which demonstrate less oxidation resistance and developed much thicker NiO external scales with subscale of Cr_2O_3-spinel-Al_2O_3, while thin Al_2O_3 external scales compatible with elevated oxidation temperature.

In 2016, a paper was published by Smith and co-workers [39], in which two superalloys ME501 and ME3, were investigated to display the effect of the novel reinforcement mechanism. The noticeable difference was the value of Z phase constituents (Ta, Nb, W, Ti, Hf), which is 13% for ME501 and 9.1 wt% for ME3. Figure 6 demonstrates the compressive creep response for ME501 and ME3 at 760 degrees Celsius for [001] direction. In order to achieve high dimensional stability in the used materials in the turbine engine disc, it is necessary to minimize these plastic strains. Figure 6(a) display the creep curves, which express that considerably decreased creep resistance of ME3 compared with ME501 at 552 MPa and 760 °C. High-angle annular dark field (HAADF) STEM with high resolution demonstrates that faults spanning both the γ and γ′ phases appeared in ME3 were nanotwins, as observed in Figure 6(d). Anyway, while many examples of separated cumulative faults can be observed, which were restricted to γ′ phases (Figure 6(e)).

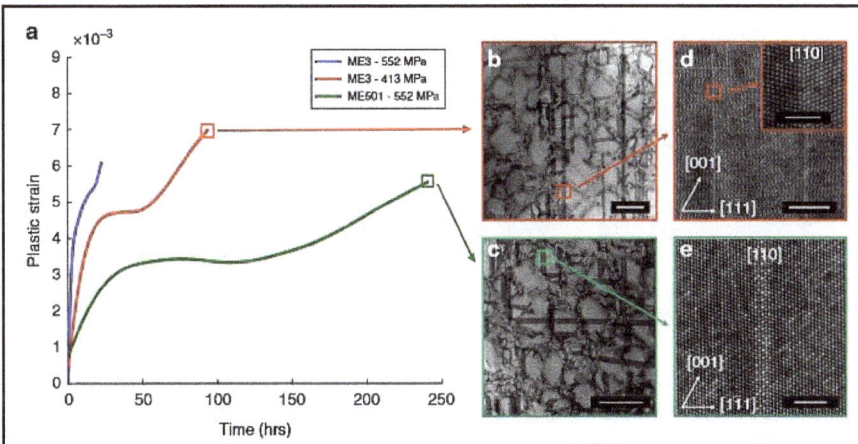

Figure 6. (a) Compression creep plots of ME501 and ME3 at 552 MPa and 760 °C. Zone axis BF-STEM image (b) ME3 and (c) ME501 crystals. HAADF–STEM images (d) a nanotwin in ME3 and (e) a SESF in ME501 (Copyright [39]).

As another attempt, in 2017, by Liu et al. [40], Incoloy A286 face-milled with cutting various agents, and fatigue life of modified surface and surface features were analyzed and measured. It has been found that cutting agents can significantly improve mechanical properties and surface integrity, and clearly effect fatigue life of workpiece. In such a way that compared with milled workpieces, polished specimens had much smaller grain size and higher hardness. Grain refinement and serious work hardening happened on all of face-milling machined surfaces.

As you can see, most of the items mentioned are based on nickel. Because nickel-based superalloys have good resistance at high temperatures. By removing grain boundaries, single crystal alloys increase resistance against creep considerably.

Likewise, abundant superalloys such as Inconel 625 [41], Allvac 718Plus [42], Haynes 282 [43], Inconel 718 [44], CMSX-4 [45], X-40 [46] and Inconel 625 [47] were utilized in jet engines for improving many properties.

Moreover, in Table 1, the chemical compositions of some utilized superalloys in jet engines are collected.

Table 1. Nominal compositions of several used superalloys in jet engines (wt. %).

Superalloy	Based on	Compositions (Weight%)										Application	Ref. (Year)
		Co	Al	Ni	C	Cr	Mo	Ti	W	Fe	Other		
ME3	Ni	20.6	3.2	50.76	0.056	12.7	3.9	3.5	2.0	-	Nb: 0.89 B: 0.033 Zr: 0.050 Mg: 0.001 Ta: 2.30	Jet engine disk	[48] (2010
PWA1484	Ni	9.92	5.69	Bal.	0.026	4.97	1.92	-	5.9	-	Ta: 8.57 Re: 2.89 Hf: 0.09	Turbine blades in jet engine	[49] (2010
ME-9	Ni	7.74	14.6	Bal.	0.63	8.18	0.95	-	2.31	-	Ta: 1.95 Re: 1.74 Hf: 0.05	Turbine blades in jet engine	[50] (2011
ME-15	Ni	7.31	15.1	Bal.	0.67	7.73	0.90	-	0.75	-	Ta: 1.97 Re: 0.46 Hf: 0.05		
ME-15	Ni	7.31	15.1	Bal.	0.67	7.73	0.90	-	0.75	-	Ta: 1.97 Re: 0.46 Hf: 0.05	Turbine blades in jet engine	[51] (2011
IN100	Ni	15.0	5.50	Bal.	0.18	10.0	3.00	4.70	-	-	B: 0.01 Zr: 0.06 V: 0.90	Turbine blades in jet engine	[52] (2011
ŻS-3DK	Ni	5.0-10.0	4.0-4.8	rest	0.07-0.12	11.0-15.0	3.5-5.0	2.5-3.5	-	< 2.0	Si < 0.4 Mn < 0.4	Jet engine exhaust flap	[53] (2013
INC 625	Ni	-	0.0	rest	0.0	22.4	9.3	0.0	-	1.00	Si: 0.18 Mn: 0.0		
ATI 718Plus	Ni	9.15	1.51	Bal.	0.020	17.88	2.69	0.76	1.03	9.72	Nb: 5.48	Hot rolled rings for the manufacture of rotating	[54] (2013

												components in jet engines	
CM247LC	Ni	9.00-9.50	5.30-5.70	Bal.	0.05-0.09	8.00-8.60	0.40-0.60	0.60-0.90	9.30-9.70	<0.25	Si < 0.03 Mn<0.10 P<0.015 S<0.015 Zr:0.007-0.22 V < 0.10 B: 0.01-0.02 Nb<0.10 Hf: 1.30-1.60 Ta:2.80-3.30	In gas turbines for electric power plant and jet engine	[55] (2013)
IN100 Billet	Ni	15.0	5.5	Bal.	0.18	10.0	3.00	4.70	-	-	V: 0.8 B: 0.014 Zr: 0.007	Slab billets	[56] (2015)
IN100 Disk	Ni	18.2	4.9	Bal.	0.07	12.1	3.22	4.22	-	-	V: 0.80 B: 0.02 Zr: 0.07	Industrial quality turbine disk	
ABD-1	Ni	10	5.8	Bal.	-	8	-	-	8.5	-	Re: 1.6 Ru: - Ta: 8.5	Jet propulsion	[57] (2016)
ABD-2	Ni	9	6.4	Bal.	-	4	-	-	7.4	-	Re: 5.6 Ru: 2.6 Ta: 5.6	Jet propulsion	
ABD-3	Ni	10	5.4	Bal.	-	13	-	-	5.0	-	Re: - Ru: - Ta: 7.0	Industrial gas turbines	
IN 792	Ni	8.90	3.40	Bal.	0.08	12.50	1.90	4.00	4.00	0.20	Ta: 4.00 Zr: 0.02 B: 0.02 Y: - Re: -		
PWA 1483	Ni	9.00	3.60	Bal.	0.07	12.00	1.90	4.20	3.80	-	Ta: 5.00 Zr: 0.03 B: 0.01 Y: - Re: -	Hot sections of gas turbines or jet engines	[58] (2017)
MCrAlY	Ni	25.0	10.0	Bal.	-	17.00	-	-	-	-	Ta: - Zr: - B: - Y: 0.40		

											Re: 1.50		
Rene 80	Ni	9.50	3.00	Bal.	0.16	14.00	4.00	5.00	4.00	-	Ta: - Zr: - B: 0.02 Y: - Re: -		
ZhS6k	Ni	4.76	5.22	Bal.	0.202	12.44	3.48	3.05	5.28	0.192	Si: 0.171 Sn: 0.164 B: 0.198	Turbine blades in former USSR DV-2 jet engine	[59] (2019
Rene 65	Ni	13.0	3.10	Bal.	-	16.0	4.00	3.70	4.00	1.00	Nb: 0.70 B: 0.016 Zr: 0.05	Different components of jet engine	[60] (2019
Inconel713C	Ni	0.04	5.93	Bal.	0.11	14.24	4.29	0.92	-	-	Nb: 2.45 Zr: 0.08 B: 0.012	Different components of jet engine	[61] (2019
Inconel 800	Ni	-	0.10	Bal.	0.02	21.5	-	0.28	-	46.3	Si: 0.48 P: 0.009 Mn: 0.47 Cu: 0.01 S: 0.004	Cooling holes in turbine blades	[62] (2020
Alloy 718	Ni	0.04	0.44	Bal.	0.43	17.7	3.00	0.93	-	21.8	Nb: 4.96 Mn:0.017 Si:0.06 B: 0.003 Cu:0.001 S: 0.004	Jet engine applications	[63] (2020
Inconel 718	Ni	0.27	0.48	53.53	0.03	18.45	3.02	0.95	-	Bal.	Nb: 5.31 Cu: 0.04 Mn: 0.06 Ta: 0.005 P: 0.007 B: 0.004 S: 0.002 Si: 0.09	In gas turbine engines	[64] (2023

3.1.2 Aircraft gas turbine engine

Environmental effects of air travel happen when aircraft engines produce heat, noise, particulates, and gas which can cause climate changes and decrease direct beam solar radiation gradually. Aircraft produce particulates and gases such as carbon dioxide, water vapour, carbon monoxide, hydro-carbon, sulphur oxide, nitrogen oxide, lead and carbon leading to reactions between these particulates and the atmosphere [65].

Superalloys: Fundamentals and Applications   Materials Research Forum LLC
Materials Research Foundations 178 (2025) 291-315   https://doi.org/10.21741/9781644903698-15

Moreover, engine turbine is one of researchers' favorite parts in aerospace industry. Aircrafts engines turbine receives fuel and combustion air from combustion chamber directly. This gas is under high pressure and it contains corrosive substances and it also has high temperature. In current engines of civilian aircrafts, the temperature can reach up to 1500°C that can be a hostile environment for most substances. Hence, suitable substances should have good strength in high temperatures. These parts are made of substances that are resistant to creep deformation in loads lower than the yield point of that substance. Because in high temperatures deformation such as creep increases. These substances must resist oxidation. Oxidation is a significant problem in high temperatures because atoms have the required energy to overcome activation energy for oxidation reactions [66].

For this reason, there is great pressure on the aerospace industry to reduce emissions as ecological targets. Because, aircraft engines need to be adapted to the environment and its functional goals, which such goals can be achieved by increasing thermal efficiency. Hence, various superalloys have been introduced for such functions [67–74].

In order to prove the importance of these contents, consider the data represented by Jithesh and co-workers [75], who evaluated the air oxidation and hot corrosion behavior of superalloy L605 at multiple temperatures. It was concluded that hot corrosion at 850 degrees Celsius and 750 degrees Celsius, L605 superalloys demonstrated heavy spallation and sputtering compared to a hot corroded superalloy of 650 degrees Celsius and air oxidized samples. Furthermore, from the thermogravimetric analysis, in comparison of all samples, a remarkable weight gain was seen for 850°C and 750°C hot corroded superalloys; this is owing to tendency of L605 superalloy to High temperature molten salt environment. It can also be concluded that after 50 cycles, a large increase in weight occurred for the 850°C hot-corroded superalloy, indicating lowest resistance to hot corrosion.

Designing superalloys with numerous targeted features is an important challenge owing to the complexity of the compositions. In 2023, Gao and co-workers [76], proposed mechanical performance at high temperature into learning of machine in order to optimize commercial superalloy (K403). This work was done in order to increase the temperature of hot parts in gas turbine engine and aircraft. Furthermore, several features containing the volume fraction of γ' precipitates, microstructure stability, processing window, density and freezing range were simultaneously optimized. Hence, ML superalloy was experimentally constructed. The creep rupture life of this superalloy compared with the commercial superalloy (K403), is improved nearly three times at 1025 °C and 975 °C. The yield strength and predicted creep rupture life at high-temperature with machine learning model has a good accommodation with experiments. Also, as observed in Figure 7., a developed creep rupture life of is observed for the ML superalloy, demonstrating new superalloy is considerably option for applications more than 1000 ∘C.

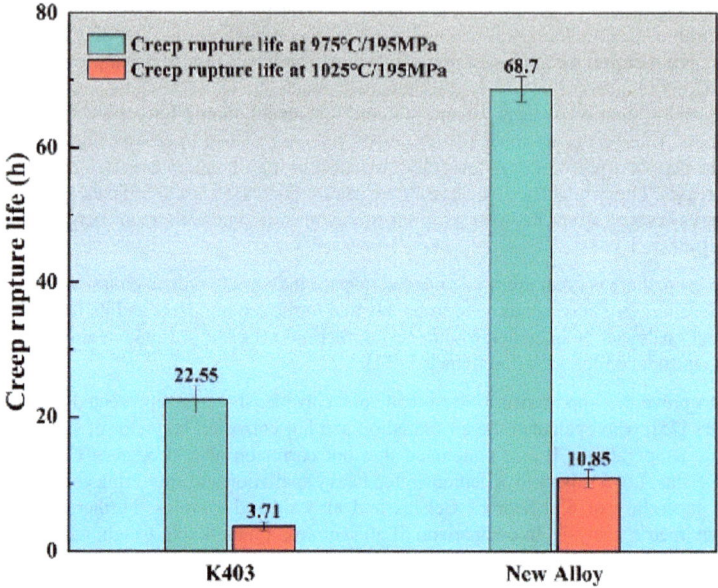

Figure 7. Creep rupture life of ML and K403 superalloys under various experiment conditions (Copyright [76]).

Likewise, abundant superalloys such as IN100 [77], ME3 [78], Nimonic alloy 90 [79], ZhS32-VI [80], alloy 718 [81], IN718 [82], IN625 [83], IN-738 [84], DZ125L [85] and GH3044 [86] were utilized in aircraft and aero-gas turbine engines for improving many properties.

Moreover, in Table 2, the chemical compositions of some utilized superalloys in aircraft and aero-gas turbine engines are collected.

Table 2. *Nominal compositions of several utilized superalloys in aircraft and aero-gas turbine engines (wt. %).*

Superalloy	Compositions (Weight%)										Application	Ref. (Year)
	Co	Al	Ni	C	Cr	Mo	Ti	W	Fe	Other		
Inconel 738 LC	8.00-9.00	3.20-3.70	rest	0.09-0.13	15.7-16.3	-	3.20-3.70	-	0.35	Nb: 0.60-1.10 Ta: 1.50-2.00 Zr: 0.03-0.08	Aircraft engines, gas turbines and turbo blowers	[87] (2010)
DZ125	10.0	5.20	Bal.	0.10	8.90	2.00	1.00	7.00	-	Ta: 3.80 Hf: 1.50 B: 0.015	Enhance of the wheel shape retention and suppress thermal damage in aircraft engine components	[88] (2016)
UDIMET 500	18.5	2.90	54.0	-	18.0	4.00	2.90	-	-	-	Gas turbine blades of an aircraft engine	[89] (2016)
IN 718	-	0.50	rest	-	19.0	3.05	0.90	-	18.5	Nb: 5.13 Mn: 0.18 Si: 0.18	Cooling holes in aero-engine and gas turbine components	[90] (2019)
MD2	5.10	5.00	Bal.	-	8.00	2.10	1.30	8.10	-	Si:0.10 Hf: 0.10 Ta:6.00	Turbine blades in aircraft engines	[91] (2020)
C-263	19.2	0.57	Bal.	0.06	20.7	5.85	1.98	-	0.03	Si: 0.40 Mn: 0.56 Cu: 0.21 S: 0.007	Turbines, power plants, and aero-engine components	[92] (2020)

GTD222	18.50-19.50	1.10-1.30	Bal.	0.08-0.12	22.20-22.80	-	2.20-2.40	1.80-2.20	-	Ta: 0.90-1.10 Nb: 0.70-0.90 Zr: 0.005-0.020 B: 0.002-0.007	Gas turbine blades or complex parts of aero-engine	[93] (2021)
GTD-111	9.50	3.00	Bal.	0.10	14.0	1.50	4.90	3.80	-	Ta: 2.80 B: 0.014 Zr: 0.03	Turbine blades of aero-engine	[94] (2021)
IC10	12.0	5.90	Bal.	0.10	7.00	1.50	-	5.00	-	Ta: 7.00 Hf: 1.50 B: 0.015	Aero engines and gas turbines	[95] (2021)
CM247LC	9.41	5.49	Bal.	0.094	8.03	0.50	0.74	9.87	-	Ta: 2.90 Hf: 1.36	Hot ends in aero-engines	[96] (2022)
GH738	13.43	1.51	Bal.	0.036	19.03	4.62	3.13	-	0.09	Mn<0.05	Critical components of aero-engines and gas turbine	[97] (2022)
XH45MBTJuBP	-	1.00	45.3	1.00	15.0	4.00	2.00	2.50	27.9	Nb: 1.30 Si: 1.00 Mn: 1.00 S: 1.00 P: 1.00	Aircraft gas turbine engine parts	[98] (2023)
DZ22B	9.50	4.90	60.64	0.140	9.00	-	1.90	12.0	-	Nb: 0.90 Hf: 1.00 B: 0.015	Blade material for aircraft engines and gas turbines	[99] (2023)
GH4169	-	0.65	Bal.	0.05	18.5	2.96	1.02	-	18.8	S: 0.002 Si: 0.12 Nb: 4.85	Turbine blades, Turbine disks, combustion chambers of aero-	[100] (2023)

											engines and gas turbines	
Nimonic 80A	1.10	1.50	Bal.	0.06	20.0	-	1.60	-	2.00	Si: 0.40 Mn: 0.20	In aircraft engines, particularly in the combustion and turbine sections	[101] (2024)

4. Summary and conclusion

Superalloys demonstrate unique thermal, mechanical and chemical features like hot hardness, hot strength, high-temperature resistance, and maintaining mechanical and chemical features at high temperatures. Most importantly, superalloys have unique thermal-resistant features and retain their strength, stiffness and toughness at a temperature higher than that of other materials. Hence, these unique features make them a suitable option for elevated temperature applications especially gas turbine components of space vehicles such as aircrafts and jets. So that the superalloys should considered as one of the most advanced man-made alloys.

In this chapter, the space application of various superalloys at different condition at elevated temperature is the center of attention. Here, the constituent elements and effect on the different features of the superalloys at elevated temperature has been discussed.

5. Future outlooks and recommendations

Nevertheless, there are reasons that show that this category of materials still needs progress and development.

- Important questions remain about the metallurgical factors governing performance, as there is no well-developed theory for these effects.
- In addition, the chemical effects as determinant the adaptability of superalloys with thermal barrier coating systems have not yet been sufficiently explained. Bond coating chemistry is not optimized yet.
- The controlling factors of the creep performance of the disc alloys is remained great extent unknown, hence, they can be an excellent subject for subsequent studies.

Acknowledgement

We express our utmost gratitude to Mohaghegh Ardabili University - IRAN, for the financial support provided. We are extremely grateful to all the researchers who collaborated in completing this chapter of the book. Finally, we are very grateful to the publications that granted us the copyright to use the images.

Superalloys: Fundamentals and Applications   Materials Research Forum LLC
Materials Research Foundations 178 (2025) 291-315   https://doi.org/10.21741/9781644903698-15

Reference

[1] R.C. Reed, The Superalloys: Fundamentals and Applications, Cambridge university press, New York, 2008.

[2] H.A. Kishawy, A. Hosseini, Machining difficult-to-cut materials, Mater. Form. Mach. Tribol. 10 (2019) 973−978. https://doi.org/10.1007/978-3-319-95966-5

[3] R. Binali, A.D. Patange, M. Kuntoğlu, T. Mikolajczyk, E. Salur, Energy saving by parametric optimization and advanced lubri-cooling techniques in the machining of composites and superalloys: A systematic review, Energies. 15 (2022) 8313. https://doi.org/10.3390/en15218313

[4] D. Srinivasan, K. Ananth, Recent advances in alloy development for metal additive manufacturing in gas turbine/aerospace applications: A review, J. Indian Inst. Sci. 102 (2022) 311−349. https://doi.org/10.1007/s41745-022-00290-4

[5] J.Y. Lee, J. An, C.K. Chua, Fundamentals and applications of 3D printing for novel materials, Appl. Mater. Today. 7 (2017) 120−133. https://doi.org/10.1016/j.apmt.2017.02.004

[6] P. Vijayakumar, S. Raja, M.A. Rusho, G.L. Balaji, Investigations on microstructure, crystallographic texture evolution, residual stress and mechanical properties of additive manufactured nickel-based superalloy for aerospace applications: Role of industrial ageing heat treatment, J. Braz. Soc. Mech. Sci. Eng. 46 (2024) 356. https://doi.org/10.1007/s40430-024-04940-9

[7] N.M. Dawood, A.M. Salim, A review on characterization, classifications, and applications of super alloys, J. Univ. Babylon Eng. Sci. (2021) 53−62.

[8] A. Behera, Superalloys, in: A. Behera (Ed.), Advanced Materials: An Introduction to Modern Materials Science, Rourkela, India, 2022, pp. 225−261. https://doi.org/10.1007/978-3-030-80359-9_7

[9] M.J. Donachie, S.J. Donachie, Superalloys: A Technical Guide, Second ed., United States of America, ASM International, 2002. https://doi.org/10.31399/asm.tb.stg2.9781627082679

[10] M.C. Hardy, M. Detrois, E.T. McDevitt, C. Argyrakis, V. Saraf, P.D. Jablonski, J. A. Hawk, R.C. Buckingham, H.S. Kitaguchi, S. Tin, Solving recent challenges for wrought Ni-base superalloys, Metall. Mater. Trans. A. 51 (2020) 2626−2650. https://doi.org/10.1007/s11661-020-05773-6

[11] N.E. Prasad, R.J.H. Wanhill, Aerospace Materials and Material Technologies, Springer, Singapore, 2017. https://doi.org/10.1007/978-981-10-2143-5

[12] M.J. Donachie, S.J. Donachie, Selection of superalloys for design, in: M. Kutz (Ed.), Mechanical Engineers' Handbook: Materials and Mechanical Design, John Wiley & Sons, 2006. https://doi.org/10.1002/0471777447.ch8

[13] S. Gialanella, A. Malandruccolo, Aerospace Alloys, Cham, Springer Cham, Switzerland, 2020. https://doi.org/10.1007/978-3-030-24440-8

[14] M.P. Brady, J. Magee, Y. Yamamoto, D. Helmick, L. Wang, Co-optimization of wrought alumina-forming austenitic stainless steel composition ranges for high-temperature creep and

oxidation/corrosion resistance, Mater. Sci. Eng. A. 590 (2014) 101−115. https://doi.org/10.1016/j.msea.2013.10.014

[15] Z. Zhang, H. Huang, Z. Zhang, Y. Wang, B. Zhu, H. Zhao, A review of the microstructure and properties of superalloys regulated by magnetic field, J. Mater. Res. Technol. 30 (2024) 9285−9317. https://doi.org/10.1016/j.jmrt.2024.05.189

[16] S. Gialanella, A. Malandruccolo, Superalloys, in: C.P. Bergmann (Ed.), Aerospace Alloys, Springer: Berlin/Heidelberg, Switzerland, 2020, pp. 267-386. https://doi.org/10.1007/978-3-030-24440-8_6

[17] M.J. Eckelman, L. Ciacci, G. Kavlak, P. Nuss, B.K. Reck, T.E. Graedel, Life cycle carbon benefits of aerospace alloy recycling, J. Clean. Prod. 80 (2014) 38−45. https://doi.org/10.1016/j.jclepro.2014.05.039

[18] M. Moniruzzaman, M.S.H. Chowdhury, D. Saha, M.M. Billah, A. Helal, R.A. Biswash, Hydrokinetic turbine technology and its prospect in Bangladesh: A, Aust. J. Eng. Innov. Tech. 4 (2022) 01−07. https://doi.org/10.34104/ajeit.022.01007

[19] E. Akca, A. Gürsel, A review on superalloys and IN718 nickel-based INCONEL superalloy, Period. Eng. Nat. Sci. 3 (2015) 15−27. https://doi.org/10.21533/pen.v3i1.43

[20] R. Batista, R. Fernanda, T. De Oliveira, M. Valério, M.B. Silva, Application of taguchi method in turning process of a superalloy NIMONIC 80A to improve the surface roughness, Int. J. Innov. Res. Eng. Manag. 2 (2015) 81−88.

[21] R. Kataria, R.P. Singh, P. Sharma, R.K. Phanden, Welding of super alloys: A review, Mater. Today Proc. 38 (2021) 265−268. https://doi.org/10.1016/j.matpr.2020.07.198

[22] M.K. Sinha, A. Pal, K. Kishore, A. Singh, Archana, H. Sansanwal, P. Sharma, Applications of sustainable techniques in machinability improvement of superalloys: A comprehensive review, Int. J. Interact. Des. Manuf. 17 (2023) 473−498. https://doi.org/10.1007/s12008-022-01053-2

[23] I.G. Akande, O.O. Oluwole, O.S.I. Fayomi, O.A. Odunlami, Overview of mechanical, microstructural, oxidation properties and high-temperature applications of superalloys, Mater. Today Proc. 43 (2021) 2222−2231. https://doi.org/10.1016/j.matpr.2020.12.523

[24] W. Xia, X. Zhao, L. Yue, Z. Zhang, Microstructural evolution and creep mechanisms in Ni-based single crystal superalloys: A review, J. Alloys Compd. 819 (2020) 152954. https://doi.org/10.1016/j.jallcom.2019.152954

[25] G.A. Rao, M. Sankaranarayana, S. Balasubramaniam, Hot isostatic pressing technology for defence and space applications, Def. Sci. J. 62 (2012) 73−80. https://doi.org/10.14429/dsj.62.372

[26] D.G. Thakur, B. Ramamoorthy, L. Vijayaraghavan, Effect of cutting parameters on the degree of work hardening and tool life during high-speed machining of Inconel 718, Int. J. Adv. Manuf. Technol. 59 (2012) 483−489. https://doi.org/10.1007/s00170-011-3529-6

[27] E.P. Ambrosio, M.R. Abdul Karim, M. Pavese, S. Biamino, C. Badini, P. Fino, Feasibility of electrochemical deposition of nickel/silicon carbide fibers composites over nickel

superalloys, Metall. Mater. Trans. A. 48 (2017) 2504−2510. https://doi.org/10.1007/s11661-017-4007-z

[28] S. Zhao, J. Dong, C. Lv, Z. Li, X. Sun, W. Zhang, Thermal mismatch effect and high-temperature tensile performance simulation of hybrid CMC and superalloy bolted joint by progressive damage analysis, Int. J. Aerospace Eng. 2020 (2020) 8739638. https://doi.org/10.1155/2020/8739638

[29] G. Barbieri, F. Cognini, C. de Crescenzo, A. Fava, M. Moncada, R. Montanari, M. Richetta, A. Varone, Process optimization in laser welding of IN792 DS superalloy, Metals. 14 (2024) 124. https://doi.org/10.3390/met14010124

[30] K.O. Lee, S.B. Lee, Modeling of materials behavior at various temperatures of hot isostatically pressed superalloys, Mater. Sci. Eng. A. 541 (2012) 81−87. https://doi.org/10.1016/j.msea.2012.02.005

[31] M. Durand-Charre, The Microstructure of Superalloys, CRC Press, London, 2017. https://doi.org/10.1201/9780203736388

[32] T.M. Pollock, S. Tin, Nickel-based superalloys for advanced turbine engines: Chemistry, microstructure and properties, J. Propuls. Power. 22 (2006) 361−374. https://doi.org/10.2514/1.18239

[33] E. Martelli, M. Girardi, P. Chiesa, Breaking 70% net electric combined cycle efficiency with CMC gas turbine blades, ASME Turbo. Expo. (2022). https://doi.org/10.1115/GT2022-81118

[34] V. Kumar, B. Kandasubramanian, Processing and design methodologies for advanced and novel thermal barrier coatings for engineering applications, Particuology. 27 (2016) 1− https://doi.org/10.1016/j.partic.2016.01.007

[35] A. Kracke, A. Allvac, Superalloys, the most successful alloy system of modern times-past, present and future, in: Proc. 7th Int. Symp. Superalloy. 718 (2010) 13−50. https://doi.org/10.7449/2010/Superalloys_2010_13_50

[36] N. Miura, Y. Kondo, Morphology of precipitates in a first stage low pressure turbine blade of a Ni-based superalloy after service and after following aging, J. of ASTM Int. 9 (2012) 1−9. https://doi.org/10.1520/JAI103463

[37] R.J. Bennett, S.W. Booth, C.M. Younes, Effect of thermal exposure on degradation of compressor blades in aero-jet engines, Corros. Eng. Sci. Technol. 50 (2015) 283−291. https://doi.org/10.1179/1743278214Y.0000000223

[38] C.K. Sudbrack, D.L. Beckett, R.A. MacKay, Effect of surface preparation on the 815° C oxidation of single-crystal nickel-based superalloys, JOM. 67 (2015) 2589−2598. https://doi.org/10.1007/s11837-015-1639-6

[39] T.M. Smith, B.D. Esser, N. Antolin, A. Carlsson, R.E.A. Williams, A. Wessman, T. Hanlon, H.L. Fraser, W. Windl, D.W. McComb, M.J. Mills, Phase transformation strengthening of high-temperature superalloys, Nat. Commun. 7 (2016) 13434. https://doi.org/10.1038/ncomms13434

Materials Research Forum LLC
https://doi.org/10.21741/9781644903698-15

[40] G. Liu, C. Huang, H. Zhu, Z. Liu, Y. Liu, C. Li, The modified surface properties and fatigue life of Incoloy A286 face-milled at different cutting parameters, Mater. Sci. Eng. A. 704 (2017) 1−9. https://doi.org/10.1016/j.msea.2017.07.072

[41] J. Liu, C. Gao, L. Shen, H. Cheng, X. Gao, X. Han, Microstructure and surface morphology of inconel 625 alloy prepared by laser melting deposition using abrasive-assisted jet electrochemical machining, Int. J. Electrochem. Sci. 13 (2018) 10654−10668. https://doi.org/10.20964/2018.11.25

[42] A. Kruk, G. Cempura, Application of analytical electron microscopy and FIB-SEM tomographic technique for phase analysis in as-cast Allvac 718Plus superalloy, Inte. J. Mater. Res. 110 (2019) 3−10. https://doi.org/10.3139/146.111678

[43] N.E. Ugodilinwa, M. Khoshdarregi, O.A. Ojo, Analysis and constitutive modeling of high strain rate deformation behavior of Haynes 282 aerospace superalloy, Mater. Today Commun. 20 (2019) 100545. https://doi.org/10.1016/j.mtcomm.2019.100545

[44] K. Khanafer, A. Eltaggaz, I. Deiab, H. Agarwal, A. Abdul-Latif, Toward sustainable micro-drilling of Inconel 718 superalloy using MQL-Nanofluid, Int. J. Adv. Manuf. Technol. 107 (2020) 3459−3469. https://doi.org/10.1007/s00170-020-05112-4

[45] M.E. Pek, A.K. Ackerman, M. Appleton, M.P. Ryan, S. Pedrazzini, Development of a novel, impurity-scavenging, corrosion-resistant coating for Ni-based superalloy CMSX-4, High Temp. Corros. Mater. 99 (2023) 3−13. https://doi.org/10.1007/s11085-022-10142-2

[46] Ł. Rakoczy, M. Grudzień-Rakoczy, R.Cygan, B. Rutkowski, T. Kargul, T. Dudziak, E. Rząd, O. Milkovič, A. Zielińska-Lipiec, Characterization of the as-cast microstructure and selected properties of the X-40 Co-based superalloy produced via lost-wax casting, Arch. Civ. Mech. Eng. 22 (2022) 143. https://doi.org/10.1007/s43452-022-00466-w

[47] L. Zhang, D. Townsend, Characteristics of quasi-static and dynamic failure of Inconel 625 superalloy under simple shear, Mater. Lett. 358 (2024) 135822. https://doi.org/10.1016/j.matlet.2023.135822

[48] J.L. Evans, Effect of surface roughness on the oxidation behavior of the Ni-base superalloy ME3, J. Mater. Eng. Perform. 19 (2010) 1001−1004. https://doi.org/10.1007/s11665-010-9605-5

[49] C.J. Pierce, A.N. Palazotto, A.H. Rosenberger, Creep and fatigue interaction in the PWA1484 single crystal nickel-base alloy, Mater. Sci. Eng. A. 527 29−30 (2010) 7484−7489. https://doi.org/10.1016/j.msea.2010.08.033

[50] Y. Amouyal, D.N. Seidman, The role of hafnium in the formation of misoriented defects in Ni-based superalloys: An atom-probe tomographic study, Acta. Mater. 59 (2011) 3321−3333. https://doi.org/10.1016/j.actamat.2011.02.006

[51] Y. Amouyal, D.N. Seidman, An atom-probe tomographic study of freckle formation in a nickel-based superalloy, Acta. Mater. 59 (2011) 6729−6742. https://doi.org/10.1016/j.actamat.2011.07.030

[52] Y.A. Nagumo, T. Jr, R. Sugiura, T. Matsuzaki, H. Takeuchi, Y. Ito, Behavior of branch cracking and the microstructural strengthening mechanism of polycrystalline Ni-base

superalloy, IN100 under creep condition, Mater. Trans. 52 (2011) 1876−1884. https://doi.org/10.2320/matertrans.M2011074

[53] A. Klimpel, A. Olejnik, D. Janicki, A. Lisiecki, Investigations of technology of repair laser welding of exhaust flaps of jet engine made of nickel superalloy, Weld. Int. 27 (2013) 331−337. https://doi.org/10.1080/09507116.2011.600039

[54] N.O. Villalón, O. Covarrubias, R. Colás, Heat treating of a 718Plus superalloy, Mater. Perform. Charact. 2 (2013) 153−162. https://doi.org/10.1520/MPC20120026

[55] A.T. Yokobori Jr, R. Sugiura, H. Takeuchi, G. Ozeki, A. Ishida, D. Kobayashi, S. Hosono, Law of fracture life for directionally solidified and poly-crystal nickel-base super alloys (CM247LC and IN100) under creep-fatigue conditions based on non-equilibrium science, Strength Fract. Complex. 8 (2013) 25−44. https://doi.org/10.3233/SFC-130154

[56] P. Fernandez-Zelaia, B.S. Adair, V.M. Barker, S.D. Antolovich, The portevin-Le Chatelier effect in the Ni-based superalloy IN100, Metall. Mater. Trans. A. 46 (2015) 5596−5609. https://doi.org/10.1007/s11661-015-3126-7

[57] R.C. Reed, Z. Zhu, A. Sato, D.J. Crudden, Isolation and testing of new single crystal superalloys using alloys-by-design method, Mater. Sci. Eng. A. 667 (2016) 261−278. https://doi.org/10.1016/j.msea.2016.04.089

[58] W.J. Nowak, Characterization of oxidized Ni-based superalloys by GD-OES, J. Anal. At. Spectrom. 32 (2017) 1730−1738. https://doi.org/10.1039/C7JA00069C

[59] J. Belan, A. Vaško, L. Kuchariková, E. Tillová, M. Chalupová, The SEM metallography analysis of vacuum cast ZhS6K superalloy turbine blade after various working hours, Manuf. Technol. 19 (2019) 727−733. https://doi.org/10.21062/ujep/362.2019/a/1213-2489/MT/19/5/727

[60] O.A. Olufayo, H. Che, V. Songmene, C. Katsari, S. Yue, Machinability of Rene 65 superalloy, Mater. 12 (2019) 2034. https://doi.org/10.3390/ma12122034

[61] Ł. Rakoczy, M. Grudzień-Rakoczy, R. Cygan, The influence of shell mold composition on the as-cast macro-and micro-structure of thin-walled IN713C superalloy castings, J. Mater. Eng. Perform. 28 (2019) 3974−3985. https://doi.org/10.1007/s11665-019-04098-9

[62] K. Venkatesan, S. Devendiran, S.C.S.R. Bhupatiraju, S. Kolluru, C. Pavan Kumar, Experimental investigation and optimization of micro-drilling parameters on Inconel 800 superalloy, Mater. Manuf. Process. 35 (2020) 1214−1227. https://doi.org/10.1080/10426914.2020.1746333

[63] T. Sonar, V. Balasubramanian, S. Malarvizhi, T. Venkateswaran, D. Sivakumar, Development of 3-Dimensional (3D) response surfaces to maximize yield strength and elongation of inter pulsed TIG welded thin high temperature alloy sheets for jet engine applications, CIRP J. Manuf. Sci. Tec. 31 (2020) 628−642. https://doi.org/10.1016/j.cirpj.2020.09.003

[64] K. Mehmood, M. Imran, L. Ali, M.A. Umer, M. Abbas, M. Saleem, Development of cost-effective microstructure and isothermal oxidation-resistant bond coats on Inconel 718 by atmospheric plasma-sprayed NiCoCrAlFe high-entropy alloy, JOM. 75 (2023) 239−247. https://doi.org/10.1007/s11837-022-05578-5

Superalloys: Fundamentals and Applications   Materials Research Forum LLC
Materials Research Foundations 178 (2025) 291-315   https://doi.org/10.21741/9781644903698-15

[65] S. Farokhi, Aircraft Propulsion: Cleaner, Leaner, and Greener, Third ed., John Wiley & Sons, 2021.

[66] A. Ashrafi, Investigating the replacement of nickel-based superalloys with niobium-based superalloys in the new generation of turbine blades in the aero-space industry, Int. J. Met. Math. Sci. 6 (2024) 22−29. https://doi.org/10.34104/ijmms.024.022029

[67] A.A. Shanyavskiy, Very-High-Cycle-Fatigue of in-service air-engine blades, compressor and turbine, Sci. China Phys. Mech. Astron. 57 (2014) 19−29. https://doi.org/10.1007/s11433-013-5364-2

[68] P. Sharma, D. Chakradhar, S. Narendranath, Effect of wire material on productivity and surface integrity of WEDM-processed Inconel 706 for aircraft application, J. Mater. Eng. Perform. 25 (2016) 3672−3681. https://doi.org/10.1007/s11665-016-2216-z

[69] F. Pashmforoush, R.D. Bagherinia, Influence of water-based copper nanofluid on wheel loading and surface roughness during grinding of Inconel 738 superalloy, J. Clean. Prod. 178 (2018) 363−372. https://doi.org/10.1016/j.jclepro.2018.01.003

[70] M. Luo, H. Li, On the machinability and surface finish of superalloy GH909 under dry cutting conditions, Mater. Res. 21 (2018) e20171086. https://doi.org/10.1590/1980-5373-mr-2017-1086

[71] C. Ming, G. Yadong, S. Yao, Q. Shuoshuo, L. Yin, Y. Yuying, Experimental study on grinding surface properties of nickel-based single crystal superalloy DD5, Int. J. Adv. Manuf. Technol. 101 (2019) 71−85. https://doi.org/10.1007/s00170-018-2839-3

[72] Y. He, Z. Zhou, P. Zou, X. Gao, K.F. Ehmann, Study of ultrasonic vibration-assisted thread turning of Inconel 718 superalloy, Adv. Mech. Eng. 11 (2019). https://doi.org/10.1177/1687814019883772

[73] A.I. Epishin, D.S. Lisovenko, Comparison of isothermal and adiabatic elasticity characteristics of the single crystal nickel-based superalloy CMSX-4 in the temperature range between room temperature and 1300 C, Mech. Solids. 58 (2023) 1587−1598. https://doi.org/10.3103/S0025654423601301

[74] D. Zhao, L. Zhou, D. Wang, H. Zeng, X. Gong, D. Shu, B. Sun, Integrated computational framework for controlling dimensional accuracy of thin-walled turbine blades during investment casting, Int. J. Adv. Manuf. Technol. 129 (2023) 1315−13 https://doi.org/10.1007/s00170-023-12319-8

[75] K. Jithesh, M. Arivarasu, An investigation on hot corrosion and oxidation behavior of cobalt-based superalloy L605 in the simulated aero-engine environment at various temperatures, Mater. Res. Express. 6 (2019) 126530. https://doi.org/10.1088/2053-1591/ab54dd

[76] J. Gao, Y. Tong, H. Zhang, L. Zhu, Q. Hu, J. Hu, S. Zhang, Machine learning assisted design of Ni-based superalloys with excellent high-temperature performance, Mater. Charact. 198 (2023) 112740. https://doi.org/10.1016/j.matchar.2023.112740

[77] W.D. Musinski, D.L. McDowell, Microstructure-sensitive probabilistic modeling of HCF crack initiation and early crack growth in Ni-base superalloy IN100 notched components, Int. J. Fatigue. 37 (2012) 41−53. https://doi.org/10.1016/j.ijfatigue.2011.09.014

[78] K.S. Chan, M.P. Enright, J.P. Moody, Development of a probabilistic methodology for predicting hot corrosion fatigue crack growth life of gas turbine engine disks, J. Eng. Gas Turbine. Power. 136 (2014) 022505. https://doi.org/10.1115/1.4025555

[79] E. Cagliyan, F. Walter, Metallurgical failure investigation of overheated brackets made of nimonic alloy 90, Pract. Metallogr. 52 (2015) 665−678. https://doi.org/10.3139/147.110251

[80] O.G. Ospennikova, M.R. Orlov, V.G. Kolodochkina, R.M. Nazarkin, Structural changes and damage of single-crystal turbine blades during life tests of an aviation gas turbine engine, Russ. Metall. (Met.). 2015 (2015) 324−331. https://doi.org/10.1134/S0036029515040114

[81] T. Sjöberg, J. Kajberg, M. Oldenburg, Calibration and validation of three fracture criteria for alloy 718 subjected to high strain rates and elevated temperatures, Eur. J. Mech. A-Solid. 71 (2018) 34−50. https://doi.org/10.1016/j.euromechsol.2018.03.010

[82] X. Shi, S. Duan, W. Yang, H. Guo, J. Guo, Solidification and segregation behaviors of superalloy IN718 at a slow cooling rate, Mater. 11 (2018) 2398. https://doi.org/10.3390/ma11122398

[83] A. Silvello, P. Cavaliere, A. Rizzo, D. Valerini, S. Dosta Parras, I. García Cano, Fatigue bending behavior of cold-sprayed nickel-based superalloy coatings, J. Therm. Spray Technol. 28 (2019) 930−938. https://doi.org/10.1007/s11666-019-00865-1

[84] A. Jokar, F. Ghadami, N. Azimzadeh, D. Salehi Doolabi, A novel approach to the protection of internal passageways of gas turbine blades by aluminide coating: Evaluation of oxidation behavior, Mater. Corros. 73 (2022) 1534−1552. https://doi.org/10.1002/maco.202213179

[85] H. Lu, J. Wang, Z. Wen, T. Liu, Y. Lian, Z. Yue, Vibration fatigue behavior and life prediction of directionally solidified superalloy based on the phase transformation theory, Eng. Fract. Mech. 282 (2023) 109184. https://doi.org/10.1016/j.engfracmech.2023.109184

[86] Z. Zhang, M. Chen, P. Yu, H. Huang, H. Li, F. Yu, Z. Zhang, Y. Niu, S. Gao, C. Wang, J. Jiang, Study of the roughness effect on the normal spectral emissivity of GH3044, Infrared Phys. Technol. 133 (2023) 104831. https://doi.org/10.1016/j.infrared.2023.104831

[87] Z. Jonsta, P. Jonsta, K. Konecna, M. Gabcova, Phase analysis of nickel superalloy Inconel 738 LC, Commun. Sci. Lett. Univ. Zilina. 12 (2010) 90−94. https://doi.org/10.26552/com.C.2010.4.90-94

[88] Z. Zhao, Y. Fu, J. Xu, Z. Zhang, Z. Liu, J. He, An investigation on high-efficiency profile grinding of directional solidified nickel-based superalloys DZ125 with electroplated CBN wheel, Int. J. Adv. Manuf. Technol. 83 (2016) 1−11. https://doi.org/10.1007/s00170-015-7550-z

[89] Z. Khan, S. Fida, F. Nisar, N. Alam, Investigation of intergranular corrosion in 2nd stage gas turbine blades of an aircraft engine, Eng. Fail. Anal. 68 (2016) 197−209. https://doi.org/10.1016/j.engfailanal.2016.05.033

[90] S. Sarfraz, E. Shehab, K. Salonitis, W. Suder, Experimental investigation of productivity, specific energy consumption, and hole quality in single-pulse, percussion, and trepanning drilling of in 718 superalloy, Energies. 12 (2019) 4610. https://doi.org/10.3390/en12244610

[91] L. Zhang, L. Zhao, R. Jiang, C. Bullough, Crystal plasticity finite-element modelling of cyclic deformation and crack initiation in a nickel-based single-crystal superalloy under low-cycle fatigue, Fatigue. Fract. Eng. Mater. Struct. 43 (2020) 1769−1783. https://doi.org/10.1111/ffe.13228

[92] F.P. Prakash, N. Jeyaprakash, M. Duraiselvam, G. Prabu, C.-H. Yang, Droplet spreading and wettability of laser textured C-263 based nickel superalloy, Surf. Coat. Technol. 397 (2020) 126055. https://doi.org/10.1016/j.surfcoat.2020.126055

[93] G. Zhu, W. Pan, R. Wang, D. Wang, D. Shu, L. Zhang, A. Dong, B. Sun, Microstructures and mechanical properties of GTD222 superalloy fabricated by selective laser melting, Mater. Sci. Eng. A. 807 (2021) 140668. https://doi.org/10.1016/j.msea.2020.140668

[94] M. Taheri, A. Halvaee, S.F. Kashani Bozorg, A. Salemi Golezani, R. Panahi Liavoli, A.A. Kashi, The effect of heat treatment on creep behavior of GTD-111 superalloy welded by pulsed Nd: YAG laser using small punch test, Eng. Fail. Anal. 122 (2021) 105255. https://doi.org/10.1016/j.engfailanal.2021.105255

[95] G. Liu, D. Du, K. Wang, Z. Pu, D. Zhang, B. Chang, Microstructure and wear behavior of IC10 directionally solidified superalloy repaired by directed energy deposition, J. Mater. Sci. Technol. 93 (2021) 71−78. https://doi.org/10.1016/j.jmst.2021.04.006

[96] X. Zhu, F. Wang, D. Ma, A. Bührig-Polaczek, Influence of Hf and MC carbide on transverse platform in single crystal grain selection of 2D grain selector, Mater. 15 (2022) 6274. https://doi.org/10.3390/ma15186274

[97] Y. Song, K. Shi, Z. He, Z. Zhang, Y. Shi, Investigation of grindability and surface integrity in creep feed grinding of GH738 alloy using different grinding wheels, Int. J. Adv. Manuf. Technol. 123 (2022) 4153−4169. https://doi.org/10.1007/s00170-022-10497-5

[98] S.N. Grigoriev, M.A. Volosova, A.A. Okunkova, Investigation of surface layer condition of SiAlON ceramic inserts and its influence on tool durability when turning nickel-based superalloy, Technol. 11 (2023) 11. https://doi.org/10.3390/technologies11010011

[99] B. Hu, W. Xie, W. Zhong, D. Zhang, X. Wang, J. Hu, Y. Wu, Y. Liu, The effect of pulling speed on the structure and properties of DZ22B superalloy blades, Coat. 13 (2023) 1225. https://doi.org/10.3390/coatings13071225

[100] W. Yu, J. Wu, Y. Li, Q. An, W. Ming, D. Chen, H. Wang, M. Chen, Investigations on surface modification of nickel-based superalloy subjected to ultrasonic surface rolling process, Int. J. Adv. Manuf. Technol. 129 (2023) 1473−1488. https://doi.org/10.1007/s00170-023-12299-9

[101] S.K. Saurabh, P. Chand, U.S. Yadav, Optimizing laser beam welding performance parameters on nimonic 80A superalloy: A study on experimentation, TGRA, and PCA, Soldag. Insp. 29 (2024) e2904. https://doi.org/10.1590/0104-9224/si29.04

Superalloys: Fundamentals and Applications Materials Research Forum LLC
Materials Research Foundations 178 (2025) 316-332 https://doi.org/10.21741/9781644903698-16

Chapter 16

Welding Materials for Superalloys

Figen Balo [1*] , Lutfu S.Sua [2]

[1] Department of METE, Engineering Faculty, Firat University, Turkey

[2] Department of Management and Marketing, Southern University and A&M College, USA

*figenbalo@gmail.com

Abstract

Some of the recent most technological productions are those that make use of superalloys, and welding these together is frequently crucial to both manufacturing and repair/remanufacturing processes. In order to successfully weld superalloys, one must have a thorough awareness of their characteristics and exercise caution while choosing the welding materials and methods to guarantee that performance requirements are satisfied. In this study, firstly, an overview of the welding materials produced for superalloys and the most effective welding material for aviation applications among the current materials were tried to be determined with a hierarchy approach.

Keywords

Welding Materials, Superalloys, Aviation Applications, Mechanical Strength

Contents

1.   Introduction

A nation's manufacturing sectors are vital to its economic expansion and rapid development. The sheet metal business, out of all the industries, is indispensable to our daily existence. Applications for sheet metal include computer panels, vehicles, airplanes, building roofs, and more. One of the most important sheet metals is superalloy. Superalloys, which are often made of aluminum and chromium, are a unique class of alloys with exceptional surface stability, resistance to creep, high

Superalloys: Fundamentals and Applications   Materials Research Forum LLC
Materials Research Foundations 178 (2025) 316-332   https://doi.org/10.21741/9781644903698-16

mechanical strength, and corrosion resistance. They are currently being used in many different industries because of their unique features. In high temperature implementations, Superalloys are alloys having good corrosion resistance and bigger mechanical strength. They have found extensive application in manufacturing due to their enhanced characteristics. For superalloy production, niobium, tungsten, molybdenum, iron, cobalt, aluminum, titanium, tantalum, zirconium, and chromium are added in small amounts to alloys, which are composed of an iron foundation, a nickel blend, a cobalt basis, or iron [1]. These materials have been selected due to their resistance to corrosion, strength at high temperatures, and endurance under challenging conditions [2]. Superalloys' unique characteristics can change based on their precise composition and processing methods. Superalloys' general grouping is given in Table 1 [3].

Table 1. Super alloys' grouping

Superalloys		
Iron-sourced	*Cobalt-sourced*	*Nickel-sourced*
Alloy 901	MAR-M918	Rene (41.95)
V-57	MP35N	Pyromet 860
H-155	Elgiloy	Udimet (720, 710, 700, 630, 520, 500,400)
Incoloy (909, 907. 903, 825, 807, 802, 601, 800)	Stellite 6B	Nimonic (PK.33, PE.16, PE.11, 942, 263, 115, 105, 90, 80A, 75)
Haynes 556	MP-159	Hastelloy (X, S, G-30, C-22)
Discaloy	L-605	Waspaloy
	Haynes 188	Inconel (X750, 718, 706, 625, 617, 601, 600, 597, 587)

Superior alloys at high temperatures are those that exhibit high tension, and high resilience to surface degradation like creep resistance, and corrosion-oxide [4-6]. Such alloys find many uses, including Chemical and petrochemical facilities, parts of medical equipment (like dental equipment and prosthesis devices, among other things.), metals processing (such as hot processing apparatus, casting dies and dies, etc.), heat processing equipment (such as conveyor belts, trays, fixtures, etc.) [7, 8], the aerospace industry, where they are used in things (like bolts, fuel chambers, vanes, gas turbine blades, etc.), nuclear power plants, where they are used in things (like valve stems, ducting, rotary controls, and springs), and steam turbine power plants, where they are used in things (like reheaters and turbine blades) [9, 10].

Superalloys play a major role in the aircraft industry and power generation. Using superalloys for various gas turbine components allows for high operating performance and temperatures. These applications under difficult conditions can be attributed to high temperatures, tear and wear, durability and strength losses, and slow resistance to oxidation and corrosion. In a nickel-sourced superalloy, at least 50% of an alloy's composition is nickel, and nickel is the main component of alloys with an iron base. Superalloys based on nickel can contain up to 10 components of alloying, including heavy/light refractive components. To improve superficies steadiness, additional alloying components like chromium and aluminum are added. To further improve their features, wrought nickel-base superalloys often contain 8% Ti, 20–10% Cr, 15–5% Co, and Al combined, along with smaller amounts of additional additives such as niobium, boron, zirconium, magnesium, molybdenum, and tungsten [11, 12]. These superalloys are made especially to resist harsh circumstances, like high temperatures, high pressures, and corrosive or oxidizing surroundings. High fracture toughness, which enables them to withstand cracking under pressure, is one of their essential characteristics. They can also bear large loads without breaking because of their strong tensile strength. Their outstanding resistance to creep, or the progressive distortion of a material under continuous load at high temperatures, is another crucial feature [13].

There are two ways to heat treat nickel-base superalloys: precipitation and resolution thermal processing. Nonbased super-alloys have proven successful in industrial applications when motor parts with intricate geometries are welded together and faulty sections are replaced during operation [14]. The process usually involves the manufacture and repair of welding of fusion operations such as laser soldering, welding of gas-tungsten-arc, and electron beam [15]. Although powerful solution fusion soldering is frequently used to enhance Ni-based superalloys, and high-intensity welding, Over time, nickel-based superalloys have grown to be a significant difficulty due to their susceptibility to porosity development in the zone, solidifying cracking, strain age cracking, and heat zone damage cracking. [16]. Therefore, these troubles restrict the restoration and development of nickel-sourced superalloys. Superalloys have been welded in a variety of ways due to their enhanced characteristics. In tungsten arc welding, an arc-sourced jointing methodology, an arc is created between a workpiece and an inert tungsten electrode. [17]. It is frequently used to guard against gases like argon and helium contaminating the atmosphere. When filler is required, it is put in the welding bowl separately. Since the electrode cannot be consumed, heat is normally transferred by tungsten arc welding to the direct-current electrode negative at a ratio of about 2/3 to 1/3 with the tungsten electrode [18]. While oxide film products, like aluminum, have issues, tungsten arc welding is employed for AC [19]. Megaalloy laser soldering has been a popular technique. When a laser is pointed at a specific spot, it can be used to perform touchless welding. The goal is to fuse the targeted region and create a soldered link [20, 21]. The combining of multiple superalloys using various welding techniques has been the subject of numerous research papers in the last decades. Nickel-based superalloys are joined using a variety of welding procedures, including laser beam, solid stage (linear friction and friction stir welding), and electron beam, gas-tungsten arc (continuous current, pulsed current). The positive characteristics of gas-tungsten arc (pulsed current) welding include a smaller heat-impacted region, a narrow fusion region, and increased stiffness at the site of weld crossing. Better ductility and yield strength were attained by the welding of the laser beam and welding of electron beam procedures as a result of the development of fine lamellar microstructure. Compared to gas-tungsten (pulsed current) arc welds, Inconel/718 electron beam welding yields welds with improved microstructures [22]. The gas-tungsten-arc of the pulsed current welding sample demonstrated properties such as decreased segregation and a notable rise in strength of tensile when compared to the continual gas-tungsten-arc of the current welding sample [23]. The gas-tungsten-arc of welding of pulsed current of Inconel718 and Inconel625 was researched by Ramkumar. Weldings are ductile under all circumstances. For these different joints, no liquation cracking or solidification flaws were seen [24]. Korrapati stated that the production of fine microstructure was accomplished through gas-tungsten-arc of pulsed current welding procedures with a higher cooling ratio and decreased heat input. More sensitive microstructure led to bigger hardness and strength in the welded specimen [25]. Research conducted by Farahani on Alloy617 revealed that gas-tungsten (pulsed current) arc welding-induced grain refinement was the cause of the material's increased toughness and impact energy [26]. Frictional heat is used to join materials in welding methods of solid-state like welding of welding of linear friction, rotary friction, and welding of friction stir. Superior mechanical qualities are achieved by producing a weld with fine grains and a dynamically recrystallized grain structure using friction stir welding. To develop the

Superalloys: Fundamentals and Applications   Materials Research Forum LLC
Materials Research Foundations 178 (2025) 316-332   https://doi.org/10.21741/9781644903698-16

skills and knowledge of researchers in the future, previous researchers have investigated their unique research discoveries.

As a result, the examined research has been condensed into the "literature review" portion that is being provided. Graneix and colleagues have examined the Nd/YAG laser's capacity to weld Haynes 188 and Hastelloy X alloys. The butt joint is used for the investigation [27]. The materials were butt welded and cut using a Tru-Laser apparatus (TRUMPH-Cel/L3000). Variations in feed rate, laser power, gas flow, and focal diameter, as well as how these affect mechanical properties and weld geometry, have all been examined. The microstructures were examined using optical microscopy in conjunction with EBSD analysis and SEM processing. Utilizing CORICO software, the welding process' most important parameter was identified. The pre-cold treating Inconel 718's impacts on the mechanical characteristics, microstructure, and welding of lasers of superalloys have been investigated by Rezaei and Moosavy [28]. Pre-cold treatment was carried out using dry ice that was approximately 30 °C in temperature. For heating, a 2000W Ronix 1101 heat gun was utilized. To record the temperature, a laser thermometer (IRG 420) was utilized. The material's microstructure was examined utilizing a scanning electron microscope (Philips XL30) and an optical microscope (Olympus BX51-M). Energy Dispersive Spectrometer and field emission SEM were utilized to thoroughly examine the microstructure. The microhardness tester (Buehler Vickers) was used to measure hardness. The samples' pre-welding temperatures were adjusted, and a laser machine (Nd:YAG pulsed) was utilized to create beads on the plaque as part of the experiment. The effects of tungsten arc welding parameters, like welding speed and input heat, on the superalloy's (Udimet-520) microstructure and mechanical characteristics were investigated by Nematzadeh et al. [29]. The components' dispersion in the weld sheet was ascertained using a quantometer, and hardness was assessed using Rockwell hardness equipment. The temperature distribution was measured using a thermocouple and a pyrometer. An optical microscope equipped and SEM analysis with a Spectrometer for Energy Dispersive were used to investigate hot cracking and microstructure. The Inconel 718's laser weldability, the correlation between welding characteristics and weld flaws, and the mechanism of weld cracking are all researched by Shinozaki et al. [30]. The CO2 laser oscillator is a low-order tri-axial cross-acting type multimode. To test the weldability, four beads with a length of 50–70 mm are utilized. To see how the shape of the bead's cross-section changed, the travel velocity was changed. In the heat-impacted region, the healed widths and boundaries of liquefied grain were measured to assess the liquation of grain boundaries. Utilizing an Energy Dispersive Electroscope, the micro-scale component on the cracks superficies is examined. Wang and his co-workers' research on the tungsten arc welding solder variables' impacts on the tensile properties, weld fraction, microstructural details, and morphological features of nickel-sourced superalloys including index for solder remelting, grooves, impulse frequency, soldering current, and welding speed. GH99 is made by tungsten arc welding and is plated to a thickness that matches the nickel foundation's butt welds [31]. On a substrate of stainless steel, Chen and his co-workers researched the weld performance of two Ni-sourced weld fillers, IN52MSS and IN52M. The welding procedure of tungsten-inert-gas was used in several passes to overlay-weld IN52MSS and IN52M upon CF8ASS. To the intermetallic development [Ni3(Mo, Nb)] micro constituents, the overlay welds' hot cracking was demonstrated to be connected. The 52MSS overlay's higher Nb and Mo contents facilitated the development of these elements at the interdendritic borders. Consequently, the 52MSS overlay's sensitivity to hot cracking was greater than the 52M overlay's [32]. In order to investigate the shift in Haynes' ductility signature after isothermal exposure from 5 to 1800 seconds at temperatures between 750

Superalloys: Fundamentals and Applications   Materials Research Forum LLC
Materials Research Foundations 178 (2025) 316-332   https://doi.org/10.21741/9781644903698-16

and 950 degrees Celsius, Hanning and his co-workers devised a Gleeble-sourced testing procedure. Microstructural components (hardening precipitates (0) and carbides of the primary and secondary types) were discovered. It was demonstrated that the mode of fracture is based on the stroke ratio, with a shift along the direction of intergranular fracture observed at stroke ratios less than 0.055 mm/s. A ductility minimal was reported at 850–800 °C. Microvoids on the grain faces indicated the presence of intergranular fracture, and continuous age-hardening reactions did not affect the ductility of Haynes with intergranular fracture [33]. Singh and his co-workers researched the various preweld thermal treatments' effects on heat-impacted area liquation cracking by using a simulation of Gleeble thermo-mechanical of the recently created Haynes cast form and constraint weldability tests. At a higher soak temperature of heat treatment, it was shown that susceptibility to cracking didn't increase with homogenization' attained a higher level, by comparison, breaking was aggravated. At the maximum thermal process temperature of 119^0C, B segregation at the borders between grains, which displayed the release from the B-C-rich precipitates' dissolution, was associated with the elevated degree of cracking [34]. Using the Varestraint weldability testing method, Alvarez et al. examined the sensitivity of investment and wrought cast Alloy718 to hot cracking. Testing on weldings of a laser beam and tungsten-inert-gas was done and the results were compared. It was demonstrated that extended center line hot cracking increased LBW's susceptibility to hot cracking and produced a "fishbone"-like cracking pattern. Concerning the material shape (worked vs. cast), pulsation, and granule extent, just a slight effect was observed. As a matter of fact, wrought material was more susceptible to heat cracking than coarse-grained cast samples [35]. Using Varestraint weldability testing, Raza and his co-workers concentrated on the impact of granule orienting on AM Alloy718's sensitivity to hot breaking. Research revealed that high-angle grain boundaries were the main locations for cracks in the heat-impacted region of the welded samples, with smaller crack extents in specimens analyzed parallel to the granule orientation elongated than in specimens analyzed transverse to the granule orientation elongated [36]. Welding of tungsten-inert-gas on additively produced (bed fusion of laser-powder, AM) Alloy718 under various circumstances of thermal treatment was studied by Raza et al. It was found that the AM as-structure circumstance had the highest propensity for breaking in the thermal-impacted region, whereas the hot-press isostatic circumstance of AM had the lowest tendency [37].

Table 2. *Literature review on superalloys [28-43]*

Process	Outputs	Researched area	Operation parameters	Research materials
Tungsten Inert gas arc welding	An inter-granular liquation film and localized liquation coarse carbides have been studied for high input heat. The utilized HX's tensile qualities are decreased by this liquation coating. The carbides dissolve back into the matrix at low input heat, especially in the heat-impacted region near the WZ.	Elongation yield strength, ultimate tensile strength,	Heat input	Hastelloy
Laser beam welding	Weld fusion zone hardness is reduced, and the length of the heat-impacted region and PMZ is reduced.	WTOP/WNECK, melted zone, partially and heat impacted region, mushy zone's length, Hardness	Temperature at which to weld before	Inconel 718

Laser beam welding	The two most important factors were laser power and focus diameter. While yield strength stays the same ductility, and tensile strength drop relative to the base metal.	ductility, tensile strength, yield strength	gas flow, laser power, focal diameter, welding speed,	Hastelloy and Haynes 188
GTAW	For the fabricating purpose, boiler equipment, the heat impacted region hardness's ideal values, weld hardness, and tensile strength were found at a gas flow ratio (15.1 l/min), velocity (9.4 cm/min), and welding current of 130A,. The most significant element is welding speed.	heat-affected zone hardness, weld hardness value, and tensile strength	flow ratio of shielding gas, welding speed, welding current,	carbon steel A516-Gr70
Machining	To be effective for machining Ni-based superalloys, modern CBN apparatus, multiple sheet carbide-coated apparatus, and ceramic-sourced apparatus are found.	cutting forces, creep strength, surface integrity,	cut depth, cutting speed, feed rate	Ni-sourced superalloy
Laser welding	When the power is at its highest (traveling at lower than 1.50 m/min), the weld grain takes on a cross-section form like a champagne glass, and when the speed is between 5 and 2,5 m/min, porosity, cracks of weld at the underfills appear and heat impacted region.	Cracks' number, important curvature radius	Power, heat input, welding speed,	Inconel 718
Gas tungsten arc welding	Hot cracking was observed at elevated input heat values; where Ti and Al concentrations are high, the hardness improved with decreasing input heat and increasing weld speed.	Ductility, hardness	welding speed, heat input	Udimet 520
XRD, Annealing, EDS, SEM	Grain size increased with prolonged warming and significant liquefaction at the granule border. Nb-ample stage precipitating during cooling, followed by an acceleration of liquid loss from the border and its breakup	All phases' mass fraction, Niobium-rich phases' cross-section, grain size	heating time, temperature	Inconel 718
EDX, XRD, FE-SEM, Laser microscope, SEM, EDM, Electronic Balance,	An oxide sheet in two layers at 600–900 CMoO3 evaporates in the external oxide film, causing gaps in the external layer (Cr2O3) and the internal layer (Al2O3).	oxide film's phase characterization, composition variation, samples' mass, surface roughness	–	Nickel-Cobalt sourced super alloy
Machining using various materials for the cutting tools	When Inconel is machined with ceramic, notch-wear happens. TIC-Alumina ceramic apparatus are favoured for great velocity cutting of Inconel/718, while uncoated carbide apparatus perform more well than coated ones.	Thermal stresses, tool wear	Feed, cutting speed, cut depth,	Inconel 718
Tungsten inert gas Welding	Tensile properties increase first and then decline with an increase in welding speed and welding current. Tensile strength decreases with an increase in impulse frequency.	Crystal morphology, heat-impacted region, Elongation, Yield strength, Tensile strength,	Impulse Frequency, Welding current, Welding speed	GRADEGH99, Ni-sourced alloy

Superalloys are high-performance alloys made to resist mechanical stress, severe temperatures, and corrosion. For example; they are made to be strong at higher temperatures, which is essential for use in the power and aerospace industries. These temperatures must also be withstood by the welding materials without compromising their integrity.

It's crucial to choose the appropriate filler metals or welding techniques for superalloys based on the particular alloy being welded and the intended application when it comes to welding materials. Superalloys' distinct characteristics and behaviors in comparison to more traditional materials must be carefully taken into account when welding them. When it comes to welding materials for superalloys, the following important characteristics and factors need to be considered [44-56];

- Post-weld heat treatment is necessary for many superalloys to reduce stress, improve ductility, and restore desired characteristics. The lifetime and integrity of the weld joint depend on understanding the proper heat treatment techniques.
- It is necessary to select a filler material that has comparable mechanical and thermal properties to the underlying superalloy. To get the best results, engineers frequently choose filler materials that complement the base material's alloying components.
- Because of problems like solidification cracking, porosity, or hot cracking, some superalloys are more challenging to weld than others. It's critical to determine the superalloy's weldability and choose the right filler materials and welding techniques.
- At high temperatures, superalloys frequently go through phase changes that might alter their mechanical characteristics. Maintaining the intended performance parameters during welding procedures requires minimizing microstructure changes.
- The coefficients of thermal expansion of various materials can differ, which could cause the welded junction to bend or retain residual stresses. These dangers can be reduced by selecting filler materials with qualities that are comparable to those of the base materials.
- Because superalloys have low thermal conductivity, heat dispersion during welding may be difficult. To prevent overheating or underheating certain sections, this calls for precise management of the welding conditions.
- For the purpose of welding and servicing, welded joints need to be ductile enough to allow for thermal expansion and contraction. Superalloys usually exhibit decreased ductility during welding, so in order to preserve toughness, the filler materials should be selected appropriately.
- At high temperatures, a lot of superalloys exhibit a strong resistance to oxidation and corrosion. It is often the case that welding materials must have identical qualities in order for the joint to function similarly in high-temperature operating settings.
- Taking distortion and heat input into account, the welding technique and welding procedures must be compatible with the properties of the superalloy.
- One of the main issues with superalloy welding is also that these materials have a high degree of segregation and strength in the cast cobalt-based nozzles, which makes them extremely sensitive to heat cracking. Exact command over the settings of the welding, including cooling ratio and heat input, is necessary to resolve this problem. The right filler material selection, preheating, and heat treatment of post-weld can also aid in improving the mechanical qualities and weldability of superalloys. To achieve strong and long-lasting weld connections while welding superalloys, it is significant to carefully obtain these parameters.

The superalloys' sheet metal business uses a variety of manufacturing techniques, including deep drawing, welding, bending, and shearing. One of the key procedures in the engineering industry is welding. The most effective technique for combining comparable and dissimilar components is welding. In order to produce high-quality weldments with desired mechanical qualities and improved serviceability, selecting affecting process factors is crucial. The choice of filler material should, for best results, match or surpass the basic superalloy's mechanical and chemical characteristics. To lessen residual stresses and prevent cracking, careful management of preheat temperatures and post-weld heat treatments may be required. Some popular superalloys and the matching welding materials are given in Table 3.

Table 3. *Some popular superalloys and matching welding materials*

Superalloy	Filler Material	Common Welding Operation
Alloy 718	ERNiFeCr-2, Alloy 718	TIG (GTAW)
Stellite	Stellite 21, Stellite 6	TIG
Rene 41	ERNiCrCo-3, Rene 41	TIG
Waspaloy	ERNiCo-3, Waspaloy	TIG
Haynes 230	ERNiMo-2, Haynes 230	MIG, TIG
Hastelloy C-276	ERNiCrMo-3, Hastelloy C-276	Stick, MIG, TIG
Inconel 718	Inconel 600, Inconel 718	MIG, TIG
Inconel 625	Inconel 82, Inconel 625	Stick, MIG, TIG

Note: TIG (Tungsten inert gas,); creates the electric arc required to melt metal using a tungsten electrode, MIG (Metal Inert Gas); a continuous wire feed method combined with semi-automation that makes long welds quicker and more effective.

Significantly, the design's welds are of high enough quality to withstand the use difficulties. In this study, after the literature research on the welding materials used for superalloys, the welding materials for superalloys used in the aviation sector were investigated and then the most performing among these materials was tried to be determined with the hierarchy approach.

2. Hierarchy Approach

Saaty invented the hierarchy approach, a potency multiple attribute decision support document with wide-ranging implementations in politics, engineering, and financials.

As per Chai and his co-workers, the AHP approach facilitates the assignment of a worth that signifies the preference degree for a certain option to every other alternative [57]. Based on a hierarchical structure, these variables can be utilized to categorize and choose between options.

It has been observed by Sonar, Jadhav, Gupta, and his co-workers that the most popular technique for assessing simulation is the hierarchy approach [58, 59]. Additionally, according to Deng and his co-workers, Tummala&Tam, and Malliga&Rajesh, a hierarchy approach has been used in supplier and vendor choice [60-62]. The hierarchy approach has recently been used in conjunction with other techniques, as described by Ahn [51]. In their respective works, Mendes et al. and Guimarães and his co-workers have utilized this novel hierarchy method [63-65].

The extra material includes a short discussion on hierarchy implementations. The AHP approach has a significant flaw that prevents it from being widely used: in order to reach a conclusion, a very high number of comparisons must be made. This restriction makes it difficult to apply to the most significant problems, which need senior executives' involvement in organizations. Consequently, an approach that minimizes the comparisons' number for each of the criteria or among attributes is given, in which comparisons are done solely between one component and overall other components, presuming consistency in decision-makers' evaluations. It is proposed that the basis be an element of ostensibly larger importance—that is, a factor whose evaluation inconsistency would be most unlikely to occur in an application of the full technique. The first step in the procedure is to arrange a decision-making dilemma as a down-upside tree with the primary target at the upper. Subgoals that support the primary aim are located at the 2nd stage. Each of the sets at each of the levels fulfills the objective of the degree to which it belongs, and each of the partial targets at the 2nd degree can be divided into goals at the 3rd stage. These incomplete goals are taken into consideration as criteria in this essay. Each of the lower-level objectives, or criteria, is satisfied at a lower level by pairing and ranking the choices. Through Saaty, binary crosschecks are applied utilizing the methodology depicted at the base measurement shown in Table 4 [66]. Each of the criteria is contrasted with each of the alternatives, and each of the criteria at a particular degree is evaluated with the relevant higher-level criterion. Ultimately, each of the 1st stage criteria is contrasted with the goal. In Table 4, a comparison is possible by putting together matrices with the identical approach and indication as those.

Table 4. Saaty's scale

Importance level	Degree
Extreme	Nine
Very strong	Seven
Strong	Five
Moderate	Three
Equal	One

The attributes' priorities are utilized as weights to compute the alternatives' priorities in each of the criteria once the criteria' priorities have been established in relation to the overall aim. Usually, inconsistencies arise when comparing several data sets. These are tolerable to a certain extent. As a result, Saaty outlined a test with a characteristic to determine if assessment incoherence is favorable and explained how to calculate an indicator of inconsistency from a range of assessments. The approach has been used and proven effective in real-world scenarios as documented by do Leal and Nunes and Leal and Vieira, as well as in the bibliometric studies as given by Guimarães and his co-workers and Mendes and his co-workers [67, 68].

2.1 Data and Variables

It is important to obtain the highest performance product by selecting the correct and appropriate welding material, taking into account the mentioned types of features together with the features of the place where the super alloy will be used. Some characteristic properties of the welding materials for superalloys are given in Table 5.

Table 5. Some welding materials' characteristic properties for superalloys.

Code	Welding Material	Welding Process	Corrosion Resistance	Hardness	Elongation	Strength of Yield	Eventual Tensile Strength
				(HRc)	(%)	(MPa)	(MPa)
W-1	Alloy 625	MIG, TIG	Excellent	30	30	310	620
W-2	Alloy 625	MIG, TIG	Excellent	40	50	690	930
W-3	Stellite 6	MIG, TIG	Excellent	45	3	550	700
W-4	Stellite 6	MIG, TIG	Excellent	65	7	900	1100
W-5	Rene 41	TIG	Good	28	15	400	700
W-6	Rene 41	TIG	Good	40	25	800	1000
W-7	Waspaloy	TIG	Good	35	12	450	700
W-8	Waspaloy	TIG	Good	45	20	700	950

W-9	Haynes 230	MIG, TIG	Good	30	20	250	460
W-10	Haynes 230	MIG, TIG	Good	40	30	570	900
W-11	Hastelloy C276	Stick, MIG, TIG	Excellent	30	30	310	580
W-12	Hastelloy C276	Stick, MIG, TIG	Excellent	40	50	600	900
W-13	Inconel 718	MIG, TIG	Excellent	30	10	500	620
W-14	Inconel 718	MIG, TIG	Excellent	40	30	1100	1200
W-15	Inconel 625	Stick, MIG, TIG	Excellent	30	30	310	620
W-16	Inconel 625	Stick, MIG, TIG	Excellent	40	50	690	930

2.2 Empirical Results

This section provides the details of the steps along with the empirical findings. This research employs an analytical hierarchy method to assess the attractiveness of various welding materials for superalloys based on a number of attributes. Table 6 provides the relative importance of each attribute based on pairwise comparison of these attributes. Comparison is made based on the 9-point scale provided in Table 4.

Table 6. Decision-matrix

Matrix	Corrosion Resistance	Hardness	Elongation	Strength of Yield	Strength of Eventual Tensile	Criteria Weights
Corrosion Resistance	1	0.33	0.25	0.5	0.5	7.92%
Hardness	3	1	0.33	0.5	0.5	13.39%
Elongation	4	3	1	3	3	42.89%
Strength of Yield	2	2	0.33	1	1	17.90%
Strength of Eventual Tensile	2	2	0.33	1	1	17.90%

Figure 1 illustrates the resulting relative priorities; *elongation* is found to be the attribute with the maximum effect on the selection of alternative materials.

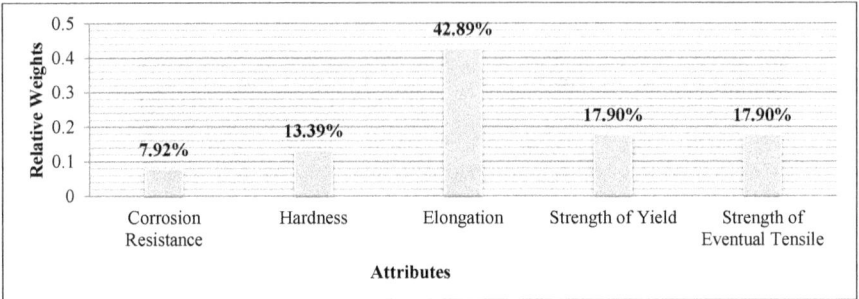

Figure 1. Relative Priorities

Since the values of different variables are subject to different scales, the values in Table 5 need to be normalized. To find the normalized values of the listed characteristics, the worths in Table 5 are divided through the total of each of the rows, resulting in the normalized worths presented in the top half of Table 7.

Table 7. Normalized and Weighted Scores of the Alternatives

	W-1	W-2	W-3	W-4	W-5	W-6	W-7	W-8	W-9	W-10	W-11	W-12	W-13	W-14	W-15	W-16
C.Resist.	0.071	0.071	0.071	0.071	0.071	0.043	0.043	0.043	0.043	0.043	0.071	0.071	0.071	0.071	0.071	0.071
Hardness	0.049	0.066	0.074	0.107	0.046	0.066	0.058	0.074	0.049	0.066	0.049	0.066	0.049	0.066	0.049	0.066
Elong.	0.073	0.121	0.007	0.017	0.036	0.061	0.029	0.049	0.049	0.073	0.073	0.121	0.024	0.073	0.073	0.121
S.of Yield	0.034	0.076	0.060	0.099	0.044	0.088	0.049	0.077	0.027	0.062	0.034	0.066	0.055	0.120	0.034	0.076
S.of E. Tens.	0.048	0.072	0.054	0.085	0.054	0.077	0.054	0.074	0.070	0.045	0.070	0.048	0.093	0.048	0.072	
Weighted Priorities																
C.Resist.	0.0057	0.005	0.005	0.005	0.005	0.003	0.003	0.003	0.003	0.003	0.005	0.005	0.005	0.005	0.005	0.005
Hardness	0.0066	0.008	0.009	0.014	0.006	0.008	0.007	0.009	0.006	0.008	0.006	0.008	0.006	0.008	0.006	0.008
Elong.	0.0312	0.052	0.003	0.007	0.015	0.026	0.012	0.020	0.020	0.031	0.031	0.052	0.010	0.031	0.031	0.052
S.of Yield	0.0061	0.013	0.010	0.017	0.007	0.015	0.008	0.013	0.004	0.011	0.006	0.011	0.009	0.021	0.006	0.013
S.of E. Tens.	0.0086	0.012	0.009	0.015	0.009	0.013	0.009	0.013	0.006	0.012	0.008	0.012	0.008	0.016	0.008	0.012
Total W. Score	0.0582	0.092	0.039	0.060	0.045	0.067	0.042	0.061	0.042	0.067	0.057	0.090	0.041	0.083	0.058	0.092

These normalized worths are then multiplied by the relative weights of each attribute to generate the weighted priorities of the decision options displayed in the bottom half of Table 7. The outputs indicate that W2 and W9 have the maximum score (0.092) among all other alternative materials. Figure 2 illustrates the resulting relative priorities of the decision alternatives along with attribute-specific values that were provided in Table 7.

Superalloys: Fundamentals and Applications   Materials Research Forum LLC
Materials Research Foundations 178 (2025) 316-332   https://doi.org/10.21741/9781644903698-16

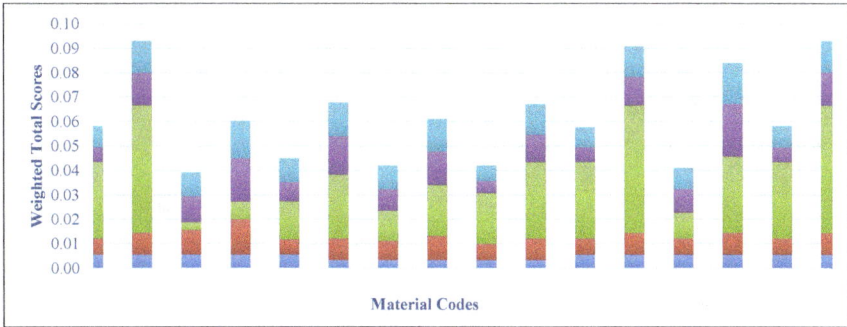

Figure 2. Weighted Scores

Conclusions

Superalloy welding materials usually contain boron, carbon, chromium, and molybdenum in certain amounts to increase the overall performance, strength, and resistance to corrosion of the weldings. They have been welded in a variety of ways due to their enhanced characteristics [16, 18–20]. Superalloy welding often uses high-strength, high-hardness welding wire with precisely measured quantities of carbon and boron. To guarantee solid and long-lasting connections between superalloy materials, various welding operations are used, including welding of electron beam, welding of gas-tungsten-arc, laser beam welding, and welding of resistance. In order to preserve performance and integrity in high-temperature and corrosive conditions, it is crucial for industries like aerospace and power generation to do research on welding materials for superalloys. Thus, component design and production experts are being forced to innovate and improve material production techniques, such as welding material, due to their growing use in aerospace applications.

As a result, it's clear that superalloy welding is reinventing the welding industry. Their various qualities and benefits are exactly what we will require in the future for things like turbines, cars, airplanes, and aerospace. Although they are limited in certain ways by the state of technology and the way they are seen in the modern world, these are readily surmountable.

References

[1] M.J. Ceslak, T.J. Headley, A.D. Romig, The welding metallurgy of Hastelloy Alloy C-4, C-22 and C-276, Metall. Mater. Trans. A. 17 (1986) 2035-2047. https://doi.org/10.1007/BF02645001

[2] O.A. Ojo, M.C. Chaturvedi, Liquation Microfissuring in the Weld Heat-Affected Zone of an Overaged Precipitation-Hardened Nickel-Base Superalloy, Metall. Mater. Trans. A 38A (2007) 356-369. https://doi.org/10.1007/s11661-006-9025-1

[3] M.J. Perricone, J.N. Dupont, M.J. Cieslak, Solidification of hastelloy alloys: An alternative interpretation, Metall. Mater. Trans. A 34A (2003) 1127-1132. https://doi.org/10.1007/s11661-003-0132-y

Materials Research Forum LLC
https://doi.org/10.21741/9781644903698-16

[4] M.H. Sun, X.J. Wang, T.D. Xia, T.Z. Liu, Welding of HP45NbTi superalloy pipe, J. Weld. Join. 8 (2006) 22-26.

[5] S. Petronic, A. Milosavljevic, The heat treatment effect on the multicomponent nickel alloys structure, FME Transaction, FME Trans. 35 (2007) 189-193.

[6] R.B. Leonard, Thermal stability of hastelloy alloy C-27, Corrosion 25 (1969) 222-228. https://doi.org/10.5006/0010-9312-25.5.222

[7] D.S. Duvall, W.A. Owczarski, Further Heat-Affected- Zone Studies in Heat Resistant Nickel Alloys, Weld. J. 46 (1997) 4235-4325.

[8] O.A. Idowu, O.A. Ojo, M.C. Chartuverdi, Effect of heat input on heat affected zone cracking in laser welded ATI Allvac 718Plus superalloy, Mater. Sci. Eng. A A454-A455 (2007) 389-397. https://doi.org/10.1016/j.msea.2006.11.054

[9] Matthews SJ 1976, Superalloys: Metallurgy and Manufacture, Baton Rouge, Claitor's Publishing Division Proc. 3rd Int. Conference on 'Superalloys' 215-226. https://doi.org/10.7449/1976/Superalloys_1976_215_226

[10] S.S. Zhao, H.J. Xu, M. Xie, Y. Zhao, Studies of mechanical properties and microstructures of argon-shielded arc welding joint of gh536 superalloy, Dalian Jiaotong Univ. 10 (2010) 47-49.

[11] T. Xu, Z.L. Gong, Y.W. Shen, H. Yang, Study on the welding procedure of heat-resistant alloy, Turbine Technol. 4 (2005) 313-314.

[12] W.F. Savage, B. Krantz, Welding research supplement, 1974 Res. Suppl. 7 292-s-302-s

[13] Saju T, Velu M, Review on welding and fracture of nickel based superalloys, Mater. Today Proc. 46, (2021) 7161-7169. https://doi.org/10.1016/j.matpr.2020.11.334

[14] E.A. Loria, Recent developments in the progress of superalloy 718, JOM 44 (1992) 33-36. https://doi.org/10.1007/BF03222252

[15] E.A. Ezugwu, A.R. Machado, I.R. Pashby, J. Wallbank, Effect of high-pressure coolant supply when machining a heat-resistant nickel-based superalloy, J. Soc. Tribologists Lub. Eng. 47 (1991) 751-757.

[16] A.R. Mashreghi, H. Monajatizadeh, M. Jahazi, S. Yue, High temperature deformation of nickel base superalloy Udimet 520, Mater. Sci. Tech 20 (2004) 161. https://doi.org/10.1179/026708304225010343

[17] H.Q. Zhang, H.Y. Zhao, H. Zhang, L.H. Li, X.A. Zhang, Analysis on the microfissuring behavior in the heat-affected zone of electron-beam welded nickel-based superalloy. J. Mater. Eng. 3 (2005) 22-23.

[18] O. Alniak, B. Fevzi, Hot forging behavior of nickel based superalloys under elevated temperatures, Mater. Des. 31 (2010) 1588-1592. https://doi.org/10.1016/j.matdes.2009.09.020

[19] K.T. Zysk, 1990 AIAA/SAE/ASME/ASEE 26th Joint Propulsion Conference, Orlando, Fla, AIAA 90-2514.

Superalloys: Fundamentals and Applications   Materials Research Forum LLC
Materials Research Foundations 178 (2025) 316-332   https://doi.org/10.21741/9781644903698-16

[20] K. Shinozaki, Welding and joining Fe and Ni-base superalloys, Jpn. Weld. Technol. J. 43 (1995) 60.

[21] W.F. Savage, C.E. Jackson, Factors influencing strain-age cracking in Inconel X-750, Weld. J. 50 (1971) 302s-303s.

[22] H. Zhang, J.K. Li, Z.W. Guan, Y.J. Liu, D.K. Qi, Q.Y. Wang, Electron beam welding of Nimonic 80A: Integrity and microstructure evaluation, Vacuum 151 (2018) 266-274. https://doi.org/10.1016/j.vacuum.2018.01.021

[23] V. Rajkumar, N. Arivazhagan, Role of pulsed current on metallurgical and mechanical properties of dissimilar metal gas tungsten arc welding of maraging steel to low alloy steel, Mater. Des. 63 (2014) 69-82. https://doi.org/10.1016/j.matdes.2014.05.055

[24] K.D. Ramkumar, W.S. Abraham, V. Viyash, N. Arivazhagan, A.M. Rabel, Investigations on the microstructure, tensile strength and high temperature corrosion behaviour of Inconel 625 and Inconel 718 dissimilar joints, J. Manuf. Process. 25 (2017) 306-322, https://doi.org/10.1016/j.jmapro.2016.12.018

[25] P.K. Korrapati, V.K. Avasarala, M. Bhushan, K. Devendranath, Ramkumar, N., Arivazhagan, N.S. Narayanan, Assessment of mechanical properties of PCGTA weldments of inconel 625, Proced. Eng. 75 (2014) 9-13 https://doi.org/10.1016/j.proeng.2013.11.002

[26] E. Farahani, M. Shamanian, F. Ashrafizadeh, A comparative study on direct and pulsed current gas tungsten arc welding of alloy 617, Growth (Lakeland). 02 (2012) 1-6.

[27] J. Graneix, J.D. Beguin, F. Pardheillan, J. Alexis, T. Masri, Optimization of Exit Diameter of Hole on Ti-6Al-4V Superalloy Using Laser Drilling. 2014 MATEC Web of Conferences 14. https://doi.org/10.1051/matecconf/20141413006

[28] M.A. Rezaie, H.N. Moosavy, The effect of pre-cold treatment on microstructure, weldability and mechanical properties in laser welding of superalloys, J. Manuf. Process. 34 (2018) 339-348. https://doi.org/10.1016/j.jmapro.2018.06.018

[29] F. Nematzadeh, M.R. Akbarpour, S. Parvizi, A.H. Kokabi, S.K. Sadrnezhaad, Effect of welding parameters on microstructure, mechanical properties and hot cracking phenomenon in Udimet 520 superalloy, Mater. Des. 36 (2012) 94-99. https://doi.org/10.1016/j.matdes.2011.10.020

[30] K. Shinozaki, H. Kuroki, X. Lou, H. Ariyoshi, M. Shirai, Effect of welding parameters on laser weldability of Inconel 718, Taylor Francis 13 (2010) 945-951. https://doi.org/10.1080/09507119909452078

[31] Q. Wang, D.L. Sun, Y. Na, Y. Zhou, X.L. Han, J. Wang, Effects of TIG welding parameters on morphology and mechanical properties of welded joint of ni-base superalloy, Procedia Eng. 10 (2011) 37-41. https://doi.org/10.1016/j.proeng.2011.04.009

[32] M. Chen, T. Wu, T. Chen, S. Jeng, L. Tsay, The comparison of cracking susceptibility of in52m and in52mss overlay welds. Metals 9 (2009) 651. https://doi.org/10.3390/met9060651

[33] F. Hanning, A. Khan, J.S. Nordenström, O. Ojo, J. Andersson, Investigation of the Effect of Short Exposure in the Temperature Range of 750-950 _C on the Ductility of Haynes® 282®

by advanced microstructural characterization. Metals. 9 (2019) 1357.
https://doi.org/10.3390/met9121357

[34] S. Singh, J. Andersson, Heat-affected-zone liquation cracking in welded cast haynes®
282®. Metals. 10 (2010) 29 https://doi.org/10.3390/met10010029

[35] P. Alvarez, L. Vázquez, N. Ruiz, P. Rodríguez, A. Magaña, A. Niklas, F. Santos,
Comparison of hot cracking susceptibility of tıg and laser beamwelded alloy 718 by
varestraint testing. Metals 2019, 9, 985. https://doi.org/10.3390/met9090985

[36] T. Raza, J. Andersson, L. Svensson, Varestraint testing of selective laser additive
manufactured alloy 718-influence of grain orientation. Metals 2019, 9, 1113.
https://doi.org/10.3390/met9101113

[37] T. Raza, K. Hurtig, G. Asala, J. Andersson, L. Svensson, O. Ojo, Influence of heat
treatments on heat affected zone cracking of gas tungsten arcwelded additivemanufactured
alloy 718. Metals 9 (2019) 881. https://doi.org/10.3390/met9080881

[38] Phanden, R. Kumar, J.C.E. Ferreira. Biogeographical and variable neighborhood search
algorithm for optimization of flexible job shop scheduling. Adv. Indust. Prod. Eng.
Springer, Singapore, 2019. 489-503. https://doi.org/10.1007/978-981-13-6412-9_48

[39] Phanden, R. Kumar, L. Kumar Saharan, J.A. Erkoyuncu. Simulation based cuckoo search
optimization algorithm for flexible job shop scheduling problem. Proc. Int. Conf. Intell. Sci.
Technol. 2018. https://doi.org/10.1145/3233740.3233752

[40] R.K. Phanden, H.I. Demir, R.D. Gupta, "Application of genetic algorithm and variable
neighborhood search to solve the facility layout planning problem in job shop production
system." 7th international conference on industrial technology and management (ICITM).
IEEE, 2018. https://doi.org/10.1109/ICITM.2018.8333959

[41] R.P. Singh, M. Tyagi, R. Kataria, Investigation of dimensional deviation in wire EDM of
M42 HSS using cryogenically treated brass wire. in: Operations Management and Systems
Engineering. Lecture Notes on Multidisciplinary Industrial Engineering. Springer,
Singapore, 2019. https://doi.org/10.1016/j.matpr.2019.08.028

[42] R. Kataria, R.P. Singh, J. Kumar, Investigation of dimensional deviation in wire EDM of
M42 HSS using cryogenically treated brass wire, 2016 AIMS Mater. Sci. 3 (2016): 1391-
1409.

[43] H. Bisht, R.P. Singh, V. Sharma, Study of impact strength in TIG welding. in: Optimization
Methods in Engineering. Lecture Notes on Multidisciplinary Industrial Engineering.
Springer, Singapore, 2021.

[44] M.B. Henderson, D. Arrell, M. Heobel, R. Larsson, Nickel-based superalloy welding
practices for ındustrial gas turbine applications m.b. henderson, Sci. Technol. Weld. Join. 9
(2004) 1-14. https://doi.org/10.1179/136217104225017099

[45] S.G.K. Manikandan, D. Sivakumar, M. Kamaraj, Microsegregation and interdendritic Laves
phase, Weld. Inconel 718 Superalloy. (2019) 47-52, https://doi.org/10.1016/B978-0-12-
818182-9.00003-7

Materials Research Forum LLC
https://doi.org/10.21741/9781644903698-16

[46] A. Benoit, S. Jobez, P. Paillard, V. Klosek, T. Baudin, Study of inconel 718 weldability using MIG CMT process, Sci. Technol. Weld. Join. 16 (2011) 477-482. https://doi.org/10.1179/1362171811Y.0000000031

[47] M.M.Z. Ahmed, B.P. Wynne, J.P. Martin, Effect of friction stir welding speed on mechanical properties and microstructure of nickel based super alloy Inconel 718, Sci. Technol. Weld. Join. 18 (2013) 680-687. https://doi.org/10.1179/1362171813Y.0000000156

[48] G.V.B. Lemos, S. Hanke, J.F. Dos Santos, L. Bergmann, A. Reguly, T.R. Strohaecker, Progress in friction stir welding of Ni alloys, Sci. Technol. Weld. Join. 22 (2017) 643-657. https://doi.org/10.1080/13621718.2017.1288953

[49] Y. Zhu, Z. Zhu, Z. Xiang, Z. Yin, Z. Wu, W. Yan, Microstructural evolution in 4Cr10Si2Mo at the 4Cr10Si2Mo/Nimonic 80A weld joint by inertia friction welding, J. Alloy. Compds. 476 (2009) 341-347 https://doi.org/10.1016/j.jallcom.2008.08.062

[50] V. Balasubramanian, V. Ravisankar, G.M. Reddy, Effect of pulsed current and post weld aging treatment on tensile properties of argon arc welded high strength aluminium alloy, Mater. Sci. Eng. A 459 (2007) 19-34. https://doi.org/10.1016/j.msea.2006.12.125

[51] S.S. Sandhu, A.S. Shahi, Metallurgical, wear and fatigue performance of Inconel 625 weld claddings, J. Mater. Process. Technol. 233 (2016) 1-8. https://doi.org/10.1016/j.jmatprotec.2016.02.010

[52] S.G.K. Manikandan, D. Sivakumar, M. Kamaraj, Physical metallurgy of alloy 718, Weld. Inconel 718 Superalloy. (2019) 1-19. https://doi.org/10.1016/B978-0-12-818182-9.00001-3

[53] S.G.K. Manikandan, D. Sivakumar, M. Kamaraj, Introduction, Weld. Inconel 718 Superalloy. (2019) xiii-xvi. 9.00014-1. https://doi.org/10.1016/B978-0-12-818182-9.00014-1

[54] H. Hänninen, P. Aaltonen, A. Brederholm, U. Ehrnsten', H. Gripenberg, A. Toivonen, J. Pitkänen, I. Virkkunen, Dissimilar metal weld joints and their performance in nuclear power plant and oil refinery conditions, VTT Tied. -Valt, Tek. Tutkimusk. (2006) 3-208.

[55] L.A. James, Fatigue-crack growth in Inconel 718 weldments at elevated temperatures. Weld. J. 57 (1978) 17-23.

[56] Welding Handbook, Weld. Handb. (1983). https://doi.org/10.1007/978-1-349-05561-6. https://doi.org/10.1007/978-1-349-05561-6

[57] Junyi Chai, James N.K. Liu, E.W.T. Ngai, Application of decision-making techniques in supplier selection: a systematic review of literature, Expert Syst. Appl. 40 (2013) 3872-3885. https://doi.org/10.1016/j.eswa.2012.12.040

[58] Ashu Gupta, K. Singh, Rajesh Verma, A critical study and comparison of manufacturing simulation softwares using analytic hierarchy process, J. Eng. Sci. Technol. 5 (2010) 108-129.

[59] A. Jadhav, R.M. Sonar, Framework for evaluation and selection of the software packages: A hybrid knowledge based system approach, J. Syst. Softw. 84 (2011) 1394-1407. https://doi.org/10.1016/j.jss.2011.03.034

[60] G. Rajesh, P. Malliga, Supplier selection based on AHP QFD methodology, Procedia Eng. 64 (2013) 1283-1292. https://doi.org/10.1016/j.proeng.2013.09.209

[61] D. Xinyang, H. Yong, D. Yong, M. Sankaran, Supplier selection using AHP methodology extended by D numbers Expert Systems with Applications, Expert Syst. Appl. 41 (2014) 156-167. https://doi.org/10.1016/j.eswa.2013.07.018

[62] M.C.Y. Tam, V.M.R. Tummala, An application of the AHP in vendor selection of a telecommunications system, Omega 29 (2001) 171-182. https://doi.org/10.1016/S0305-0483(00)00039-6

[63] S.A. Byeong, The analytic hierarchy process with interval preference statements, Omega 67 (2017) 177-185. https://doi.org/10.1016/j.omega.2016.05.004

[64] A.M. Guimaraes, J.E. Leal, P. Mendes, Discrete-event simulation software selection for manufacturing based on the maturity model, Comput. Ind. 103 (2018) 14-27. https://doi.org/10.1016/j.compind.2018.09.005

[65] P. Mendes, J.E. Leal, A.M.T. Thomé, A maturity model for demand-driven supply chains in the consumer product goods industry, Int. J. Prod. Econ. 179 (2016) 153-165. https://doi.org/10.1016/j.ijpe.2016.06.004

[66] T.L. Saaty, Theory and Applications of the Analytic Network Process: Decision Making With Benefits, Opportunities, Costs and Risks, RWS Publications, Pittsburg, USA, 2013. https://doi.org/10.1007/978-1-4614-7279-7

[67] V.N. Vieira, J.E. Leal, Performance criteria for Liquid Bulk Storage Terminals (LBST), using AHP, in: A. Leiras, C. González-Calderón, I. de Brito Junior, S. Villa, H. Yoshizaki (Eds.), Operations Management for Social Good. POMS 2018. Springer Proceedings in Business and Economics, Springer, Cham, 2020, pp. 421-429. https://doi.org/10.1007/978-3-030-23816-2_41

[68] G.S. Nunes, J.E. Leal, Decision-making method for facility location for offshore logistic operation based on AHP, in: A. Leiras, C. González-Calderón, I. de Brito Junior, S. Villa, H. Yoshizaki (Eds.), Operations Management for Social Good. POMS 2018. Springer Proceedings in Business and Economics, Springer, Cham, 2020, pp. 721-729. https://doi.org/10.1007/978-3-030-23816-2_71

Superalloys: Fundamentals and Applications   Materials Research Forum LLC
Materials Research Foundations 178 (2025) 333-363   https://doi.org/10.21741/9781644903698-17

Chapter 17

Single Crystal Superalloys

Annada Mishra, Pooja Mohapatra, Shreelata Behera, Saleja Sahoo, Lipsa Shubhadarshinee,
Bigyan Ranjan Jali, Aruna Kumar Barick*, and Priyaranjan Mohapatra

Department of Chemistry, Veer Surendra Sai University of Technology, Siddhi Vihar, Burla,
Sambalpur 768018, Odisha, India

* akbarick_chem@vssut.ac.in, akbarick@gmail.com

Abstract

Advanced materials such as single crystal superalloys (SCSs) are made especially for high temperature uses, including turbine engines in the aerospace and electricity generation companies. Their single crystalline shape, which distinguishes their microstructures, confers superior mechanical behaviours, including remarkable fatigue strength and creep resistance. As a result, components can continue to function under severe mechanical and thermal stresses, significantly increasing their longevity and efficiency. The creation of these superalloys requires thorough control over processing methods, including directed solidification, which maximizes crystal orientation alignments. The ability to create complicated shapes is made possible by recent advancements in additive manufacturing, i.e., 3D printing, which expands the range of possible uses for these materials. In order to meet the demands of contemporary turbine technology, Ni-based SCSs (Ni-SXs) are essential due to their key performance characteristics, such as oxidation resistance and high-temperature stability. These superalloys will be crucial for the development of turbine design and operation as the need for cleaner and more effective energy sources. Continuous investigation endeavors to improve the compositions and processing techniques of superalloys in order to further extend the performance features. Future high temperature applications depend on the ongoing development of Ni-SXs, which open the door to the development of next-generation energy and aerospace systems.

Keywords

Single Crystal Superalloys, Ni-SXs, Microstructure, Properties, Deformation Mechanism, Applications

Contents

1.   Introduction

The super alloys are the compounds preferred for component functioning under the load at temperatures above or more than 540 °C, owing to their incredible capacity to sustain endurance, tolerate creep and fatigue, and withstand oxidation at such high temperatures [1–4]. Turbine blades, which typically perform at elevated temperatures and endless centrifugal forces, are among the most widely recognized purposes to superalloys [5,6]. Superalloys are typically put into three distinct groups: Fe-based, Ni-based, and Co-based alloys. Given that, the Ni-SXs are more reasonably priced than Co-based superalloys as well as possess much better extreme temperature qualities than the Fe-based superalloys. Ni-SXs constitute one of the most frequently implemented among these three types of superalloys [7–9]. Initially developed superalloys evolved out of Fe, Co, Ni, and Cr; such alloys can usually be employed, while Ni-based superalloys are the most contemporary [10]. Nevertheless, the vast majority of Fe-Ni-Cr alloys have been classified as superalloys developed from Fe. Eventually, the wrought alloys were further strengthened by precipitates; which initially occurred via micro-adding carbon in the alloy framework, leading to the precipitation of the carbide particles [11]. Countless generations of SCSs have been originated following the initially developed SCSs i.e., PWA1480, emerged in the beginning of the eighties. In order to support the frequent development of the complicated aircraft engines, SCSs have been

Superalloys: Fundamentals and Applications   Materials Research Forum LLC
Materials Research Foundations 178 (2025) 333-363   https://doi.org/10.21741/9781644903698-17

upgraded up to the fifth generation [12]. The use of Ni-SXs is increasing in turbines, specifically in turbine blades, due to their exceptional durability in complicated and aggressive conditions, where centrifugal forces and thermal tensions fall behind the combustor [13–15].

The elevated conditions, along with stresses imposed upon turbine blades, would eventually cause the happening of creep deformation alongside diminish their service lifespan [16–18]. Hence, one of the utmost objective of formulating novel eras of Ni-SXs was to strengthen their creep resistance at higher temperatures [19]. Quality assurance, engineering methods, and materials science are used to generate Ni-SXs. A suitable composition must be chosen, a homogenous melt must be produced by vacuum induction melting or arc remelting, and contaminations must be avoided by regulating temperature and environment. Single crystals are grown using the directional solidification technique, also known as the Bridgman method, in which a temperature gradient is maintained by heating the mold. Heat treatments, which frequently include solutionizing and aging, enhance microstructures. Components are shaped using investment casting or additive manufacturing processes and finely machined to ensure stringent tolerances. The resistance to oxidation and fatigue is improved by different surface treatments like coating or shot peening. Material integrity is guaranteed by rigid testing such as TEM and SEM, which includes mechanical, microstructural, and non-destructive tests. A combination of materials science, engineering methods, and quality control is needed to produce high-performance SCSs that can be used in gas turbines and jet engines. Careful management is required at each phase to ensure the necessary qualities. Tensile, compression, and creep properties are crucial elements in assessing the effectiveness and longevity of the SCSs, particularly in challenging conditions like gas turbines.

2.   Composition and Microstructure of Ni-SXs

2.1   Microstructure of Ni-SXs

These alloys are composed of single crystal of metal, means that they lack grain boundaries. This eliminates the potential for grain boundary degradation and improves the mechanical strength at elevated temperature. It consists of both γ and γ' phases. The γ phase is the primary matrix phase of the alloy, which is made up of solid solution of Ni or Co that provides the general structural support. The γ' phase is a precipitate phase often consisting of compounds like $Ni_3(Al, Ti)$. These precipitates are dispersed within the γ matrix. Both γ and γ' phases have face centred cubic (FCC) crystal structure but with different atomic arrangements due to the presence of different elements. Ni-SXs primarily possess a dual phase microstructure, which consists of an ordered L12 structured γ' phase alongside an FCC structured γ phase. This dual phase combined framework, which is formed by γ' precipitation by means of post-casting thermal treatment, is the alloys 'service' microstructure. The γ matrix exists in slender parallel channels between the γ' cuboids, whereas the precipitates of γ' are cuboidal in form having {100} facets and are all oriented in the directions of the single crystals. The interactions between the two phases are fully cohesive [4]. The strength and creep resistance of the alloys that are derived from the high-volume fraction in alloy are enhanced due to the high ordered structure of γ' [20]. There is a high possibility of presence of γ–γ' eutectic framework for SCSs. The created solid solution γ phase of Ni-SXs exist in same FCC crystal structure of original Ni [19]. Macroscopically, the compound is a single crystal owing to its γ–γ' eutectic framework and randomly assigned crystallographic arrangement; nevertheless, microscopically, it is polycrystalline, causing degradation of its efficiency at high temperature

[21]. About 300–500 nm in size, the γ' cubic precipitates are uniformly distributed in the γ matrix and the γ/γ' contact is located along {001} planes [22]. Misfit engineering has optimized the constituents of more than ten components in Ni-SXs to control the volume fraction, overall mechanical properties, and precipitate morphology. W and Re are more specifically added to improve the γ matrix phase. Al, Ta, and Ti have stabilized γ' precipitates [23].

2.2 Compositions of SCSs

The compositions of SCSs are carefully engineered, often incorporating elements like Co, Cr, Mo, W, and Re in addition to Ni. These elements enhance various properties such as, oxidation resistance, strength, and thermal stability. Table 1 represents the composition of various Ni-SXs.

Table 1: Compositions of Some Typical Ni-SXs (wt.%) [1,12,24–27].

Generation	Alloys	Ni	Co	Cr	Mo	W	Al	Ti	Ta	Re	Ru	Others
First	PWA1480	Bal.	5.0	10.0	–	4.0	5.0	1.5	12.0	–	–	0.015C
	CMSX-2	Bal.	4.6	8.0	0.6	8.0	5.6	1.0	9.0	–	–	0.015C
	SRR99	Bal.	5.0	8.0	–	10.0	5.5	2.2	12.0	–	–	0.015C
	NASAIR100	Bal.	–	8.5	1.1	10.1	5.8	1.2	3.3	–	–	0.015C
Second	PWA1484	Bal.	10.0	5.0	2.0	6.0	5.6	–	9.0	3.0	–	0.1Hf
	CMSX-4	Bal.	8.0	7.0	2.0	5.0	6.2	–	7.0	3.0	–	0.2Hf
	ReneN5	Bal.	9.0	6.5	0.6	6.0	5.6	1.0	6.5	3.0	–	0.1Hf
Third	ReneN6	Bal.	12.5	4.2	4.4	6.0	5.7	–	7.2	5.4	–	0.05C, 0.004B
	CMSX-10	Bal.	3.0	2.0	0.4	5.0	5.7	0.2	8.0	6.0	–	0.03Hf
	TMS-75	Bal.	12.0	3.0	2.0	6.0	6.0	–	6.0	5.0	–	0.1Hf
Fourth	PWA1497	Bal.	16.5	2.0	2.0	6.0	5.5	–	8.2	5.9	3.0	0.15Hf, 0.03C
	MC-NG	Bal.	–	4.0	1.0	5.0	6.0	0.5	5.0	–	–	0.1Hf
	EPM-102	Bal.	16.5	2.0	2.0	6.0	5.5	–	8.3	5.9	3.0	0.15Hf, 0.03C
Fifth	TMS-162	Bal.	5.8	3.0	3.9	5.8	5.8	–	5.6	4.9	6.0	0.09Hf
	TMS-196	Bal.	5.6	4.5	2.4	5.0	5.6	–	5.6	6.4	5.0	0.1 Hf
Sixth	TMS-238	Bal.	6.5	4.6	1.1	4.0	5.9	–	7.6	6.4	5.0	0.1 Hf

3. Preparation and Characterization of Ni-SXs

Each of these steps must be carefully controlled to produce high-quality SCSs suitable for demanding applications, as shown in Figure 1.

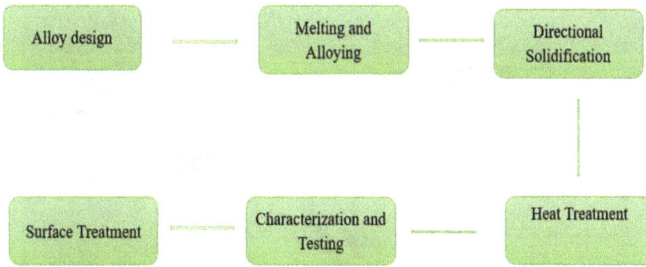

Figure 1: Flow Chart for Preparation and Characterization of SCSs.

4. Properties of Ni-SXs

4.1 Creep Properties

Since the primary factors leading to the reduction of creep resistance involve the advancement of the γ' phase's volume fraction, size, morphology, and distribution, as well as the topologically close pack (TCP) phases formation caused by heat exposure, this related creep property is an additional excellent method for calculating the microstructural reduction. Zhang *et al.* studied the creep life expectancy of DD6 alloy at 140 MPa following various amounts of heat exposure at 980 °C and 1070 °C [28]. From studies, it was observed that none of the research specimens exhibited the formation of a distinct TCP phase. The structures displayed an apparent rafting framework at 1070 °C for 1000 h, in contrast to a practically cuboidal morphology at 980 °C for 1000 h [28,29]. It was studied that the creep life significantly declines with higher duration of heat exposure up to 400 h, beyond which it becomes an extremely smaller increase in creep life with increased thermal exposure time. The primary reason for this fact is that the thermal exposure time (more than 400 h) proved to be an ideal heat treatment approach for the phase's optimum development in accordance with the overestimated characteristics. Nevertheless, over multiple periods of heat exposure at 1070 °C, the related creep life gradually decreases because of the significant deterioration of the γ and γ' phases [28,29].

Figure 2 depicts the creep tests that Cheng *et al.* carried out on the subjected specimens that include TCP phases of CMSX-4 alloy [30]. In this case, at 1050 °C for 2000 h, more TCP phases developed than at 950 °C for 2000 h, which indicates that the creep characteristic decreases as it is exposed to higher temperature or the duration is increased. The creep life clearly demonstrates an even greater divergence between 1000 h and 2000 h thermal exposure times when conducted at higher temperatures. This is mostly due to the notable rise in the TCP formation at 1050 °C for 2000 h [28].

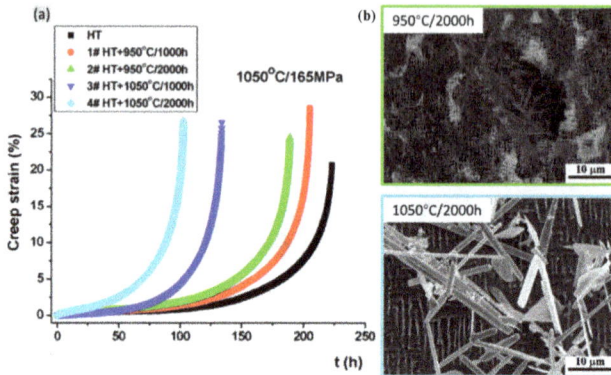

Figure 2: *Representation of (a) Creep Characteristics of Specimens in the Heated Condition and After Thermal Exposure of CMSX-4 Alloy at Various Temperatures and Times and (b) Corresponding Microstructures Following Thermal Exposure at 950 and 1050 °C for 2000 h at 165 MPa [30].*

4.2 Compression Properties

Table 2 depicts the alloy's compression features [34]. The alloy's yield strength improves as temperature rises from room temperature (RT) to 760 °C. The yield strength reduces as temperature rises from 760 °C to 980 °C. That is to say that the alloy exhibits the same aberrant yield behaviour as the other Ni-SXs. The Bertnollide type of Ni_3Al with the L12 structure [31] leads to an unusual strength-temperature relationship, meaning that the alloy strength does not progressively decrease as the temperature rises. Thermally induced cross-slip of dislocations from {111} to {110} planes have been proposed as the explanation for the temperature-dependent rise in γ' strength [32]. For the portions of the dislocations that are still in the {111} plane, the cross-slipped dislocation segments serve as the pinning centres. Consequently, when temperature rises, the yield strength increases due to the transition from a glissile to a sessile dislocation. The sole possibly movable component of the dislocations at low temperatures may be these pinning centres, which are located in the glide plane [33]. At 760 °C, the alloy's compressive strength achieves its maximum and then starts to decline as the temperature rises.

The microstructures of the alloy during compression at RT are displayed in Figure 3. The volume fraction declines to nearly about 52%, the γ' phase size develops to about 0.58 μm, the cubic γ' phase becomes irregular, and the pattern of distribution is not homogeneous. The alloy's compression microstructures at multiple temperatures are shown in Figure 4. The size and volume fraction of the γ' phase in the alloy following compression at multiple temperatures are shown in Figure 5. At ambient temperature, the γ' phase's volume fraction is not significantly changed, but its size is changed. In addition, when the alloy shape becomes erratic, a more unequal distribution may be seen. The γ' phase's size drops to near 0.44 μm, volume fraction elevates to 62%, and shape becomes more regular as the temperature elevates to 760 °C. The strength at elevated temperature of the precipitated strengthened alloy is closely correlated with the γ' phase vol.%. These results additionally show that, at 760 °C, the alloy has the highest possible compression strength. It appears that the volume fraction and size of the γ' phase elevate significantly in contrast with those at 950 °C, proving that the alloy still has good compression strength when the temperature is up to 980 °C. However, the volume fraction of the γ' phase gradually drops, and the size distribution widens more inconsistently short after compression at 980 °C [34].

Table 2: *Compressive Properties of the Alloys at Different Temperature [34].*

	RT	650 °C	760 °C	850 °C	900 °C	950 °C	980 °C
$\sigma_{0.2}$ (MPa)	935.6±9.3	958.8±24.2	1058.3±26.2	833.4±19.0	771.1±17.6	697.1±2.4	635.0±7.9

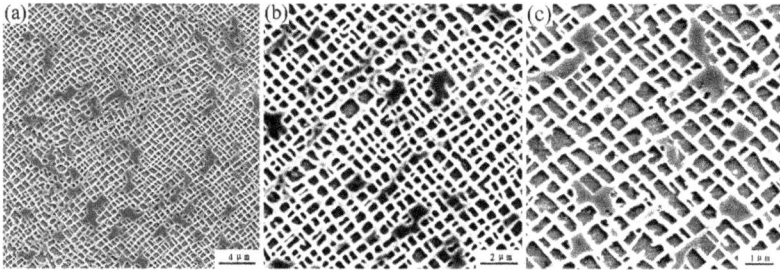

Figure 3: *Microstructure of Alloy Compressed at Ambient Temperature [34].*

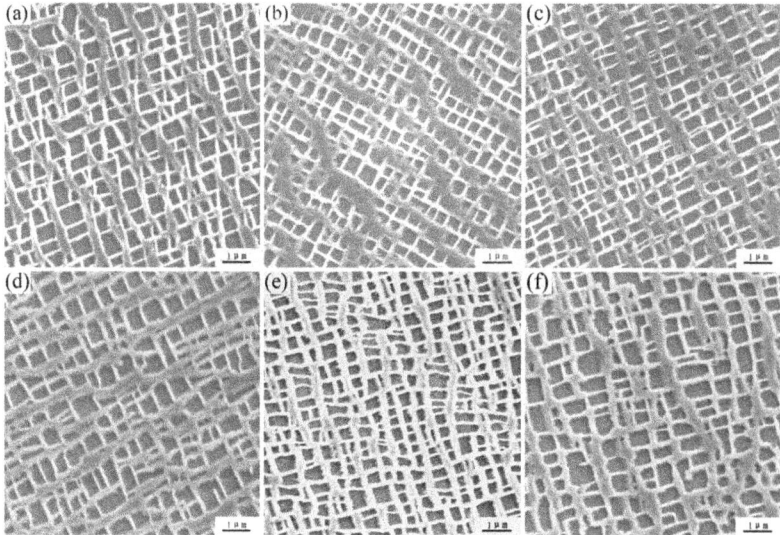

Figure 4: *Microstructures of the Alloy During Compression at Various Temperatures: (a) 650, (b) 760, (c) 850, (d) 900, (e) 950, and (f) 980 °C [34].*

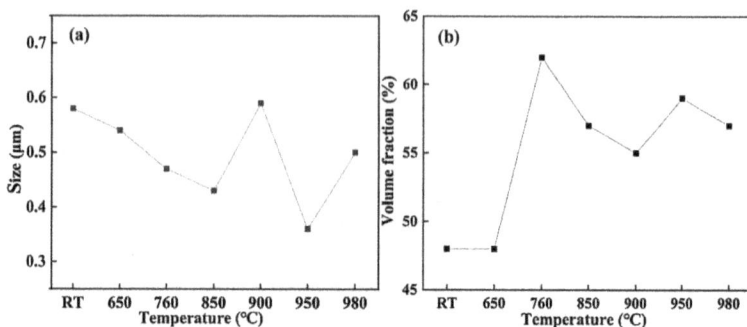

Figure 5: Modification in Dimensions and Volume Fraction of the γ': (a) Changes in Proportion to the γ' Phase and (b) Changes in Volume Fraction of the γ' Phase [34].

4.3 Tensile Properties

Figure 6 and Table 3 illustrates the tensile stress-strain curves and test results of three SCSs at different temperatures [35]. The stress-strain graphs indicate a rapid increase before to attain the yield point, followed by varying behaviours thereafter. Significant strain hardening occurs during the tensile process at relatively low temperatures of 650 and 760 °C. At 650 °C, the stress-strain curves of alloys exhibit a marked decline upon reaching the ultimate fracture strength, signifying the occurrence of typical brittle fracture, which correlates with a reduced percentage of area under the curve. When the temperature exceeds 850 °C, the stress-strain curve of the alloy predominantly indicates an initial decrease in stress, followed by a subsequent increase in post-yielding. When the tensile temperature is below 850 °C, atomic mobility within the alloy is restricted, resulting in a reduced number of active slip systems during deformation. When the tensile temperature exceeds 850 °C, atomic movement accelerates, potentially activating additional slip mechanisms during deformation. The alloy exhibits both octahedral and hexahedral slip, with numerous slip systems resulting in pronounced deformation softening [36]. The tensile strength of the three superalloys does not diminish progressively with rising temperature due to the typical relationship between flow stress and the temperature of the Ni_3Al phase with an L12 structure [37,38]. As the temperature rises to 760 °C, the enhanced strength of the γ' phase can be attributed to the thermally induced cross-slip of dislocations from the octahedral {111} planes to the cubic {001} planes. Thermal activation enhances the cross-slip of $a/2$ (011) dislocations from the {111} plane to the {001} plane; however, the Peierls force for $a/2$ (011) dislocations on the {100} plane is substantial, hence restricting their movement on the {100} plane. A substantial quantity of dislocations is confined between the {111} and {001} planes, hence enhancing the strength of the superalloy [38–40]. The tensile test findings of the three alloys indicate that the variations in alloying elements minimally influence tensile strength at lower temperatures. However, as the tensile temperature exceeds 900 °C, the strength diminishes rapidly. Superalloys can be enhanced through many techniques, including work hardening, solution strengthening, and precipitation strengthening [34,41,42].

Table 3: *Results of the Tensile Tests of SCSs at Various Temperatures [35].*

Compositions	Parameters	650 °C	760 °C	850 °C	950 °C	980 °C
0Re Alloy	$R_{p0.2}$ (MPa)	920±42	1075±41	848±6	653±13	604±11
	R_m (MPa)	1088±90	1299±23	1074±13	810±7	750±15
	A (%)	7.6±0.4	19.5±2.5	26.5±0.7	28.0±2	31.8±4.8
1.5Re Alloy	$R_{p0.2}$ (MPa)	938±36	1017±30	829±6	606±2	548±8
	R_m (MPa)	1196±27	1283±6	1013±1	745±5	671±7
	A (%)	6.0±1.5	17.0±1.0	30.0±2.0	40.0±0.6	42.0±4.1
3Re Alloy	$R_{p0.2}$ (MPa)	898±13	938±22.5	844±12	669±18	634±6
	R_m (MPa)	1110±25	1245±30	1068±9	843±16	757±18
	A (%)	9.5±3.5	19.0±7.0	25.0±0.4	25.0±1.0	22.0±2.2

$R_{p0.2}$ (MPa): Yield Strength, R_m (MPa): Tensile Strength, and A (%): Elongation at Break

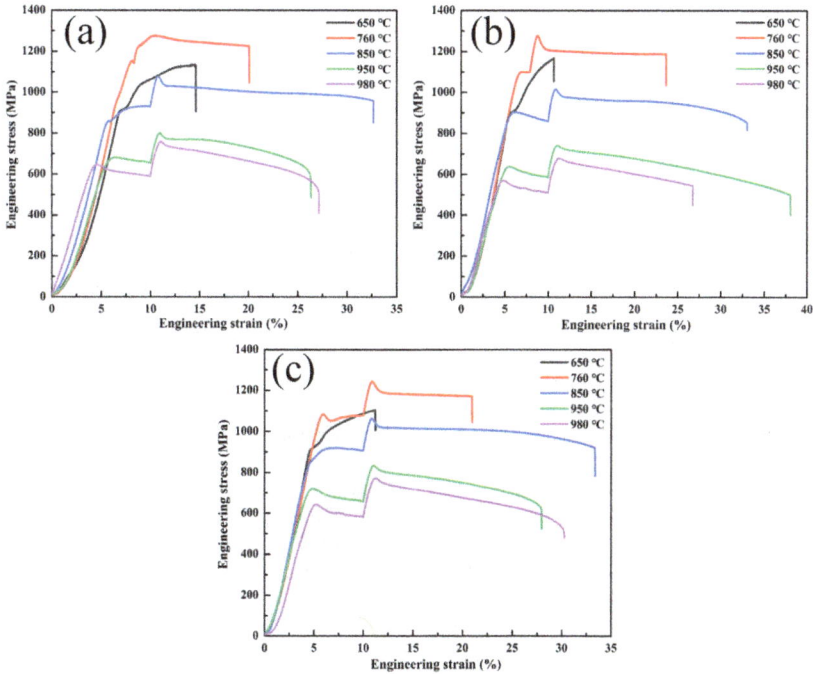

Figure 6: *Tensile Stress-Strain Curves of SCSs at Various Temperatures: (a) 0Re, (b) 1.5Re, and (c) 3Re Alloys [35].*

5.   Mechanisms of Ni-SXs

5.1   Creep Mechanism

The rate at which the original microstructures degrade i.e., highly responsive to the external factors like temperature and stress on the alloy, determines the creep behaviour. Naturally, a greater temperature and stress correspond to a quicker rate of alloy deterioration and a shorter creep life [43]. Larger creep strain (ε) can result from increasing stresses over time, as shown in Figure 7(a). Furthermore, Figure 7(b) and (c) demonstrates that larger stresses also result in higher rates of creep strain (ε). Conversely, the same tension might result in more deformation at higher temperatures. It is interesting to note that the stress-strain curves exhibit diverse tendencies under various external conditions, which are a result of various dislocation motions and interactions.

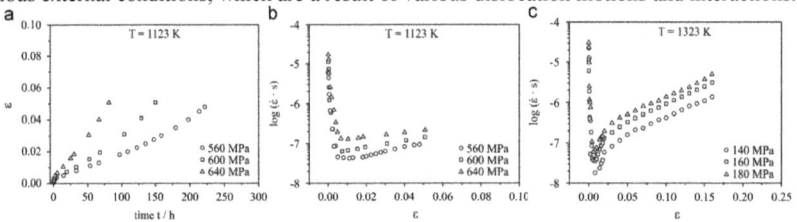

Figure 7: *Series of Creep Curves in a Ni-SX at Varying Temperatures and Applied Forces. (**a**) Stress Versus Time Creep Curves for Different Stresses at 1123 K and Creep Rate Versus Strain Curves in a Log-Linear Plot for Different Stresses at (**b**) 1123 K, and (**c**) 1323 K [43].*

The creep mechanisms of Ni-SXs are essentially separated into three kinds of behaviour as per the applied temperatures: lower temperature creep about 750 °C, mid-temperature creep about 950 °C, and higher temperature creep above 1100 °C, as displayed in Figure 8–10. Temperatures have the greatest influence on the dislocation movements. At low temperatures, particularly around 750 °C, dislocations are trapped inside γ matrix, and at low stresses, dislocation motion is inactive with regard to the stable γ/γ' microstructure. Creep deformation occurs as external stress rises, based on the movements and interactions of dislocations [44,45]. It can be seen in Figure 8(a) that the noticeable creep deformation only arises up to a higher stress, indicating a threshold stress for lower temperature creep [19]. If the stress threshold for particle cutting processes is surpassed, significant main creep occurs [47,48]. As seen in Figure 8(b) that the dislocations sweep over the γ/γ' microstructure to generate the anti-phase boundaries (APBs) and subsequently tear the γ' precipitates [49,50]. Under high pressures, the shearing of γ' phase is considered an indication of primary creep [51]. Under these conditions, the primary driving factor behind the creep deformation is external stress, and as more creep strains accumulate, the rate of creep strain increases initially before declining. Dislocations have a brief incubation phase prior to major creep, during which they spread over previously dislocation free zones [52]. Dislocations percolate via the cross-slip behaviour along the γ/γ' microstructure for the significant creep deformation. Dislocations first proliferate from dislocation sources during the cross-slip process, and they subsequently glide preferentially in γ channels that are normal to the applied stress. The dislocation density rapidly rises during this incubation time as creep strain increases, indicating a higher strain rate. It will negatively restrict the dislocation motion to follow the dense dislocations in γ channels.

Superalloys: Fundamentals and Applications   Materials Research Forum LLC
Materials Research Foundations 178 (2025) 333-363   https://doi.org/10.21741/9781644903698-17

Therefore, following a brief incubation period, the strain rate was seen to be decreased [53]. As a result, the first creep gives rise to a subsequent one that exhibits the almost constant creep strain rate. There is very little creep deformation accumulated at this stage because of the low creep strain rate in the secondary creep [47]. Ultimately, the creep strain rate increases once again until the alloy's final rupture, as it approaches a threshold. Lower strains can result in the noticeable creep deformation at comparatively high temperatures (950 °C), as seen by a comparison with the low temperature (750 °C) creep. In a similar line, higher applied stress can promote alloy creep at the intermediate temperatures by exhibiting a rapid rise in the strain rate. Prominent creep deformation for lower temperature creep is due to the dislocation of γ' precipitates, shearing at high loads. However, the mid-temperature creep involves a completely separate creep process. On the other hand, it appears that the activation energy for dislocation gliding is lowered at the relatively high temperature of about 950 °C, resulting in more dislocation motions occurring in the matrix [43]. Conversely, the reduced applied stress at mid-temperature creep limits the majority of the dislocation motions inside the matrix. Furthermore, dislocations cannot pass through the γ' precipitates unless a very high degree of creep strain is present. A portion of the softening effects that raise the strain rate can be attributed to the increased dislocation activity [46]. The alloy is softening because of a slow coarsening of the γ' precipitates, which also occurs at these high temperatures. Similar to this, a hardening impact is indicated by the detrimental restraint on the dislocation motions caused by an increasing dislocation density [46]. As with the tertiary creep, the mid-temperature creep of an alloy often exhibits a continual softening process. Without exhibiting a steady-state regime, the strain softening behaviour is linked to a proportionality of creep strain rate and cumulative creep strain [55]. As seen in Figure 9, the creep strain rate steadily rises with creep strain in tertiary creep until ultimate rupture. Ultimately, the rafted structure development that was seen as the directional coarsening behaviour of γ' precipitates is linked to creep deformation at high temperatures at over 1100 °C [54]. As seen in Figure 10(a), there are often no noticeable incubation durations under such conditions because of the extremely high exposure temperatures [56,57]. Low stresses result in the dislocation motions to be limited inside the matrix, allowing the dislocations to ascend and glide across the γ' precipitates. Figure 10(a) illustrates the creep strength rapidly degrades with a rapid rise in strain rate [19]. Figure 10(b) shows that the strain rate achieves a small value, as it approaches the dynamic equilibrium between the generation of new dislocations (strain hardening) and the annihilation of dislocations (recovery), and it remains almost constant, indicating secondary creep. In addition, the creep accounts for the majority of the creep life in this stage. Complex dislocation networks are formed because of the enhanced dislocation activity in the matrix, as displayed in Figure 11(a). The dislocation networks that arise can ease the unequal stress. The length of the γ channels for dislocation climbing is significantly extended with creation of rafted structures. Dislocations of the opposite sign in distinct slip planes might approach one another to react and annihilate one another based on the dislocation rise. In this case, as Figure 11(b) illustrates, the dislocation motions cause the created dislocation networks to gradually get denser. Creep fractures are always originated in areas where stresses are concentrated, and ultimate failure is exceptionally controlled [49]. Furthermore, because of the altered triaxiality of the stress state close to the fracture surface, large stresses in fracture zone are enough to force the rafts to reorient [57].

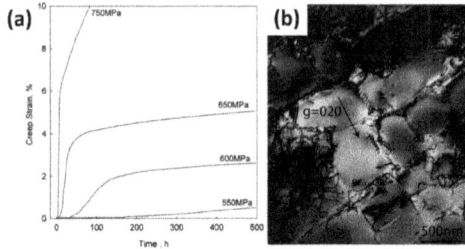

Figure 8: A Ni-SXs with Low Temperature and High Stress Creep. *(a)* Creep Curves Show Strain Versus Time for Various Applied Stressors and *(b)* Shearing Processes in the γ' Precipitates Indicate the Beginning of Primary Creep at high Stress and Low Temperature Creep [19].

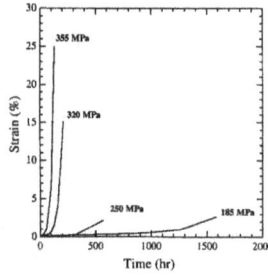

Figure 9: Creep Curves Representing Strain against Time at 950 °C for Ni-SXs under Varying Applied Stresses for samples orientated within 10^0 of [001] [46].

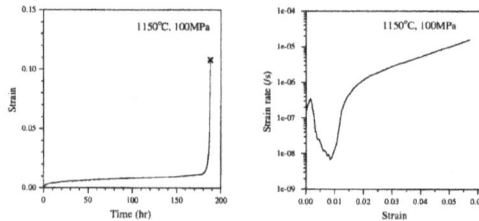

Figure 10: Creep Curves with Low Stress and High Temperature [48].

Superalloys: Fundamentals and Applications   Materials Research Forum LLC
Materials Research Foundations 178 (2025) 333-363   https://doi.org/10.21741/9781644903698-17

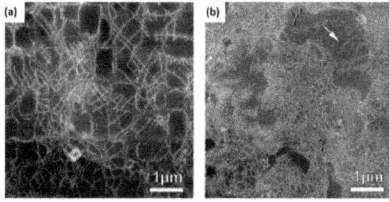

Figure 11: Development of a Ni-SX's Dislocation Networks at High Temperature and Low Stress Creep (1293 K and 160 MPa) [106].

5.2 Coarsening Mechanism

The Lifshitz-Slyozov-Wagner (LSW) hypothesis was initially put out by Lifshitz and Slyozof [58] and then by Wagner [59] to quantitatively explain the precipitate coarsening kinetics. Diffusion-controlled coarsening depends on the time exponents in the precipitate coarsening kinetics, which is a notable characteristic of the developed theory. The LSW theory offered an analytical equation based on the Gibbs–Thomson equation: $\langle rn \rangle = Kt$, with the exponent of $n = 3$. In this equation, t is the related time, K is the coarsening constant, and $\langle r \rangle$ is the average precipitates radius. Nonetheless, the LSW theory presumed a diluted system with a tiny volume proportion of precipitates and was created based on binary systems. The development of several models, including Modified LSW (MLSW) [60], Davies-Nash-Stevens and Lifshitz-Slyozov encounter modified (LSEM) [61], Brailsford-Wynblatt (BW) [62], and others, considered high precipitate volume fraction. Philippe and Voorhees (PV) have suggested a novel model that includes the precipitate volume fraction and the multi-component impact [63]. However, several investigations have successfully extended the LSW theory to multi-component superalloys, maintaining its status as the most widely recognized explanation to explain the coarsening behaviour. The trans-interface diffusion controlled (TIDC) model was another coarsening model developed that took the view that the coarsening process was controlled by diffusion through the interface. This model considers the interfacial width and the exponent of n = 2 in the given power-law function, but it does not account for the influence of vol.%. Recently, multi-component superalloys have also been effectively treated using the TIDC theory [64–66]. The two processes that regulate coarsening, matrix-diffusion controlled and interfacial-diffusion controlled, typically arise concurrently and engage in competition within a solitary coarsening process [67]. The coarsening-controlled process may be identified by analyzing the particle size distributions (PSDs) in various coarsening phases. In a Ni-Al-Cr-Re alloy, Figure 12 displays both the experimentally measured PSDs and the PSD predictions for LSW and TIDC theory. The lower temperature first represents the stage of the alloy before heat treatment, while the higher temperature and longer aging time represent the Re containing alloy [65]. Since the PSDs are substantially closer to the prediction of the LSW theory, it is suggested that matrix diffusion primarily controls the coarsening throughout the heat treatment process. However, as thermal exposure times increase, the interfacial diffusion gradually becomes more and more dominant in controlling coarsening [28].

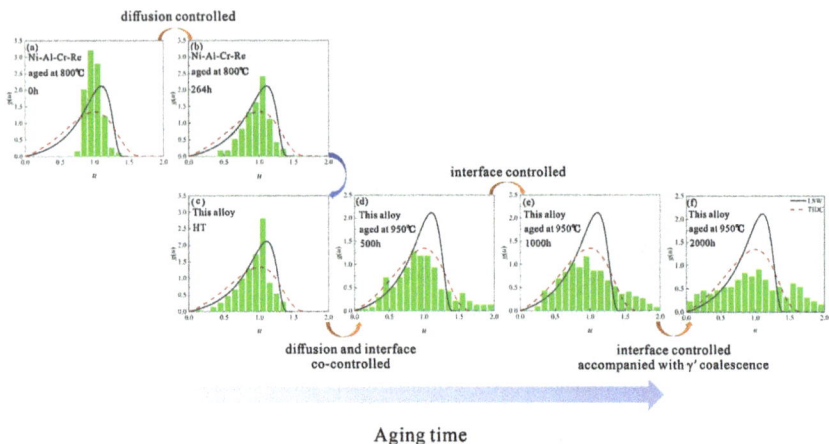

Figure 12: PSDs for the Alloy under Investigation for the (c) HT State and Aged for (d) 500 h, (e) 1000 h, and (f) 2000 h at 950 °C, as well as PSDs for Ni-Al-Cr-Re Aged for [65]. (Where u is the normalized size of γ.)

5.3 Twin Formation Mechanism

5.3.1 Annealing Mechanism

In metals like Ni-SXs that have an FCC structure and a stacking fault energy of less than 0.15 J/m^2 show annealing twins. When twins are annealed, the grain boundary that separates them from the parent grain is usually long and straight. The process of recrystallization and grain development produces these annealing twins. The twin boundary accepts the (111) plane as a symmetrical plane and has a disorientation of 60° with the ⟨111⟩ axis [68,69]. These results align with the twin's crystallographic orientation connection, as shown in Figure 13(a) [73]. A typical coherent twin structure is shown by the identical atomic structure and arrangement on both sides of the twin border, as shown in Figure 13(b) [73]. In recent years, grain boundary engineering has focused a lot of effort on improving the intergranular characteristics through coherent architectures with low-energy interfaces [70]. Lin *et al.* discussed that in the completely recrystallized microstructure, twin grains were seen to develop along three lines [71]. The total grain boundary energy decreases with each additional twin, despite the fact that the total grain border area grows. Grain boundary migration is always present along the twin limits of Σ3 according to the 'growth accident' hypothesis [72]. The frequency of appearance of Σ3 twin borders increases with the growth of recrystallized grains. The recrystallized grains are likewise nucleated at the Σ3 twin boundary created at the triple junction [73]. This agrees with the transmission electron microscope (TEM) and electron backscatter diffraction (EBSD) data, which show several Σ3 twin boundaries in the recrystallized grain structure in the Ni-SXs caused by oxidation. It is inferred that the Σ3 twin boundary contributes significantly to the nucleation and development of recrystallization as well as lowering the energy of the grain boundary during the process [73].

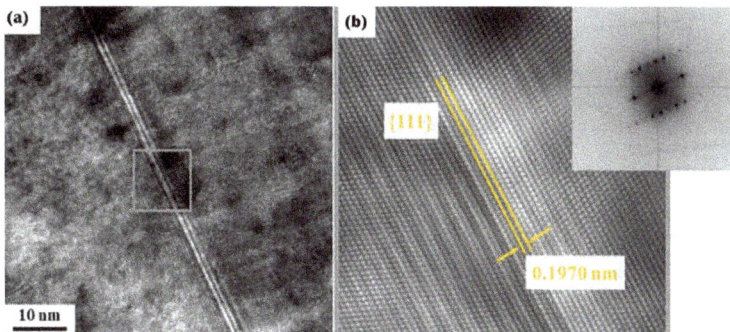

Figure 13: *Microstructures Near the Oxidation Layer and (**a**) Twin Boundary HR-TEM Image, and (**b**) Inverse Fast Fourier analysis and Quick Image of the Fourier Transform of (**b**) [73].*

5.4   Deformation Mechanism

The work shows how ultrasonic surface rolling (USR) changes the slip behaviour of materials and how heat exposure after USR treatment makes sheared γ' segments rougher in some places. Initially, cubic γ' phases coherently embed within the γ matrix, as illustrated in figure 14(a). However, mild plastic deformation at the sublayer leads to the shear transitions with insufficient shear step length, resulting in inadequate γ terminals or channel necking, as illustrated in Figure 14(b). This small channel has the ability to slow down mobile dislocations, but it cannot completely stop them within specified periods. The USR treatment creates three series of slip traces (STs) on the upper surface that has been deformed a lot. These STs have the right length of shear steps to allow for terminal forms, which leads to multiple cross-shaped phases, as shown in Fig. 14(c). The γ_s'/γ_t' interfaces, which are crossed by the γ terminal and γ' segments, have parallel lattice striations that are filled with superlattice extrinsic stacking fault (SESF) at the nanoscale level because of stacking fault reordering. The serration interface has a lower energy threshold for dislocation invasions than the main γ'/γ interface. This means that some dislocations can stay in the γ' segments. The negative misfit gets worse as the strip deforms more, which helps the interfacial energy build up and lowers the augmented lattice distortion energy. The work hardening of γ channels is mostly controlled by dislocation strengthening. On the other hand, γ' hardening is caused by dislocation proliferation, planar fault contributions, and increases in net system energy. However, premature terminals at the sub-layer significantly diminish impingement processes at the γ_s'/γ_t' interfaces, resulting in reduced nano-hardness compared to the uppermost layer. Due to reduced diffusion energy dissipation, thermal exposure enhances dislocation invasions and Co/Cr solute directional diffusion into γ' segments at the γ_s'/γ_t' interface on the upper surface of the USR sample, as shown in Figure 14(d). After that, they are probably used to change the γ' morphologies to make coarsening easier through interface movements or coalescences. This leads to ST eliminations and changes to the advanced stages of coarsening, as shown in figure 14(e) [22]. When there is, a positive lattice mismatch in Ni-SXs, the triaxial compressive stress changes the shape from triangles to half-circles and it occurs at γ' phase [74]. The coarsening can greatly weaken the alloy's mechanical properties because it breaks down interfacial coherency, leading to lower nano-hardness and compressive residual stress (CRS) amplitudes. Due to early γ terminals, the free dislocation tends to move within the channel interior and more locally penetrates into γ' segments that are perpendicular to the Burgers vector. This causes less coarsening and higher CRS values than the top surface. Even when the temperature is kept at 650 °C for long enough, as shown in Figure 14(f), it is not enough to make the γ'/γ structure coarser at the substrate or severe plastic deformation (SPD) layer away from the STs [22].

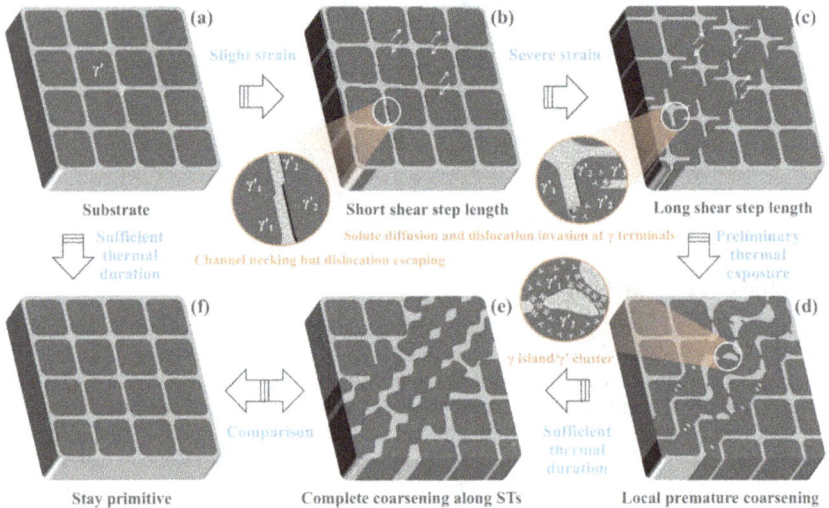

Figure 14: *Schematic Representation of the γ'/γ Microstructure of the Following Samples: (**a**) AR (as-received) Sample; (**b**) USR Sample at Sub-layer; (**c**) USR Sample at Topmost Layer; (**d**) USR Sample at Topmost Layer with Initial Thermal Duration; (**e**) USR Sample at Topmost Layer with Adequate Thermal Duration; and (**f**) AR Sample with Adequate Thermal Duration for Evaluation [22].*

6. Applications of Ni-SXs

6.1 Applications in Power and Aerospace

Although the SCSs are categorized as superalloys with a Co, Fe, and Ni basis, they can also be categorized as aerospace and industrial gas turbine (IGT) alloys depending on their intended use. The type of stress applied on these two alloys makes a noticeable effect. Turbine components in aeronautical applications are made from superalloys. While the fundamental operating load distribution of an aero engine's turbine blade and that of a ground-based turbine are comparable, flying maneuvers, frequent on-off cycles, and other factors subject the aero engine's turbine blade to extra stresses. Once they survive creep loads and high-temperature oxidation, superalloys utilized in stationary turbines have to work for an extended period up to 10,000 h [75-79]. The alloys utilized in IGT applications can withstand extended exposure to oxidative conditions at high temperatures. Therefore, action must be made to increase the material's resistance to oxidation.

Superalloys: Fundamentals and Applications   Materials Research Forum LLC
Materials Research Foundations 178 (2025) 333-363   https://doi.org/10.21741/9781644903698-17

Compared to aerospace alloys for high-pressure turbine (HPT) applications have a lower Cr content percentage and an improved refractory metal content than the Ni-based superalloys for IGT applications, which have higher Cr concentrations. Specifically, the future aerospace alloys for use in gas turbine blades have a reduced Cr content and rigorously avoid compositions that might cause β-NiAl and other phases like TCP to develop, which would hinder the performance of the component [80]. By altering the γ/γ' interface properties, new-generation alloys are able to stabilize the γ' precipitates for extended periods at elevated temperatures, which leads to improve mechanical performance. The notable enhancement in mechanical behaviour is the reason for the popularity of Re and Ru additions. Grain boundary engineering for purposes, such as engine shafts and turbine disks and SCSs for turbine blade applications are the two fields of research on the development of Ni-SXs. At working temperatures, most Ni-SXs exhibit creep-brittleness. Cast superalloys, on the other hand, are creep-ductile. As a result, wrought alloys are usually employed in IGT applications; however, they are only suitable for disc applications, such as gas turbines utilized in HPT aero engines. There are very few superalloys designed specifically for turbine disc applications out of all those produced for gas turbine applications. In order to minimize inter-stage working fluid leaks, turbine blades generally need to withstand wear damage. Gas turbine blades with abrasive tips can be effectively treated by laser-assisted surface deposition of a carbon-enriched superalloy onto the blade's tip. Diffusion at the interface between the abrasive superalloy coat and blade face must also be considered since the composition of the substrate and coat varies. According to Suzuki et al., they were able to create a SCSs that was Re-free but had high creep and oxidation resistance, on par with compositions including Re. The comparable creep resistance of alloys without Re content can be attributed to the solution treatments and higher refractory contents, followed by somewhat long aging at higher temperature. The production of TCP is inhibited throughout the aging process by adding more refractories than the next generation Ni-SXs and keeping lower concentration of sigma-forming elements like Mo and Cr along with the absence of Fe [11]. After a solution annealing procedure, Nazmy et al. claimed to create a SCSs using HIP of CMSX-4 alloy at 160 MPa. It is stated that the HIP process avoids internal tensions and chemical inhomogeneity. According to Hu et al., a superalloy based on Ni and enhanced with Pt is suitable for mending gas turbine blades. In addition, inclusion of Si when combined with other factors like oxidation and thermal stress, enhance the resistance of material to erosion in high-pressure turbine uses [81,82]. According to Segletes et al., filler wire of an appropriate composition must be deposited over the damaged region using welding procedures appropriate for high-performance superalloys in order to repair damaged steeples on gas turbine disks using the weld build-up approach. Direct aging temperatures are those that are reached by preheating in proximity to the aging temperature but below the solution temperature [11].

6.2   Jet Propulsion

An ideal alloy is probably going to have a proportion of the γ' phase between 0.60 and 0.70; applying this condition significantly shrinks the dataset to 16,700. A relative density of less than 9.0 is reported for all. By carefully selecting the lattice misfit, one can further reduce the size of the dataset. For instance, 443, 638, and 992 alloys are available when needed an alloy with a relative density < 8.5 and a tolerance of metal-d levels (M_d) < 0:9615 than 286, 317, and 348 alloys, which might also be required. The alloy with the highest anticipated creep resistance is negatively misfitting with a formulation of Ni–4Cr–10Co–3Re–3W–7Al–5Ta (wt.%) and a creep merit index (CMI) of 5:4 1015 m^{2}s. Analyzing the alloy compositions in these significantly smaller datasets is interesting. It is discovered that a variety of Cr composition alloys exist, most likely as a result of Cr's notable solubility in Ni lattice, as observed in Ni–Cr binary phase diagram. It should be noted that neither density nor stability have been given any criteria. As the resistance to oxidation, corrosion, and sulphidization is significantly enhanced at high Cr content, it is expected that the creep behaviour can vary noticeably with Cr concentration. Interestingly, the negatively misfitting alloys are projected to have a creep resistance around the theoretical limit over quite a wide range of Cr concentrations. For the zero and negative lattice misfitting categories, respectively, the creep merit indices for the alloys that are expected to act best resulting from both Cr and Re concentrations. The negatively misfitting alloys are expected to perform better than the zero misfitting ones, especially for medium to high Cr concentrations. Our calculations suggest that the creep performance will be degraded at low to mid-range Cr levels if one demands a relative density of 8.75 or below. These results suggest that alloys with negative misfit and Cr concentrations between 7 and 9 wt.% are promising. According to the calculations, Cr content of 8% wt.% may be the optimum that may be accepted before noticeably increasing TCP stability, since the anticipated M_d number grows noticeably at higher concentrations. One may wonder if this can be improved, given that the computations were done with a 1 wt.% resolution only. An extra degree of optimization has been done to show that this is feasible, but at significant computational expense. This needed an extra 114 calculations per alloys in the original dataset of 2661. There are several methods to query the finished dataset. For instance, the alloy with 3 wt.% Re, which is expected to work with best in creep can be Ni–8Cr–10Co–3Re–8.5W–5.9Al–8.5Ta (wt.%). This alloy has an expected CMI of 8:2 1015 m^2 s and a relative density of 8.88. According to the prediction, it should outperform CMSX-4, the widely utilized second-generation alloy with 3 wt.% Re, outlasting its life at 900 °C and 390 MPa by around 100 h. It is viable to do additional calculations; for instance, the alloy with the lowest Re content is reported to be Ni–8Cr–10Co–1.6Re–8.5W–5.8Al–8.5Ta (wt.%), with a relative density of 8.83 and an expected CMI of 7:2 1015 m^2 s. This alloy is projected to tally the creep performance of CMSX-4 and is approximately 1.6/3 or 53% less expensive due to its lower Re content. These two alloys are referred to as Alloy ABD-1 and Alloy ABD-2, respectively. With an anticipated ratio of CMI/density that is higher than CMSX-4, the alloy Ni–8Cr–10Co–2.1Re–8.3W–6.1Al–8.2Ta (wt.%), with a relative density of 8.78 and a forecasted CMI of 7:5 1015 m^2 s, should provide a favourable balance between creep performance and cost [83].

6.3   Industrial Gas Turbine

Historically, jet engines have created SCSs, while industrial gas turbines (IGTs), which produce electricity, have created very few. As a result, the IGT community has a propensity to employ aerospace alloys for their applications, such as CMSX-4 and Rene N5, even if custom alloys can be preferable in certain cases. Because aero engine fuels are often cleaner than IGT fuels, it is likely that more focus has to be positioned on providing stronger corrosion resistance than has traditionally been necessary. Furthermore, the blades are far bigger than those of the aero engine, which means that (i) the price of the raw elements additions is a significant concern, and (ii) the alloy's castability should be optimized because it is well known that freckling improves with increasing casting size. The alloy design issue is shown from the viewpoint of an industrial gas turbine. The alloys like Rene N5, PWA 1484, MC2, etc. have a notable creep strength because of their Al content, among other reasons, but do not have enough Cr to provide sufficient corrosion resistance for IGT applications. While alloys like IN792 and IN738LC are not intended to be SCSs, they do have enough corrosion resistance but lack strength. It has been suggested that new grades of SCSs with greater Cr levels to improve corrosion resistance may be designed for these uses,

Superalloys: Fundamentals and Applications
Materials Research Foundations 178 (2025) 333-363

Materials Research Forum LLC
https://doi.org/10.21741/9781644903698-17

matching the Al content of the aero engine alloys. For IGT applications, new grades of SCSs have been designed using these techniques. The following requirements are considered: (1) the Re concentration is zero, as in these applications, castability could be improved and costs could be reduced (while maintaining an appropriately longer component lifetime); (2) the Cr concentration could be significantly more than the aero engine alloys that ranges 12–20 wt.%, due to the requirement for corrosion resistance; (3) the γ' fraction was kept in the range of 0.5–0.6, slightly lower than for the aero engine alloy, to maintain a good combination of fatigue, creep resistance, and microstructural stability; and (iv) the lattice misfit should be in the range of $\delta = (-8+2.5) \times 10^{-4}$ [83]. Higher-level Cr alloys are needed to satisfy the corrosion resistance requirements for IGTs operating in the temperature regimes prone to hot-corrosions, as well as to enable the use of 'flexible' and 'dirtier fuels' with higher S and V contents. IGTs are used in many different industries, including industrial facilities, oil and gas production, maritime propulsion, and power generation. A substantial overhaul is not necessary for some power producing turbines, which may operate for more than 20,000 h. Improvements in creep strength and oxidation resistance are also necessary for the most recent generation of IGT, which is intended to function at temperatures far greater than those of its elder counterparts. The more recent versions of IGT buckets have a temperature range of around 1100 °C at the tip to 700 °C in the shank area. The alloy should form a protective alumina scale in the hotter regions and a chromia scale in the colder regions. The low Cr content of the SCSs, which makes them prone to corrosion, prevents IGT from using them in propulsion turbines. Many efforts have been made to develop alloys for IGT applications that would meet the critical requirements of high amounts of Cr, enhanced strength, and resistance to oxidation and corrosion. Low Mo and Ti levels are also necessary for the alloying approach because of their negative impact on hot-corrosion resistance, in addition to greater Cr values. Additionally, because of their high elemental costs, alloying additives like Ru and Re are undesirable. High Cr concentrations are seen in the majority of directionally solidified (DS) or SCS created for IGT applications. With an excellent mix of mechanical and environmental behaviour, Reed *et al.* [15, 84] and Sato *et al.* [85] have created a promising SCS called STAL-15, which contains 15 wt.% Cr, 5 wt.% Co, 1 wt.% Mo, 3.7 wt.% W, 4.55 wt.% Al, 8 wt.% Ta, 0.1 wt.% Hf, 0.03 wt.% Ce, and 0.25 wt.% Si were the composition of the alloy after it was exposed to an alumina forming at 1000 °C. The lack of Ti and doping with 0.25 wt.% Si improved the capacity of alumina to form scales. Re is an attractive economic benefit needed for larger IGT blade applications, yet it is absent from this alloy. The IGT buckets have a maximum length of 90 cm and a maximum weight of 18 kg each. The investment casting of these enormous buckets in single-crystal form presents major financial and technological constraints. Consequently, the majority of alloying and processing work has been focused on making sections that can be cast as DS components [86].

7.   Factors Affecting Performances of SCSs

7.1   Stress Corrosion Cracking (SCC)

Aero engine hot section components, which are frequently made of products of radical combustion, work in harsh conditions with high temperatures and loads. These combustion products are a combination of small impurities in the fuel, alkaline oxides, and partly oxidized corrosive gases. Contaminants cause detrimental chemical alterations on component surfaces throughout the gas turbine's open cycle, resulting in diminished mechanical behaviour, particularly fracture

Superalloys: Fundamentals and Applications   Materials Research Forum LLC
Materials Research Foundations 178 (2025) 333-363   https://doi.org/10.21741/9781644903698-17

toughness, at both surface and subsurface levels [87]. Yoo *et al.* discussed about the potential for initiation of SCC because of passivating components like Cr depletion and short-range order creation [88]. Superalloy systems with tertiary components, such as Alloy 600, are vulnerable to corrosive conditions, prolonged thermal aging, and frequent residual stress from welding. Although they have better corrosion properties than 316 L steels, the grain morphology alters largely from the fusion zone throughout the heat-affected zone (HAZ) due to an improvement in the density of Σ3 boundaries, a reduction in misorientation, and creation of Cr carbides at higher angle grain boundaries, which increases the susceptibility of SCC in the HAZ [84]. Lu *et al.* described a mathematical examination of the impact of stress on oxidation, as shown in Equation 1. They found a relationship between a vacancy concentration and the type of load applied.

$$R_f = e^{\left(\frac{\sigma \Omega_{metal}(R_{PB}-1)}{K_b T}\right)}$$ Equation 1

It is well known that the vacancy concentration directly influences the tension in the oxide film. At the interfaces of oxide layers with the substrate and environment, the kind of stress, such as tensile or compression, can influence the cation vacancy concentration, a phenomenon known as stress-assisted oxidative failure [90]. The aforementioned formula has R_f (enhancement factor on stress-induced oxidation rate), Ω (metal ion volume), k_b constant, R_{PB} (Boltzmann's ratio of molar volume of scale to metal), T (absolute temperature), and σ (stress) [11,92]. When polycrystalline materials are exposed to a corrosive environment, intergranular SCC, which occurs when oxygen penetrates the material, usually leads to the failure. Complex oxides are then formed because of the depletion of elements that separate the grain boundaries. On the other hand, in single crystals, the SCC without grain boundaries, makes the crack tip more vulnerable to chemical alterations, which encourages the crack to continuously grow. Brooking *et al.* reported that the CMSX-4 SCS was claimed to have undergone a 3-point bending test in low-temperature hot corrosion settings (500–700 °C) by means of cyclic deposition of alkaline sulphates. Apart from the effect of corrosive medium radicals on SCC, tungsten isolated to the γ-γ' interface is expected to cause SCC only because of strain localizations in a corrosive condition [11].

7.2. Effect of Oxygen Partial Pressure on Oxidation of SCSs

When working fluid has an oxygen content of 15 vol.% and a volumetric oxygen content of 8.54%, the flue gas released from the turbine's combustion chamber attains an adiabatic temperature of 927 °C. However, later on appropriate cooling systems are installed to reach the temperature up to 1427 °C [91]. The working fluid's adiabatic temperature or the amount of oxygen left in the flue gas can both rise in response to an increase in oxygen concentration. An elevation in turbine intake temperatures requires materials with superior metallurgical thresholds, while a rise in oxygen levels will require enhanced oxidation resistance measures. According to Li *et al.*, a higher oxygen level in the examining condition tends to form spinels and transition metal oxides, which decrease the mechanical stability of the alumina layer at 1100 °C. This temperature corresponds to the gas turbine's operating range for single crystals that are [001] oriented and have a dendritic orientation deviation of 15° [100]. According to Huczkowski *et al.*, the oxidation of alloy 625 is dependent on the working fluid's flow characteristics. They also emphasized how the thickness of the boundary layer affects the amount of Cr lost because of oxidation at various points along the aerodynamic

Superalloys: Fundamentals and Applications   Materials Research Forum LLC
Materials Research Foundations 178 (2025) 333-363   https://doi.org/10.21741/9781644903698-17

profile [92]. When changing the working fluid flow rate or raising the oxygen content in the test environment, the partial pressure of oxygen is the key factor that determines how the oxidation precedence order is affected. Al is more prone to oxidation at low partial pressures of oxygen, as may be shown from Ellingham diagrams [94]. According to Dalton's partial pressure law, the partial pressure of oxygen in the surrounding air should be around 0.2 atm. Fe, Cu, and Ni will take order at these partial pressure levels and temperature ranges of 800–1000 °C. The majority of oxygen reacts with Ni in Ni-SXs, creating NiO at the surface. Ishizaki *et al.* have documented the impact of Ellingham phase diagrams' higher-pressure extension for Si_3N_4 and AlN [95]. Given that the working pressure at the turbine entrance is substantially greater than the ambient environments, an understanding of the oxidation kinetics of Ni-SXs under higher pressure is necessary. Three main circumstances lead to oxidation of the SCSs. The first occurs when a motor is started, and the second occurs when the engine is switched down, lowering the working pressure from about 20–40 bar to ambient level. The third cause arises when the flue gas's oxygen concentration rises over the ideal level [11].

7.3   HIP Technology on Crack Propagation

This study investigates the impact of HIP treatment on crack propagation in Ni-SXs. Fatigue crack propagation is traditionally divided into mode I and mode II, perpendicular to the load direction and parallel to the crystallographic planes, respectively [99]. Mode II occurs when the temperature rises to an intermediate temperature [100]. Early crack growth from micropores, residual eutectics, or carbides typically occurs in mode I cracking, which is governed by maximum normal stress [96,97,101]. Mode II only occurs in the final stage of crack propagation [98,101,102]. The fatigue crack propagation modes of alloys AS-A and AS-HIP-SA were significantly different at 760 °C. Cracks are prone to rapid propagation near micropores and residual eutectics [96,97]. In alloy AS-A, microcracks rapidly propagate perpendicular to the stress axis along micropores and residual eutectics, creating interconnected microcracks. These areas lead to fatigue fractures dominated by mode I, which is illustrated in Figure 15(1-2). The fatigue fracture in AS-A is characterized by mode I, where cracks grow along the interdendritic regions influenced by stress concentrations around the micropores and residual eutectics, as illustrated in Figure 15(1-3). In alloy AS-HIP-SA, HIP treatment significantly reduces the micropores and residual eutectics in the interdendritic region, decreasing stress concentrations and preventing rapid crack propagation in these areas, as illustrated in Figure 15(2-1). Figure 15(2-2) explains that the microcracks in AS-HIP-SA do not propagate further through the interdendritic regions after entering the matrix, as the reduced stress concentrations hinder further crack growth. In the final stage, cracks in AS-HIP-SA propagate along slip planes, with primary and secondary cracks following dense slip bands. This leads to a fatigue fracture dominated by mode II, which is illustrated in Figure 15(2-3) [104]. From all these, it can be seen that HIP treatment in AS-HIP-SA changes the crack propagation mode from mode I to mode II, enhancing fatigue life by reducing the impact of micropores and residual eutectics [105].

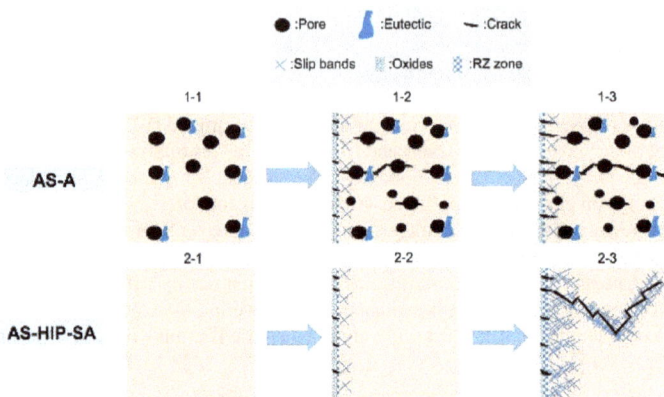

Figure 15: Schematic Diagram of Crack Propagations in AS-A and AS-HIP-SA Alloys of LCF (Low cycle- fatigue) at 760 °C [105].

Conclusions

Ni-SXs, ideal for aerospace and power generation, are shaped by their composition and preparation methods. The use of alloying elements like Cr, Co, and Mo enhances the oxidation resistance and strength. Advanced techniques like directional solidification and investment casting minimize defects. Research is ongoing to optimize compositions and processing methods, enhancing performance under extreme conditions. Ni-SXs are essential for high-temperature applications, especially in aerospace and power generation. They have excellent creep resistance due to their microstructural stability, minimizing grain boundary sliding and enhancing deformation resistance. Their unique crystallographic orientation and strong intermetallic phases contribute to high compressive strength, allowing them to withstand operational stresses effectively. These superalloys also show high yield strength and elongation in tensile testing, crucial for dynamic loading applications. Their ability to maintain mechanical strength at high temperatures without significant reduction in ductility ensures reliability in service conditions. The combination of superior creep, compression, and tensile properties makes them essential for advanced high-performance applications. Further research and development are needed to enhance these properties for next-generation applications. Ni-SXs are crucial for high-temperature applications due to their behaviour influenced by dislocation motion. The interaction of dislocations with γ' precipitates enhances creep resistance, while power-law creep dominates at high temperatures. Coarsening, or Ostwald ageing, occurs during high-temperature exposure, reducing precipitate density and influencing strength. Controlling temperature and time is essential for maintaining desired microstructure and mechanical properties. Annealing relieves internal stresses and enhances microstructural stability, allowing for dislocation recovery and recrystallization. Proper annealing treatments improve ductility and reduce residual stresses. Deformation mechanisms involve increasing and decreasing of dislocations, with anisotropic behaviour dependent on crystallographic orientation. The interaction between dislocations and precipitates can lead to the complex deformation pathways, including super dislocations and channeling. Understanding these mechanisms is essential for designing and optimizing Ni-SXs. Advances in processing techniques and alloy development aim to mitigate undesirable effects, enhance creep and deformation resistance, and ensure component reliability and performance in demanding environments. The interaction of stress corrosion cracking, oxidation behaviour under varying oxygen partial pressures, and the effects of HIP significantly influences the performance of Ni-SXs. The SCC is a critical failure mode for superalloys, especially in high-stress, corrosive environments. Factors such as microstructure, residual stresses, and environmental conditions influence the susceptibility of these materials to SCC. Effective alloying and heat treatment strategies are crucial to mitigate SCC risk by refining microstructural features and reducing susceptibility to environmental attack. The oxidation behaviour of Ni-SXs is significantly affected by the partial pressure of oxygen. Higher oxygen levels can lead to rapid oxidation and detrimental scale formation, affecting the material's longevity and mechanical properties. Optimizing the oxidation environment can enhance protective oxide layer formation, improving resistance to high-temperature oxidation. Understanding the relationship between oxygen partial pressure and oxidation kinetics is essential for developing superalloys with improved performance in oxidizing environments. HIP technology enhances the mechanical properties of Ni-SXs by reducing porosity, improving density, and promoting better homogenization of the microstructure. This comprehensive understanding aids in the development of materials that meet the demanding requirements of advanced engineering applications, ensuring reliability and durability in extreme environments.

Reference

[1] H. Long, S. Mao, Y. Liu, Z. Zhang, X. Han, Microstructural and compositional design of Ni-based single crystalline superalloys – A review, J. Alloys Compd. 743 (2018) 203–220. https://doi.org/10.1016/j.jallcom.2018.01.224

[2] M.J. Donachie, S.J. Donachie, Superalloys A Technical Guide, second ed., ASM International, Ohio, 2002

[3] P.W. Auburtin, T. Wang, S.L. Cockcroft, A. Mitchell, Freckle formation and freckle criterion in superalloy castings, Metall. Mater. Trans. B 31 (2000) 801–811. https://doi.org/10.1007/s11663-000-0117-9

[4] H. Fecht, D. Furrer, Processing of nickel-base superalloys for turbine engine disc applications, Adv. Eng. Mater. 2 (2000) 777–787. https://doi.org/10.1002/1527-2648(200012)2:12%3C777::AID-ADEM777%3E3.0.CO;2-R

[5] Y. Kiyak, B. Fedelich, T. May, A. Pfennig, Simulation of crack growth under low cycle fatigue at high temperature in a single crystal superalloy, Eng. Fract. Mech. 75 (2008) 2418–2443. https://doi.org/10.1016/j.engfracmech.2007.08.002

[6] J.D. Mattingly, Elements of Gas Turbine Propulsion, first ed., McGraw-Hill, New York 1996

[7] B. Zhang, R. Wang, D. Hu, K. Jiang, X. Hao, J. Mao, F. Jing, Damage-based low-cycle fatigue lifetime prediction of nickel-based single-crystal superalloy considering anisotropy and dwell types, Fatigue Fract. Eng. Mater. Struct. 43 (2020) 2956–2965. https://doi.org/10.1111/ffe.13345

[8] M.N. Babu, C.K. Mukhopadhyay, G. Sasikala, B.S. Dutt, S. Venugopal, S.K. Albert, A.K. Bhaduri, T. Jayakumar, Fatigue crack growth characterisation of RAFM steel using acoustic

emission technique, Procedia Eng. 55 (2013) 722–726.
https://doi.org/10.1016/j.proeng.2013.03.321

[9] B.S. Dutt, G. Shanthi, G. Sasikala, M.N. Babu, S. Venugopal, S.K. Albert, A.K. Bhaduri, T. Jayakumar, Effect of nitrogen addition and test temperatures on elastic-plastic fracture toughness of SS 316 LN, Procedia Eng. 86 (2014) 302–307. https://doi.org/10.1016/j.proeng.2014.11.042

[10] S. Gialanella, A. Malandruccolo, *Aerospace Alloys*, Cham, Switzerland, 2020

[11] G. Gudivada, A.K. Pandey, Recent developments in nickel-based superalloys for gas turbine applications, J. Alloys Compd. 963 (2023). https://doi.org/10.1016/j.jallcom.2023.171128

[12] M. Huang, J. Zhu, An overview of rhenium effect in single-crystal superalloy, Rare Met. 35 (2016) 127–139. https://doi.org/10.1007/s12598-015-0597-z

[13] J. Xu, X. Zhao, Q. Yue, W. Xia, H. Duan, Y. Gu, Z., Zhang, A morphological control strategy of γ′ precipitates in nickel-based single-crystal superalloys: an aging design, fundamental principle, and evolutionary simulation, Mater. Today Nano. 22 (2023) 100335. https://doi.org/10.1016/j.mtnano.2023.100335.

[14] T.M. Pollock, S. Tin, Nickel-based superalloys for advanced turbine engines: chemistry, microstructure and properties, J. Propuls. Power. 22 (2006) 361–374. https://doi.org/10.2514/1.18239

[15] A. Sato, Y.L. Chiu, R.C. Reed, Oxidation of nickel-based single-crystal superalloys for industrial gas turbine applications, Acta Mater. 59 (2011) 225–240. https://doi.org/10.1016/j.actamat.2010.09.027

[16] M. Durand-Charre (Ed.), The Microstructure of Superalloys, first ed., CRC Press, Florida, 2017. https://doi.org/10.1201/9780203736388

[17] Y. Koizumi, Z. Jianxin, T. Kobayashi, T. Yokokawa, H. Harada, Y. Aoki, M. Arai, Development of next generation Ni-base single crystal superalloys containing ruthenium, J. Japan Inst. Metals. 67 (2003) 468–471. https://doi.org/10.2320/jinstmet1952.67.9_468

[18] A.F. Giamei, Development of single crystal superalloys: a brief history, AM&P Tech. Articles, 171 (2013) 26–30. https://doi.org/10.31399/asm.amp.2013-09.p026

[19] W. Xia, X. Zhao, L. Yue, Z. Zhang, Microstructural evolution and creep mechanisms in Ni-based single crystal superalloys: A review, J. Alloys Compd. 819 (2020). https://doi.org/10.1016/j.jallcom.2019.152954

[20] W. Xia, X. Zhao, L. Yue, Z. Zhang, A review of composition evolution in Ni-based single crystal superalloys, J. Mater. Sci. Technol. 44 (2020) 76–95. https://doi.org/10.1016/j.jmst.2020.01.026

[21] Y. Li, X. Liang, Y. Yu, D. Wang, F. Lin, Review on additive manufacturing of single-crystal nickel-based superalloys, Chin. J. Mech. Eng. Addit. Manuf. Front. 1 (2022). https://doi.org/10.1016/j.cjmeam.2022.100019

[22] H. Chen, S. Sun, F. Tian, D. Min, L. Liu, L. Li, Deformation mechanism of Ni-based single crystal superalloy under ultrasonic surface rolling and subsequent thermal exposure, J. Surf. Coat. Technol. (2024). https://doi.org/10.1016/j.surfcoat.2024.131369

[23] X. Wu, S.K. Makineni, C.H. Liebscher, G. Dehm, J. Mianroodi. Rezaei, P. Shanthraj, B. Svendsen, D. Bürger, G. Eggeler, D. Raabe, B. Gault, Unveiling the Re effect in Ni-based single crystal superalloys, Nat. Commun. 11 (2020) 389. https://doi.org/10.1038/s41467-019-14062-9

[24] J.X. Zhang, H. Harada, Y. Ro, Y. Koizumi, T. Kobayashi, Thermomechanical fatigue mechanism in a modern single crystal nickel base superalloy TMS-82, Acta Mater. 56 (2008) 2975–2987. https://doi.org/10.1016/j.actamat.2008.02.035

[25] K. Kawagishi, A.C. Yeh, T. Yokokawa, T. Kobayashi, Y. Koizumi, H. Harada, Development of an oxidation-resistant high-strength sixth-generation single-crystal superalloy TMS-238, Superalloys. 9 (2012) 189–195. https://doi.org/10.1002/9781118516430.ch21

[26] G.L. Erickson, The development and application of CMSX-10, Superalloys. 1996; (1996) 35–44. https://doi.org/10.7449/1996/superalloys_1996_35_44

[27] A. Volek, F. Pyczak, R.F. Singer, H. Mughrabi, Partitioning of Re between γ and γ′ phase in nickel-base superalloys, Scripta Mater. 52 (2005) 141–145. https://doi.org/10.1016/j.scriptamat.2004.09.013

[28] J. Zhang, F. Lu, L. Li, An overview of thermal exposure on microstructural degradation and mechanical properties in Ni-based single crystal superalloys, Materials. 16 (2023) 1787. https://doi.org/10.3390/ma16051787

[29] H.P. Jin, J.R. Li, S.Z. Liu, Stress rupture properties of the second-generation single crystal superalloy DD6 after high temperature exposure, Mater. Sci. Forum. 546 (2007) 1249–1252. https://doi.org/10.4028/www.scientific.net/MSF.546-549.1249

[30] K. Cheng, C. Jo, T. Jin, Z. Hu, Precipitation behavior of μ phase and creep rupture in single crystal superalloy CMSX-4, J. Alloys Compd. 509 (2011) 7078–7086. https://doi.org/10.1016/j.jallcom.2011.04.001

[31] C.P. Liu, X.N. Zhang, L. Ge, S.H. Liu, C.Y. Wang, T. Yu, Y.F. Zhang, Z. Zhang, Effect of rhenium and ruthenium on the deformation and fracture mechanism in nickel-based model single crystal superalloys during the in-situ tensile at room temperature, Mater. Sci. Eng. A. 682 (2017) 90–97. https://doi.org/10.1016/j.msea.2016.10.107

[32] X. Xiong, P. Dai, D. Quan, Z. Wang, Q. Zhang, Z. Yue, Intermediate temperature brittleness and directional coarsening behavior of nickel-based single-crystal superalloy DD6, Mater. Des. 86 (2015) 482–486. https://doi.org/10.1016/j.matdes.2015.07.063

[33] A. Sengupta, S.K. Putatunda, L. Bartosiewicz, J. Hangas, P.J. Nailos, M. Peputapeck, F.E. Alberts, Tensile behavior of a new single-crystal nickel-based superalloy (CMSX-4) at room and elevated temperatures, J. Mater. Eng. Perform. 3 (1994) 73–81. https://doi.org/10.1007/BF02654502

[34] Z. Shang, X. Wei, D. Song, J. Zou, S. Liang, G. Liu, L. Nie, X. Gong, Microstructure and mechanical properties of a new nickel-based single crystal superalloy, J. Mater. Res. Technol. 9 (2020) 11641–11649. https://doi.org/10.1016/j.jmrt.2020.08.032

[35] Z. Shang, H. Niu, X. Wei, D. Song, J. Zou, G. Liu, S. Liang, L. Nie, X. Gong, Microstructure and tensile behavior of nickel-based single crystal superalloys with different

Re contents, J. Mater. Res. Technol. 18 (2022) 2458–2469.
https://doi.org/10.1016/j.jmrt.2022.03.149

[36] H. Zhang, P. Li, X. Gong, T. Wang, L. Li, Y. Liu, Q. Wang, Tensile properties, strain rate sensitivity and failure mechanism of single crystal superalloys CMSX-4, Mater. Sci. Eng. A. 782 (2020). https://doi.org/10.1016/j.msea.2020.139105

[37] D.P. Pope, S.S. Ezz, Mechanical properties of Ni_3Al and nickel-base alloys with high volume fraction of γ', Int. Met. Rev. 29 (1984) 136–167.
https://doi.org/10.1179/imtr.1984.29.1.136

[38] N. Sun, L. Zhang, Z. Li, A. Shan, Effect of heat treatment on microstructure and high-temperature deformation behavior of a low rhenium-containing single crystal nickel-based superalloy, Mater. Sci. Eng. A. 606 (2014) 417–425.
https://doi.org/10.1016/j.msea.2014.03.093

[39] F.R. Nabarro, M.S. Duesbery (Eds.), Dislocations in Solids, Volume 11, first ed., Elsevier, Amsterdam, 2002.

[40] L.N. Wang, Y. Liu, J.J. Yu, Y. Xu, X.F. Sun, H.R. Guan, Z.Q. Hu, Orientation and temperature dependence of yielding and deformation behavior of a nickel-base single crystal superalloy, Mater. Sci. Eng. A. 505 (2009) 144–150.
https://doi.org/10.1016/j.msea.2008.12.039

[41] R.C. Reed, The superalloys: Fundamentals and Applications, first ed., Cambridge University Press, Cambridge, 2008

[42] Q. Ding, H. Bei, X. Zhao, Y. Gao, Z. Zhang, Processing, microstructures and mechanical properties of a Ni-based single crystal superalloy, Crystals. 10 (2020) 572. https://doi.org/10.3390/cryst10070572

[43] P. Wollgramm, H. Buck, K. Neuking, A.B. Parsa, S. Schuwalow, J. Rogal, R. Drautz, G. Eggeler, On the role of Re in the stress and temperature dependence of creep of Ni-base single crystal superalloys, Mater. Sci. Eng. A. 628 (2015) 382–395.
https://doi.org/10.1016/j.msea.2015.01.010

[44] R. Wu, S. Sandfeld, Insights from a minimal model of dislocation-assisted rafting in single crystal nickel-based superalloys, Scripta Mater. 123 (2016) 42–45.
https://doi.org/10.1016/j.scriptamat.2016.05.032

[45] R. Wu, S. Sandfeld, A dislocation dynamics-assisted phase field model for nickel-based superalloys: The role of initial dislocation density and external stress during creep, J. Alloys Compd. 703 (2017) 389–395. https://doi.org/10.1016/j.jallcom.2017.01.335

[46] N. Matan, D.C. Cox, P. Carter, M.A. Rist, C.M. Rae, R.C. Reed, Creep of CMSX-4 superalloy single crystals: effects of misorientation and temperature, Acta Mater. 47 (1999) 1549–1563. https://doi.org/10.1016/S1359-6454(99)00029-4

[47] V. Sass, U. Glatzel, M. Feller-Kniepmeier, Creep anisotropy in the monocrystalline nickel-base superalloy CMSX-4, Superalloys. 96 (1996) 283–290.
https://doi.org/10.7449/1996/superalloys_1996_283_290

[48] R.C. Reed, N. Matan, D.C Cox, M.A Rist, C.M Rae, Creep of CMSX-4 superalloy single crystals: effects of rafting at high temperature, Acta Mater. 47 (1999) 3367–3381. https://doi.org/10.1016/S1359-6454(99)00217-7

[49] P. Zhang, Y. Yuan, S.C. Shen, B. Li, R.H. Zhu, G.X. Yang, X.L. Song, Tensile deformation mechanisms at various temperatures in a new directionally solidified Ni-base superalloy, J. Alloys Compd. 694 (2017) 502–509. https://doi.org/10.1016/j.jallcom.2016.09.303

[50] P. Geng, W. Li, X. Zhang, Y. Deng, H. Kou, J. Ma, J. Shao, L. Chen, X. Wu, A theoretical model for yield strength anomaly of Ni-base superalloys at elevated temperature, J. Alloys Compd. 706 (2017) 340–343. https://doi.org/10.1016/j.jallcom.2017.02.262

[51] G.R. Leverant, B.H. Kear, The mechanism of creep in γ' precipitation-hardened nickel-base alloys at intermediate temperatures, Metall. Mater. Trans. B. 1 (1970) 491–498. https://doi.org/10.1007/BF02811560

[52] V. Sass, U. Glatzel, M. Feller-Kniepmeier, Anisotropic creep properties of the nickel-base superalloy CMSX-4, Acta Mater. 44 (1996) 1967–1977. https://doi.org/10.1016/1359-6454(95)00315-0

[53] F.R. Nabarro, Rafting in superalloys, Metall. Mater. Trans. A. 27 (1996) 513–530. https://doi.org/10.1007/BF02648942.

[54] T.M. Pollock, A.S. Argon, Creep resistance of CMSX-3 nickel base superalloy single crystals, Acta Metall. Mater. 40 (1992) 1–30. https://doi.org/10.1016/0956-7151(92)90195-K

[55] L.M. Pan, I. Scheibli, M.B. Henderson, B.A. Shollock, M. McLean, Asymmetric creep deformation of a single crystal superalloy, Acta Metall. Mater. 43 (1995) 1375–1384. https://doi.org/10.1016/0956-7151(94)00383-S

[56] P. Caron, High γ'solvus new generation nickel-based superalloys for single crystal turbine blade applications, Superalloys. 2000 (2000) 737–746. https://doi.org/10.7449/2000/superalloys_2000_737_746

[57] R.C. Reed, D.C. Cox, C.M. Rae, Damage accumulation during creep deformation of a single crystal superalloy at 1150 °C, Mater. Sci. Eng. A. 448 (2007) 88–96. https://doi.org/10.1016/j.msea.2006.11.101

[58] I.M. Lifshitz, V.V. Slyozov, The kinetics of precipitation from supersaturated solid solutions, J. Phys. Chem. Solids. 19 (1961) 35–50. https://doi.org/10.1016/0022-3697(61)90054-3

[59] C. Wagner, Theorie der Alterung von Niederschlägen durch Umlösen (Ostwald-Reifung, J. Electrochem. Soc. 65 (1961) 581–591. https://doi.org/10.1002/bbpc.19610650704

[60] A.J. Ardell, The effect of volume fraction on particle coarsening: theoretical considerations, Acta Metall. 20 (1972) 61–71. https://doi.org/10.1016/0001-6160(72)90114-9

[61] C.K. Davies, P. Nash, R.N. Stevens, Precipitation in Ni-Co-Al alloys: part 1 continuous precipitation, J. Mater. Sci. 15 (1980) 1521–1532. https://doi.org/10.1007/BF00752134

[62] A.D. Brailsford, P. Wynblatt, The dependence of Ostwald ripening kinetics on particle volume fraction, Acta Metall. 27 (1979) 489–497. https://doi.org/10.1016/0001-6160(79)90041-5

[63] T. Philippe, P.W. Voorhees, Ostwald ripening in multicomponent alloys, Acta Mater. 61 (2013) 4237–4244. https://doi.org/10.1016/j.actamat.2013.03.049Get rights and content

[64] J. Tiley, G.B. Viswanathan, R. Srinivasan, R. Banerjee, D.M. Dimiduk, H.L. Fraser, Coarsening kinetics of γ′ precipitates in the commercial nickel base superalloy René 88 DT, Acta Mater. 57(8) (2009) 2538–2549. https://doi.org/10.1016/j.actamat.2009.02.010

[65] J. Zhang, L. Liu, T. Huang, J. Chen, K. Cao, X. Liu, J. Zhang, H. Fu, Coarsening kinetics of γ′ precipitates in a Re-containing Ni-based single crystal superalloy during long-term aging, J. Mater. Sci. Technol. 62 (2021) 1–10. https://doi.org/10.1016/j.jmst.2020.05.034

[66] F. Lu, S. Antonov, S. Lu, J. Zhang, L. Li, D. Wang, J. Zhang, Q. Feng, Unveiling the Re effect on long-term coarsening behaviors of γ′ precipitates in Ni-based single crystal superalloys, Acta Mater. 233 (2022). https://doi.org/10.1016/j.actamat.2022.117979

[67] W. Sun, Kinetics for coarsening co-controlled by diffusion and a reversible interface reaction, Acta Mater. 55 (2007) 313–320. https://doi.org/10.1016/j.actamat.2006.07.045

[68] D.L. Olmsted, S.M. Foiles, E.A. Holm, Survey of computed grain boundary properties in face-centered cubic metals: I. Grain boundary energy, Acta Mater. 57 (2009) 3694–3703. https://doi.org/10.1016/j.actamat.2009.04.007

[69] C. Baruffi, C. Brandl, On the structure of (111) twist grain boundaries in diamond: atomistic simulations with Tersoff-type interatomic potentials, Acta Mater. 215 (2021). https://doi.org/10.1016/j.actamat.2021.117055

[70] R.D. Moore, T. Beecroft, G.S. Rohrer, C.M. Barr, E.R. Homer, K. Hattar, B.L. Boyce, F. Abdeljawad, The grain boundary stiffness and its impact on equilibrium shapes and boundary migration: analysis of the Σ5, 7, 9, and 11 boundaries in Ni, Acta Mater. 218 (2021). https://doi.org/10.1016/j.actamat.2021.117220

[71] B. Lin, Y. Jin, C.M. Hefferan, S.F. Li, J. Lind, R.M. Suter, M. Bernacki, N. Bozzolo, A.D. Rollett, G.S. Rohrer, Observation of annealing twin nucleation at triple lines in nickel during grain growth, Acta Mater. 99 (2015) 63–68. https://doi.org/10.1016/j.actamat.2015.07.041

[72] Y. Jin, B. Lin, M. Bernacki, G.S. Rohrer, A.D. Rollett, N. Bozzolo, Annealing twin development during recrystallization and grain growth in pure nickel, Mater. Sci. Eng. A. 597 (2014) 295–303. https://doi.org/10.1016/j.msea.2014.01.018

[73] Q. Xiao, Y. Xu, X. Liu, Y. Wang, W. Zhang, Oxidation-induced recrystallization and damage mechanism of a Ni-based single-crystal superalloy during creep, Mater. Charact. 195 (2023). https://doi.org/10.1016/j.matchar.2022.112465

[74] K. Arora, K. Kishida, K. Tanaka, H. Inui, Effects of lattice misfit on plastic deformation behavior of single-crystalline micropillars of Ni-based superalloys, Acta Materialia 138 (2017) 119–130. https://doi.org/10.1016/j.actamat.2017.07.044

[75] Q. Han, Y. Gu, J. Huang, L. Wang, K.W. Low, Q. Feng, Y. Yin, R. Setchi, Selective laser melting of Hastelloy X nanocomposite: effects of TiC reinforcement on crack elimination and strength improvement, Compos. Part B Eng. 202 (2020). https://doi.org/10.1016/j.compositesb.2020.108442

[76] C. Körner, M. Ramsperger, C. Meid, D. Bürger, P. Wollgramm, M. Bartsch, G. Eggeler, Microstructure and mechanical properties of CMSX-4 single crystals prepared by additive manufacturing, Metall. Mater. Trans. A. 49 (2018) 3781–3792. https://doi.org/10.1007/s11661-018-4762-5

[77] K. Wang, D. Du, G. Liu, Z. Pu, B. Chang, J. Ju, Microstructure and mechanical properties of high chromium nickel-based superalloy fabricated by laser metal deposition, Mater. Sci. Eng. A. 780 (2020). https://doi.org/10.1016/j.msea.2020.139185

[78] X. Yu, X. Lin, F. Liu, L. Wang, Y. Tang, J. Li, S. Zhang, W. Huang, Influence of post-heat-treatment on the microstructure and fracture toughness properties of Inconel 718 fabricated with laser directed energy deposition additive manufacturing, Mater. Sci. Eng. A. 798 (2020). https://doi.org/10.1016/j.msea.2020.140092

[79] S.Y. Lee, Y.L. Lu, P.K. Liaw, L.J. Chen, S.A. Thompson, J.W. Blust, P.F. Browning, A.K. Bhattacharya, J.M. Aurrecoechea, D.L. Klarstrom, Tensile-hold low-cycle-fatigue properties of solid-solution-strengthened superalloys at elevated temperatures, Mater. Sci. Eng. A. 504 (2009) 64–72. https://doi.org/10.1016/j.msea.2008.10.030

[80] F. Masuyama, Low-alloyed steel grades for boilers in ultra-supercritical power plants, In: Materials for ultra-supercritical and advanced ultra-supercritical power plants; Woodhead Publ. (2017) 53–76. https://doi.org/10.1016/B978-0-08-100552-1.00002-6

[81] Y. Hu, Honeywell International Inc., Platinum-modified nickel-based superalloys, methods of repairing turbine engine components, and turbine engine components. US Patent 8,652,650, 2014.

[82] Y. Hu, Honeywell International Inc., Nickel-based superalloys, repaired turbine engine components, and methods for repairing turbine components. US Patent App. 12/171,538, 2010

[83] R.C. Reed, T. Tao, N. Warnken, Alloys-by-design: application to nickel-based single crystal superalloys, Acta Materialia. 57 (2009) 5898–5913. https://doi.org/10.1016/j.actamat.2009.08.018

[84] R.C. Reed, J.J. Moverare, A. Sato, F. Karlsson, M. Hasselqvist, A new single crystal superalloy for power generation applications, Superalloys 2012. 2012

[85] A. Sato, J.J. Moverare, M. Hasselqvist, R.C. Reed, On the mechanical behavior of a new single-crystal superalloy for industrial gas turbine applications, Metall. Mater. Trans. A. 43 (2012) 2302–2315. https://doi.org/10.1007/s11661-011-0995-2

[86] R. Darolia, Development of strong, oxidation and corrosion resistant nickel-based superalloys: critical review of challenges, progress and prospects, In Effect of porosity and eutectics on the high-temperature low-cycle fatigue performance of a nickel-base single-crystal superalloy t. Mater. Rev. 64 (2019) 355–380. https://doi.org/10.1080/09506608.2018.1516713

[87] R.P. Gangloff, Probabilistic fracture mechanics simulation of stress corrosion cracking using accelerated laboratory testing and multi-scale modelling, Corrosion. 72 (2016) 862–880. https://doi.org/10.5006/1920

[88] S.C. Yoo, K.J. Choi, T. Kim, S.H. Kim, J.Y. Kim, J.H. Kim, Microstructural evolution and stress-corrosion-cracking behavior of thermally aged Ni-Cr-Fe alloy, Corros. Sci. 111 (2016) 39–51. https://doi.org/10.1016/j.corsci.2016.04.051

[89] W. Wu, T.P. Cheng, W.H. Hsu, Stress corrosion behaviour of nickel superalloy weldments, Sci. Technol. Weld. Join. 5 (2000) 45–1718. https://doi.org/10.1179/stw.2000.5.1.45

[90] M. Sarvghad, S. Bell, R. Raud, T.A. Steinberg, G. Will, Stress assisted oxidative failure of Inconel 601 for thermal energy storage, Solar Energy Mater. Solar Cells. 159 (2017) 510–517. https://doi.org/10.1016/j.solmat.2016.10.008

[91] C.Y. Liu, G. Chen, N. Sipöcz, M. Assadi, X.S. Bai, Characteristics of oxy-fuel combustion in gas turbines, Appl. Energy. 89 (2012) 387–394. https://doi.org/10.1016/j.apenergy.2011.08.004

[92] Z. Lu, T. Shoji, S. Yamazaki, K. Ogawa, Characterization of microstructure, local deformation and microchemistry in Alloy 600 heat-affected zone and stress corrosion cracking in high temperature water, Corros. Sci. 58 (2012) 211–228. https://doi.org/10.1016/j.corsci.2012.01.029

[93] M. Li, P. Wang, Y.Q. Yang, Y.Z. Yang, H.Q. Pei, Z.X. Wen, Z.F. Yue, Oxidation behavior of a nickel-based single crystal superalloy at 1100° C under different oxygen concentration. J. Mater. Sci. 57 (2022) 3822–3841. https://doi.org/10.1007/s10853-022-06885-7

[94] P. Stratton, Ellingham diagrams–their use and misuse, Int. Heat Treat. Surf. Eng. 7 (2013) 70–73. https://doi.org/10.1179/1749514813Z.00000000053

[95] K. Ishizaki, Phase diagrams under high total gas pressures—Ellingham diagrams for hot isostatic press processes, Acta Metall. Mater. 38 (1990) 2059–2066. https://doi.org/10.1016/0956-7151(90)90073-P

[96] B. Ruttert, C. Meid, L.M. Roncery, I. Lopez-Galilea, M. Bartsch, W. Theisen, Effect of porosity and eutectics on the high-temperature low-cycle fatigue performance of a nickel-base single-crystal superalloy, Scripta Mater. 155 (2018) 139–143. https://doi.org/10.1016/j.scriptamat.2018.06.036

[97] Y. Liu, M. Kang, Y. Wu, M. Wang, M. Li, J. Yu, G. Hao, and J. Wang, Crack formation and microstructure-sensitive propagation in low cycle fatigue of a polycrystalline nickel-based superalloy with different heat treatments, Int. J. Fatigue. 108 (2018) 79–89. https://doi.org/10.1016/j.ijfatigue.2017.10.012

[98] L. Rémy, M. Geuffrard, A. Alam, A. Köster, E. Fleury, Effects of microstructure in high temperature fatigue: Lifetime to crack initiation of a single crystal superalloy in high temperature low cycle fatigue, Int. J. Fatigue 57 (2013) 37–49. https://doi.org/10.1016/j.ijfatigue.2012.10.013

[99] B.A. Miller, R.J. Shipley, R.J. Parrington, D.P. Dennies (Eds.), ASM Handbook: Failure Analysis and Prevention, ASM Int. 2021

[100] C. Busse, F. Palmert, B. Sjödin, P. Almroth, D. Gustafsson, K. Simonsson, D. Leidermark, Evaluation of the crystallographic fatigue crack growth rate in a single-crystal nickel-base superalloy, Int. J. Fatigue. 127 (2019) 259–267. https://doi.org/10.1016/j.ijfatigue.2019.05.023

Superalloys: Fundamentals and Applications   Materials Research Forum LLC
Materials Research Foundations 178 (2025) 333-363   https://doi.org/10.21741/9781644903698-17

[101] J. Zhang, Y.Y. Guo, M. Zhang, Z.Y. Yang, Y.S. Luo, Low-cycle fatigue and creep-fatigue behaviors of a second-generation nickel-based single-crystal superalloy at 760°C, Acta Metall. Sin. 33 (2020) 1423–1432. https://doi.org/10.1007/s40195-020-01056-6

[102] S. Yandt, X.J. Wu, N. Tsuno, A. Sato, Cyclic dwell fatigue behaviour of single crystal Ni-base superalloys with/without rhenium, Superalloys 2012. (2012) 501–508. https://doi.org/10.1002/9781118516430.ch55

[103] A. Paraschiv, G. Matache, C. Puscasu, the effect of heat treatment on the homogenization of CMSX-4 single-crystal Ni-based superalloy, Transp. Res. Procedia. 29 (2018) 303–311. https://doi.org/10.1016/j.trpro.2018.02.027

[104] J.C. Stinville, P.G. Callahan, M.A. Charpagne, M.P. Echlin, V. Valle, T.M. Pollock, Direct measurements of slip irreversibility in a nickel-based superalloy using high resolution digital image correlation, Acta Mater. 186 (2020) 172–189. https://doi.org/10.1016/j.actamat.2019.12.009

[105] S. He, L. Li, Y. Zhao, W. An, F. Lu, J. Zhang, S. Lu, J. Cormier, Q. Feng, Low-cycle fatigue behavior of a solution-treated and HIPped nickel-based single-crystal superalloy at 760°C, Mater. Sci. Eng. A. 881 (2023). https://doi.org/10.1016/j.msea.2023.145369

[106] L.A. Jacome, P. Nörtershäuser, J.K. Heyer, A. Lahni, J. Frenzel, A. Dlouhy, C. Somsen, G. Eggeler, High-temperature and low-stress creep anisotropy of single-crystal superalloys, Acta Mater. 61 (2013) 2926-2943. https://doi.org/10.1016/j.actamat.2013.01.052

Keyword Index

www.ingramcontent.com/pod-product-compliance
Lightning Source LLC
Chambersburg PA
CBHW071319210326
41597CB00015B/1283